Lecture Notes in Artificial Intelligence 8777

Subseries of Lecture Notes in Computer Science

T0213850

Sašo Džeroski   Panče Panov
Dragi Kocev   Ljupčo Todorovski (Eds.)

Discovery Science

17th International Conference, DS 2014
Bled, Slovenia, October 8-10, 2014
Proceedings

 Springer

Volume Editors

Sašo Džeroski
Panče Panov
Dragi Kocev
Jožef Stefan Institute
Department of Knowledge Technologies
Jamova cesta 39, 1000 Ljubljana, Slovenia
E-mail: {saso.dzeroski,pance.panov,dragi.kocev}@ijs.si

Ljupčo Todorovski
University of Ljubljana
Faculty of Administration
Gosarjeva 5, 1000 Ljubljana, Slovenia
E-mail: ljupco.todorovski@ijs.si

ISSN 0302-9743   e-ISSN 1611-3349
ISBN 978-3-319-11811-6   e-ISBN 978-3-319-11812-3
DOI 10.1007/978-3-319-11812-3
Springer Cham Heidelberg New York Dordrecht London

Library of Congress Control Number: 2014949190

LNCS Sublibrary: SL 7 – Artificial Intelligence

Typesetting: Camera-ready by author, data conversion by Scientific Publishing Services, Chennai, India

Printed on acid-free paper

Springer is part of Springer Science+Business Media (www.springer.com)

Preface

This year's International Conference on Discovery Science (DS) was the 17th event in this series. Like in previous years, the conference was co-located with the International Conference on Algorithmic Learning Theory (ALT), which is already in its 25th year. Starting in 2001, ALT/DS is one of the longest-running series of co-located events in computer science. The unique combination of recent advances in the development and analysis of methods for discovering scientific knowledge, coming from machine learning, data mining, and intelligent data analysis, as well as their application in various scientific domains, on the one hand, with the algorithmic advances in machine learning theory, on the other hand, makes every instance of this joint event unique and attractive.

This volume contains the papers presented at the 17th International Conference on Discovery Science, while the papers of the 25th International Conference on Algorithmic Learning Theory are published by Springer in a companion volume (LNCS Vol. 8776). We had the pleasure of selecting contributions from 62 submissions by 178 authors from 22 countries. Each submission was reviewed by at least three Program Committee members. The program chairs eventually decided to accept 30 papers, yielding an acceptance rate of 48%. The program also included three invited talks and two tutorials. In the joint DS/ALT invited talk, Zoubin Ghahramani gave a presentation on "Building an Automated Statistician." DS participants also had the opportunity to attend the ALT invited talk on "Cellular Tree Classifiers", which was given by Luc Devroye. The two tutorial speakers were Anuška Ferligoj ("Social Network Analysis") and Eyke Hüllermeier ("Online Preference Learning and Ranking").

This year, both conferences were held in Bled, Slovenia, and were organized by the Jožef Stefan Institute (JSI) and the University of Ljubljana. We are very grateful to the Department of Knowledge Technologies (and the project MAESTRA) at JSI for sponsoring the conferences and providing administrative support. In particular, we thank the local arrangement chair, Mili Bauer, and her team, Tina Anžič, Nikola Simidjievski, and Jurica Levatić from JSI for their efforts in organizing the two conferences. We would like to thank the Office of Naval Research Global for the generous financial support provided under ONRG GRANT N62909-14-1-C195.

We would also like to thank all authors of submitted papers, the Program Committee members, and the additional reviewers for their efforts in evaluating the submitted papers, as well as the invited speakers and tutorial presenters. We are grateful to Sandra Zilles, Peter Auer, Alexander Clark and Thomas Zeugmann for ensuring a smooth coordination with ALT, Nikola Simidjievski for putting up and maintaining our website, and Andrei Voronkov for making

EasyChair freely available. Finally, special thanks go to the Discovery Science Steering Committee, in particular to its past and current chairs, Einoshin Suzuki and Akihiro Yamamoto, for entrusting us with the organization of the scientific program of this prestigious conference.

July 2014

Sašo Džeroski
Pance Panov
Dragi Kocev
Ljupčo Todorovski

Organization

ALT/DS General Chair

Ljupčo Todorovski   University of Ljubljana, Slovenia

Program Chair

Sašo Džeroski   Jožef Stefan Institute, Slovenia

Program Co-chairs

Panče Panov   Jožef Stefan Institute, Slovenia
Dragi Kocev   Jožef Stefan Institute, Slovenia

Local Organization Team

Mili Bauer   Jožef Stefan Institute, Slovenia
Tina Anžič   Jožef Stefan Institute, Slovenia
Nikola Simidjievski   Jožef Stefan Institute, Slovenia
Jurica Levatić   Jožef Stefan Institute, Slovenia

Program Committee

Albert Bifet   HUAWEI Noahs Ark Lab, Hong Kong and
           University of Waikato, New Zealand
Hendrik Blockeel   KU Leuven, Belgium
Ivan Bratko   University of Ljubljana, Slovenia
Michelangelo Ceci   Università degli Studi di Bari Aldo Moro, Italy
Simon Colton   Goldsmiths College, University of London, UK
Bruno Cremilleux   Université de Caen, France
Luc De Raedt   KU Leuven, Belgium
Ivica Dimitrovski   Ss. Cyril and Methodius University, Macedonia
Tapio Elomaa   Tampere University of Technology, Finland
Bogdan Filipič   Jožef Stefan Institute, Slovenia
Peter Flach   University of Bristol, UK
Johannes Fürnkranz   TU Darmstadt, Germany
Mohamed Gaber   University of Portsmouth, UK
Joao Gama   University of Porto and INESC Porto, Portugal
Dragan Gamberger   Rudjer Boskovic Institute, Croatia

Yolanda Gil   University of Southern California, Information
                             Sciences Institute, Marina del Rey, CA, USA
Makoto Haraguchi   Hokkaido University, Japan
Kouichi Hirata   Kyushu Institute of Technology, Japan
Jaakko Hollmén   Aalto University, Finland
Geoffrey Holmes   University of Waikato, New Zealand
Alipio Jorge   University of Porto and INESC Porto, Portugal
Kristian Kersting   Technical University of Dortmund and
                             Fraunhofer IAIS, Bonn, Germany
Masahiro Kimura   Ryukoku University, Japan
Ross King   University of Manchester, UK
Stefan Kramer   Johannes Gutenberg University Mainz,
                             Germany
Nada Lavrač   Jožef Stefan Institute, Slovenia
Philippe Lenca   Telecom Bretagne, France
Gjorgji Madjarov   Ss. Cyril and Methodius University, Macedonia
Donato Malerba   Università degli Studi di Bari Aldo Moro, Italy
Dunja Mladenic   Jožef Stefan Institute, Slovenia
Stephen Muggleton   Imperial College, London, UK
Zoran Obradovic   Temple University, Philadelphia, PA, USA
Bernhard Pfahringer   University of Waikato, New Zealand
Marko Robnik-Sikonja   University of Ljubljana, Slovenia
Juho Rousu   Aalto University, Finland
Kazumi Saito   University of Shizuoka, Japan
Ivica Slavkov   Centre for Genomic Regulation, Barcelona,
                             Spain
Tomislav Šmuc   Rudjer Bosković Institute, Croatia
Larisa Soldatova   Brunel University London, UK
Einoshin Suzuki   Kyushu University, Japan
Maguelonne Teisseire   Cemagref - UMR Tetis, Montpellier, France
Grigorios Tsoumakas   Aristotle University, Thessalonki, Greece
Takashi Washio   ISIR, Osaka University, Japan
Min-Ling Zhang   Southeast University, Nanjing, China
Zhi-Hua Zhou   Nanjing University, China
Indré Žliobaité   Aalto University, Finland
Blaž Zupan   University of Ljubljana, Slovenia

Additional Reviewers

Mariam Adedoyin-Olowe   João Gomes
Reem Al-Otaibi   Chao Han
Darko Aleksovski   Sheng-Jun Huang
Jérôme Azé   Inaki Inza
Behrouz Bababki   Tetsuji Kuboyama
Janez Brank   Meelis Kull

Vladimir Kuzmanovski
Fabiana Pasqua Lanotte
Carlos Martin-Dancausa
Louise Millard
Alexandra Moraru
Benjamin Negrevergne
Inna Novalija
Hai Phan
Gianvito Pio
Marc Plantevit
Pascal Poncelet
John Puentes
Hani Ragab-Hassen
Dusan Ramljak

Shoumik Roychoudhury
Hiroshi Sakamoto
Rui Sarmento
Francesco Serafino
Nikola Simidjievski
Jasmina Smailović
Arnaud Soulet
Jovan Tanevski
Aneta Trajanov
Niall Twomey
Alexey Uversky
João Vinagre
Bernard Ženko
Marinka Žitnik

Invited Talks
(Abstracts)

Building an Automated Statistician

Zoubin Ghahramani

Department of Engineering,
University of Cambridge,
Trumpington Street
Cambridge CB2 1PZ, UK
zoubin@eng.cam.ac.uk

Abstract. We live in an era of abundant data and there is an increasing need for methods to automate data analysis and statistics. I will describe the "Automated Statistician", a project which aims to automate the exploratory analysis and modelling of data. Our approach starts by defining a large space of related probabilistic models via a grammar over models, and then uses Bayesian marginal likelihood computations to search over this space for one or a few good models of the data. The aim is to find models which have both good predictive performance, and are somewhat interpretable. Our initial work has focused on the learning of unknown nonparametric regression functions, and on learning models of time series data, both using Gaussian processes. Once a good model has been found, the Automated Statistician generates a natural language summary of the analysis, producing a 10-15 page report with plots and tables describing the analysis. I will discuss challenges such as: how to trade off predictive performance and interpretability, how to translate complex statistical concepts into natural language text that is understandable by a numerate non-statistician, and how to integrate model checking. This is joint work with James Lloyd and David Duvenaud (Cambridge) and Roger Grosse and Josh Tenenbaum (MIT).

References

1. Duvenaud, D., Lloyd, J.R., Grosse, R., Tenenbaum, J.B., Ghahramani, Z.: Structure Discovery in Nonparametric Regression through Compositional Kernel Search. In: ICML 2013 (2013), http://arxiv.org/pdf/1302.4922.pdf
2. Lloyd, J.R., Duvenaud, D., Grosse, R., Tenenbaum, J.B., Ghahramani, Z.: Automatic Construction and Natural-language Description of Nonparametric Regression Models. In: Twenty-Eighth AAAI Conference on Artificial Intelligence, AAAI 2014 (2014), http://arxiv.org/pdf/1402.4304.pdf

Social Network Analysis

Anuška Ferligoj

Faculty of Social Sciences,
University of Ljubljana
anuska.ferligoj@fdv.uni-lj.si

Abstract. Social network analysis has attracted considerable interest from the social and behavioral science communities in recent decades. Much of this interest can be attributed to the focus of social network analysis on relationship among units, and on the patterns of these relationships. Social network analysis is a rapidly expanding and changing field with a broad range of approaches, methods, models and substantive applications. In the talk, special attention will be given to:

1. General introduction to social network analysis:
   - What are social networks?
   - Data collection issues.
   - Basic network concepts: network representation; types of networks; size and density.
   - Walks and paths in networks: length and value of path; the shortest path, k-neighbours; acyclic networks.
   - Connectivity: weakly, strongly and bi-connected components; contraction; extraction.
2. Overview of tasks and corresponding methods:
   - Network/node properties: centrality (degree, closeness, betweenness); hubs and authorities.
   - Cohesion: triads, cliques, cores, islands.
   - Partitioning: blockmodeling (direct and indirect approaches; structural, regular equivalence; generalised blockmodeling); clustering.
   - Statistical models.
3. Software for social network analysis (UCINET, PAJEK, ...)

References

1. Batagelj, V., Doreian, P., Ferligoj, A., Kejžar, N.: Understanding Large Temporal Networks and Spatial Networks: Exploration, Pattern Searching, Visualization and Network Evolution. Wiley, Chichester (2014)
2. Carrington, P.J., Scott, J., Wasserman, S. (eds.): Models and Methods in Social Network Analysis. Cambridge University Press, Cambridge (2005)
3. Doreian, P., Batagelj, V., Ferligoj, A.: Generalized Blockmodeling. Cambridge University Press, Cambridge (2005)
4. de Nooy, W., Mrvar, A., Batagelj, V.: Exploratory Social Network Analysis with Pajek, Revised and expanded second edition. Cambridge University Press, Cambridge (2011)

5. Scott, J., Carrington, P.J. (eds.): The SAGE Handbook of Social Network Analysis. Sage, London (2011)
6. Wasserman, S., Faust, K.: Social Network Analysis: Methods and Applications. Cambridge University Press, Cambridge (1994)

Cellular Tree Classifiers[*]

Gérard Biau[1,2] and Luc Devroye[3]

[1] Sorbonne Universités, UPMC Univ Paris 06, France
[2] Institut universitaire de France
[3] McGill University, Canada

Abstract. Suppose that binary classification is done by a tree method in which the leaves of a tree correspond to a partition of d-space. Within a partition, a majority vote is used. Suppose furthermore that this tree must be constructed recursively by implementing just two functions, so that the construction can be carried out in parallel by using "cells": first of all, given input data, a cell must decide whether it will become a leaf or an internal node in the tree. Secondly, if it decides on an internal node, it must decide how to partition the space linearly. Data are then split into two parts and sent downstream to two new independent cells. We discuss the design and properties of such classifiers.

---

[*] The full paper can be found in Peter Auer, Alexander Clark, Sandra Zilles, and Thomas Zeugmann, Proceedings of the 25th International Confer- ence on Algorithmic Learning Theory (ALT-14), Lecture Notes in Computer Science Vol. 8776, Springer, 2014.

Online Preference Learning and Ranking[*]

Eyke Hüllermeier

Department of Computer Science
University of Paderborn, Germany
eyke@upb.de

Abstract. A primary goal of this tutorial is to survey the field of preference learning [7], which has recently emerged as a new branch of machine learning, in its current stage of development. Starting with a systematic overview of different types of preference learning problems, methods to tackle these problems, and metrics for evaluating the performance of preference models induced from data, the presentation will focus on theoretical and algorithmic aspects of ranking problems [6, 8, 10]. In particular, recent approaches to preference-based online learning with bandit algorithms will be covered in some depth [12, 13, 11, 2, 9, 1, 3–5, 14].

References

1. Ailon, N., Karnin, Z., Joachims, T.: Reducing dueling bandits to cardinal bandits. In: Proc. ICML, Beijing, China (2014)
2. Busa-Fekete, R., Szörényi, B., Weng, P., Cheng, W., Hüllermeier, E.: Top-k selection based on adaptive sampling of noisy preferences. In: Proc. ICML, Atlanta, USA (2013)
3. Busa-Fekete, R., Hüllermeier, E., Szörényi, B.: Preference-based rank elicitation using statistical models: The case of Mallows. In: Proc. ICML, Beijing, China (2014)
4. Busa-Fekete, R., Szörényi, B., Hüllermeier, E.: PAC rank elicitation through adaptive sampling of stochastic pairwise preferences. In: Proc. AAAI (2014)
5. Busa-Fekete, R., Hüllermeier, E.: A survey of preference-based online learning with bandit algorithms. In: Proc. ALT, Bled, Slovenia (2014)
6. Cohen, W.W., Schapire, R.E., Singer, Y.: Learning to order things. Journal of Artificial Intelligence Research 10 (1999)
7. Fürnkranz, J., Hüllermeier, E. (eds.): Preference Learning. Springer (2011)
8. Fürnkranz, J., Hüllermeier, E., Vanderlooy, S.: Binary decomposition methods for multipartite ranking. In: Buntine, W., Grobelnik, M., Mladenić, D., Shawe-Taylor, J. (eds.) ECML PKDD 2009, Part I. LNCS, vol. 5781, pp. 359–374. Springer, Heidelberg (2009)

---

[*] The full version of this paper can be found in Peter Auer, Alexander Clark, Sandra Zilles, and Thomas Zeugmann, Proceedings of the 25th International Confer- ence on Algorithmic Learning Theory (ALT-14), Lecture Notes in Computer Science Vol. 8776, Springer, 2014.

 9. Urvoy, T., Clerot, F., Féraud, R., Naamane, S.: Generic exploration and k-armed voting bandits. In: Proc. ICML, Atlanta, USA (2013)
10. Vembu, S., Gärtner, T.: Label ranking: A survey. In: Fürnkranz, J., Hüllermeier, E. (eds.) Preference Learning. Springer (2011)
11. Yue, Y., Broder, J., Kleinberg, R., Joachims, T.: The K-armed dueling bandits problem. Journal of Computer and System Sciences 78(5), 1538–1556 (2012)
12. Yue, Y., Joachims, T.: Interactively optimizing information retrieval systems as a dueling bandits problem. In: Proc. ICML, Montreal, Canada (2009)
13. Yue, Y., Joachims, T.: Beat the mean bandit. In: Proc. ICML, Bellevue, Washington, USA (2011)
14. Zoghi, M., Whiteson, S., Munos, R., de Rijke, M.: Relative upper confidence bound for the k-armed dueling bandit problem. In: Proc. ICML, Beijing (2014)

Table of Contents

Explaining Mixture Models through Semantic Pattern Mining and Banded Matrix Visualization

Prem Raj Adhikari[1], Anže Vavpetič[2], Jan Kralj[2], Nada Lavrač[2],
and Jaakko Hollmén[1]

[1] Helsinki Institute for Information Technology HIIT and Department of Information
and Computer Science, Aalto University School of Science,
PO Box 15400, FI-00076 Aalto, Espoo, Finland
{prem.adhikari,jaakko.hollmen}@aalto.fi
[2] Jožef Stefan Institute and Jožef Stefan International Postgraduate School,
Jamova 39, 1000 Ljubljana, Slovenia
{anze.vavpetic,jan.kralj,nada.lavrac}@ijs.si

Abstract. Semi-automated data analysis is possible for the end user
if data analysis processes are supported by easily accessible tools and
methodologies for pattern/model construction, explanation, and exploration. The proposed three–part methodology for multiresolution 0–1
data analysis consists of data clustering with mixture models, extraction of rules from clusters, as well as data, cluster, and rule visualization
using banded matrices. The results of the three-part process—clusters,
rules from clusters, and banded structure of the data matrix—are finally
merged in a unified visual banded matrix display. The incorporation
of multiresolution data is enabled by the supporting ontology, describing
the relationships between the different resolutions, which is used as background knowledge in the semantic pattern mining process of descriptive
rule induction. The presented experimental use case highlights the use-
fulness of the proposed methodology for analyzing complex DNA copy
number amplification data, studied in previous research, for which we
provide new insights in terms of induced semantic patterns and clus-
ter/pattern visualization.

Keywords: Mixture Models, Semantic Pattern Mining, Pattern Visualization.

1 Introduction

In data analysis, the analyst aims at finding novel ways to summarize the data to
become easily understandable [6]. The interpretation aspect is especially valued
among application specialists who may not understand the data analysis process
itself. Semi-automated data analysis is hence possible for the end user if data analysis processes are supported by easily accessible tools and methodologies for pat-
tern/model construction, explanation and exploration. This work draws together

S. Džeroski et al. (Eds.): DS 2014, LNAI 8777, pp. 1–12, 2014.

different approaches developed in our previous research, leading to a new three-part data analysis methodology. Its utility is illustrated in a case study concerning the analysis of DNA copy number amplifications represented as a 0–1 (binary) dataset [12]. In our previous work we have successfully clustered this data using mixture models [13, 16]. Furthermore, in [8], we have learned linguistic names for the patterns that coincide with the natural structure in the data, enabling domain experts to refer to clusters or the patterns extracted from these clusters, with their names. In [7] we report that frequent itemsets describing the clusters, or extracted from the 'one cluster at a time' clustered data are markedly different than those extracted from the whole dataset. The whole set of about 100 DNA amplification patterns identified from the data have been reported in [13].

With the aim of better explaining the initial mixture model based clusters, in this work we consider the cluster labels as class labels in descriptive rule learning [14], using a semantic pattern mining approach [20]. This work proposes a crossover of unsupervised methods of probabilistic clustering with supervised methods of subgroup discovery to determine the specific chromosomal locations, which are responsible for specific types of cancers. Determining the chromosomal locations and their relation to certain cancers is important to study and understand pathogenesis of cancer. It also provides necessary information to select the optimal target for cancer therapy on individual level [10]. We also enrich the data with additional background knowledge in different forms such as pre–discovered patterns as well as taxonomies of features in multiresolution data, cancer genes, and chromosome fragile sites. The background knowledge enables the analysis of data at multiple resolutions. This work reports the results of the Hedwig semantic pattern mining algorithm [19] performing semantic subgroup discovery, using the incorporated background knowledge.

While a methodology, consisting of clustering and semantic pattern mining, has been suggested in our previous work [9, 20], we have now for the first time addressed the task of explaining sub-symbolic mixture model patterns (clusters of instances) using symbolic rules. In this work, we propose this two-step approach to be enhanced through pattern comparison by their visualization on the plots resulting from banded matrices visualization [5]. Using colored overlays on the banded patient–chromosome matrix (induced from the original data), the mixture model clusters are first visualized, followed by visualizing the sets of patterns (i.e., subgroups) induced by semantic pattern mining.

Matrix visualization is a very popular method for information mining [1] and banded matrix visualization provides new means for data and pattern exploration, visualization and comparison. The addition of visualization helps to determine if the clustering results are plausible or awry. It also helps to identify the similarities and differences between clusters with respect to the amplification patterns. Moreover, an important contribution of this work is the data analytics task addressed, i.e., the problem of explaining chromosomal amplification in cancer patients of 73 different cancer types where data features are represented in multiple resolutions. Data is generated in multiple resolutions (different dimensionality) if a phenomenon is measured in different levels of detail.

The main contributions of this work are as follows. We propose a three-part methodology for data analysis. First, we cluster the data. Second, we extract semantic patterns (rules) from the clusters, using an ontology of relationships between the different resolutions of the multiresolution data [15]. Finally, we integrate the results in a visual display, illustrating the clusters and the identified rules by visualizing them over the banded matrix structure.

2   Methodology

A pipeline of algorithmic steps forming the proposed three–part methodology is outlined in Figure 1. The methodology starts with a set of experimental data (*Load Data*) and background knowledge and facts (*Load Background Knowledge*) as shown in the Figure 1. Next, both a mixture model (*Compute Mixture Model*) and a banded matrix (*Compute Banded Matrix*) are induced independently from the data in Sections 3.1 and 3.2, respectively. The mixture model is then applied to the original data, to obtain a clustering of the data (*Apply Mixture Model*). The banded structure enables the visualization of the resulting clusters in Section 3.2 (*Banded matrix cluster visualization*). Semantic pattern mining is used in Section 3.3 to describe the clusters in terms of the background knowledge (*Semantic pattern mining*). Finally, all three models (the mixture model, the banded matrix and the patterns) are joined in Section 3.4 to produce the final visualization (*Banded Matrix Rule Visualization*).

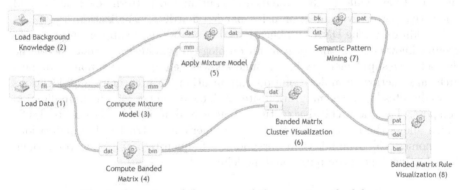

Fig. 1. Overview of the proposed three-part methodology

2.1   Experimental Data

The dataset under study describes DNA copy number amplifications in 4,590 cancer patients. The data describes 4,590 patients as data instances, with attributes being chromosomal locations indicating aberrations in the genome. These aberrations are described as 1's (amplification) and 0's (no amplification). Authors in [12] describe the amplification dataset in detail. In this paper, we consider datasets in different resolutions of chromosomal regions as defined by International System of Cytogenetic Nomenclature (ISCN) [15].

Given the complexity of the multiresolution data, we were forced to reduce the complexity of the learning setting to a simpler setting, allowing us to develop and test the proposed methodology. To this end, we have reduced the size of the dataset: from the initial set of instances describing 4,590 patients, each belonging to one of the 73 different cancer types, we have focused on 34 most frequent cancer types only, as there were small numbers of instances available for many of the rare cancer types, thus reducing the dataset from 4,590 instances to a 4,104 instances dataset. In addition, in the experiments we have focused on a single chromosome (chromosome 1), using as input to step 2 of the proposed methodology the data clusters obtained at the 393 locations granularity level using a mixture modeling approach [13]. Inferencing and density estimation from entire data would produce degenerate results because of the curse of large dimensionality. When chromosome 1 is extracted from the data, some cancer patients show no amplifications in any bands of the chromosome 1. We have removed such samples without amplifications (zero vectors) because we are interested in the amplifications and their relation to cancers, not their absence. This reduces the sample size of chromosome 1 from 4,104 to 407. Similar experiments can be performed for each chromosome in such a way that every sample of data is properly utilized. While this data reduction may be an over-simplification, finding relevant patterns in this dataset is a huge challenge, given the fact that even individual cancer types are known to consist of cancer sub-types which have not yet been explained in the medical literature. The proposed methodology may prove, in future work, to become a cornerstone in developing means through which such sub-types could be discovered, using automated pattern construction and innovative pattern visualization using banded matrices visualization.

In addition to the DNA amplifications dataset, we used supplementary background knowledge in the form of an ontology to enhance the analysis of the dataset. The supplementary background knowledge used are taxonomies of hierarchical structure of multiresolution amplification data, chromosomal locations of fragile sites [3], virus integration sites [21], cancer genes [4], and amplification hotspots. The hierarchical structure of multiresolution data is due to ISCN which allows the exact description of all numeric and structural amplifications in genomes [15]. Amplification hotspots are frequently amplified chromosomal loci identified using computational modeling [12].

2.2 Mixture Model Clustering

Mixture models are probabilistic models for modeling complex distributions by a mixture or weighted sum of simple distributions through a decomposition of the probability density function into a set of component distributions [11]. Since the dataset of our interest is a 0–1 data, we use multivariate Bernoulli distributions as component distributions to model the data. Mathematically, this can be expressed as

$$p(\boldsymbol{x}|\boldsymbol{\Theta}) = \sum_{j=1}^{J} \pi_j P(\mathbf{x} \mid \theta_j) = \sum_{j=1}^{J} \pi_j \prod_{i=1}^{d} \theta_{ji}^{x_i} (1 - \theta_{ji})^{1-x_i}. \tag{1}$$

Here, $j = 1, 2, \ldots, J$ indexes the component distributions and $i = 1, 2, \ldots, d$ indexes the dimensionality of the data. π_j defines the mixing proportions or mixing coefficients determining the weight for each of the J component distributions. Mixing proportions satisfy the properties of convex combination such as: $\pi_j \geq 0$ and $\sum_{J=1}^{J} \pi_j = 1$. Individual parameters θ_{ji} determine the probability that a random variable in the jth component in the ith dimension takes the value 1. x_i denotes the data point such that $x_i \in \{0, 1\}$. Therefore, the parameters of mixture models can be represented as: $\boldsymbol{\Theta} = \{J, \{\pi_j, \Theta_j\}_{j=1}^{J}\}$.

Expectation maximization (EM) algorithm can be used to learn the maximum likelihood parameters of the mixture model if the number of component distributions are known in advance [2]. Whereas the mixture model is merely a way to represent the probability distribution of the data, the model can be used in clustering the data into (hard) partitions, or subsets of data instances. We can achieve this by allocating individual data vectors to mixture model components that maximize the posterior probability of that data vector.

2.3   Semantic Pattern Mining

Existing semantic subgroup discovery algorithms are either specialized for a specific domain [17] or adapted from systems that do not take into the account the hierarchical structure of background knowledge [18]. The general purpose Hedwig system overcomes these limitations by using domain ontologies to structure the search space and formulate generalized hypotheses. Semantic subgroup discovery, as addressed by the Hedwig system, results in relational descriptive rules. Hedwig uses ontologies as background knowledge and training examples in the form of Resource Description Framework (RDF) triples. Formally, we define the semantic data mining task addressed in the current contribution as follows.

Given:
- set of training examples in empirical data expressed as RDF triples
- domain knowledge in the form of ontologies, and
- an object-to-ontology mapping which associates each object from the RDF triplets with appropriate ontological concepts.

Find:
- a hypothesis (a predictive model or a set of descriptive patterns), expressed by domain ontology terms, explaining the given empirical data.

The Hedwig system automatically parses the RDF triples (a graph) for the subClassOf hierarchy, as well as any other user-defined binary relations. Hedwig also defines a namespace of classes and relations for specifying the training examples to which the input must adhere. The rules generated by Hedwig system using beam search are repetitively specialized and induced as discussed in [20].

The significance of the findings is tested using the Fisher's exact test. To cope with the multiple–hypothesis testing problem, we use Holm–Bonferroni direct adjustment method with $\alpha = 0.05$.

2.4   Visualization Using Banded Matrices

Consider a $n \times m$ binary matrix M and two permutations, κ and π of the first n and m integers. Matrix M_κ^π, defined as $(M_\kappa^\pi)_{i,j} = M_{\kappa(i),\pi(j)}$, is constructed by applying the permutations π and κ on the rows and columns of M. If, for some pair of permutations π and κ, matrix M_κ^π has the following property:

- For each row i of the matrix, the column indices for which the value in the matrix is 1 appear consecutively, i.e., on indices $a_i, a_i + 1, \ldots, b_i$,
- For each i, we have $a_i \leq a_{i+1}$ and $b_i \leq b_{i+1}$,

then the matrix M is *fully banded* [5]. The motivation behind using banded matrices to exposes the clustered structure of the underlying data through the banded structure. We use barycentric method used to extract banded matrix in [5] to find the banded structure of a matrix. The core idea of the method is the calculation of barycenters for each matrix row, which are defined as

$$Barycenter(i) = \frac{\sum_{j=1}^m j \cdot M_{ij}}{\sum_{j=1}^m M_{ij}}.$$

The barycenters of each matrix row are best understood as centers of gravity of a stick divided into m sections corresponding to the row entries. An entry of 1 denotes a weight on that section. One step of the `barycentric` method now: calculate the barycenters for each matrix row and sort the matrix rows in order of increasing barycenters. In this way, the method calculates the best possible permutation of rows that exposes the banded structure of the input matrix. It does not, however, find any permutation of columns. In our application, neighboring columns of a matrix represent chromosome regions that are in physical proximity to one another, the goal is to only find the optimal row permutation while not permuting the matrix columns.

The image of the banded structure can then be overlaid with a visualization of clusters, as described in Section 2.2. Because the rows of the matrix represent instances, highlighting one set of instances (one cluster) means highlighting several matrix rows. If the discovered clusters are exposed by the matrix structure, we can expect that several adjacent matrix rows will be highlighted, forming a wide band. Furthermore, all the clusters can be simultaneously highlighted because each sample belongs to one and only one cluster. The horizontal colored overlay of the clusters in Figure 3 can also be supplemented with another colored vertical overlay of the rules explaining the clusters as discussed in Section 2.3. If an important chromosome region is discovered for the characterization of a cluster, we highlight the corresponding column. In the case of composite rules of the type $\boxed{\text{Rule 1: Cluster3(X)} \leftarrow \text{1q43-44(X)} \wedge \text{1q12(X)}}$, both chromosomal regions 1q43 − −44 and 1q12 are understood as equally important and are therefore both highlighted. If a chromosome band appears in more than one rule, this is visualized by a stronger highlight of the corresponding matrix column.

3 Experiments and Results

3.1 Clusters from Mixture Models

We used the mixture models trained in our earlier contribution [13]. Through a model selection procedure documented in [16], the number of components for modeling the chromosome 1 had been chosen to be $J = 6$, denoting presence of six clusters in the data.

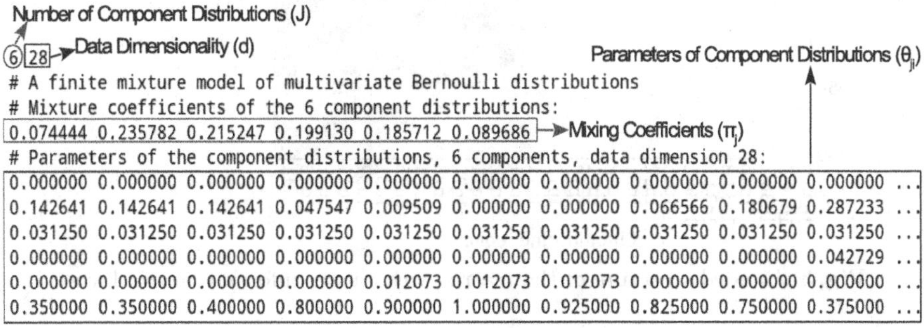

```
Number of Component Distributions (J)

 6  28  →Data Dimensionality (d)                    Parameters of Component Distributions (θ_ji)
# A finite mixture model of multivariate Bernoulli distributions
# Mixture coefficients of the 6 component distributions:
 0.074444 0.235782 0.215247 0.199130 0.185712 0.089686 →Mixing Coefficients (π_j)
# Parameters of the component distributions, 6 components, data dimension 28:
 0.000000 0.000000 0.000000 0.000000 0.000000 0.000000 0.000000 0.000000 0.000000 0.000000 ...
 0.142641 0.142641 0.142641 0.047547 0.009509 0.000000 0.000000 0.066566 0.180679 0.287233 ...
 0.031250 0.031250 0.031250 0.031250 0.031250 0.031250 0.031250 0.031250 0.031250 0.031250 ...
 0.000000 0.000000 0.000000 0.000000 0.000000 0.000000 0.000000 0.000000 0.000000 0.042729 ...
 0.000000 0.000000 0.000000 0.000000 0.012073 0.012073 0.012073 0.000000 0.000000 0.000000 ...
 0.350000 0.350000 0.400000 0.800000 0.900000 1.000000 0.925000 0.825000 0.750000 0.375000 ...
```

Fig. 2. Mixture model for chromosome 1. Figure shows first 10 dimensions of the total 28 for clarity.

Figure 2 shows a visual illustration of the mixture model for chromosome 1. In the figure, the first line denotes the number of components (J) in the mixture model and the data dimensionality (d). The lines beginning with # are comments and can be ignored. The fourth line shows the parameters of component distributions (π_j) which are six probability values summing to 1. Similarly, the last six lines of the figure denote the parameters of the component distributions, θ_{ji}. Figure 2 does not visualize and summarize the data as it consists of many numbers and probability values. Therefore, we use banded matrix for visualization as discussed in Section 2.4. Here, we focus on hard clustering of the samples of chromosomal amplification data using the mixture model depicted in Figure 2. The dataset is partitioned into six different clusters allocating data vectors to the component densities that maximize the probability of data. The number of samples in each cluster are the following: Cluster 1→30, Cluster 2→96, Cluster 3→88, Cluster 4→81, Cluster 5→75, and Cluster 6→37.

3.2 Cluster Visualization Using Banded Matrices

We used the barycentric method, described in Section 2.4, to extract the banded structure in the data. The black color indicates ones and white color denotes zeros in the data. The banded data was then overlaid with different colors for

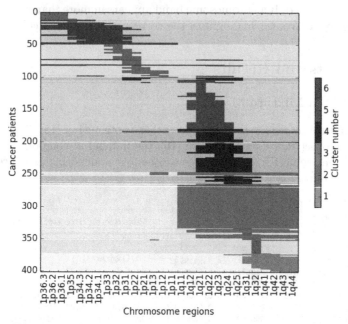

Fig. 3. Banded structure of the matrix with cluster information overlay

the 6 clusters, discovered in Section 3.1, as shown in the Figure 3. By expos-
ing the banded structure of the matrix, Figure 3 allows a clear visualization of
the clusters discovered in the data. Figures 3 shows that each cluster captures
amplifications in some specific regions of the genome. The figure captures a phenomenon that the left part of the figure showing chromosomal regions beginning
with $1p$ shows a comparatively smaller number of amplifications whereas the
right part of the figure showing chromosomal regions beginning with $1q$ (q–arm)
shows a higher number of amplifications.

The Figure 3 also shows that cluster 1 is characterized by pronounced amplifications in the end of the q–arm (regions 1q32–q44) of chromosome 1. The
figure also shows that samples in the second cluster contain sporadic amplifications spread across both p and q–arms in different regions of chromosome 1. This
cluster does not carry much information and contains cancer samples that do
not show discriminating amplifications in chromosomes as the values of random
variables are near 0.5. It is the only cluster that was split into many separate
matrix regions. In contrast, cluster 3 portrays marked amplifications in regions
1q11–44. Cluster 4 shows amplifications in regions 1q21–25. Similarly, cluster
5 is denoted by amplifications in 1q21–25. Cluster 6 is defined by pronounced
amplifications in the p-arm of chromosome 1. The visualization with banded
matrices in Figure 3 also draws a distinction between clusters each cluster which
upon first viewing show no obvious difference to the human eye when looking at
the probabilities of the mixture model shown in the Figure 2.

3.3 Rules Induced Through Semantic Pattern Mining

Using the method described in Section 2.3, we induced subgroup descriptions for each cluster as the target class [19]. For a selected cluster, all the other clusters represent the negative training examples, which resembles a one-versus-all approach in multiclass classification. In our experiments, we consider only the rules without negations, as we are interested in the presence of amplifications characterizing the clusters (and thereby the specific cancers), while the absence of amplifications normally characterizes the absence of cancers not their presence [10]. We focus our discussion only on the results pertaining to cluster 3 because of the space constraints.

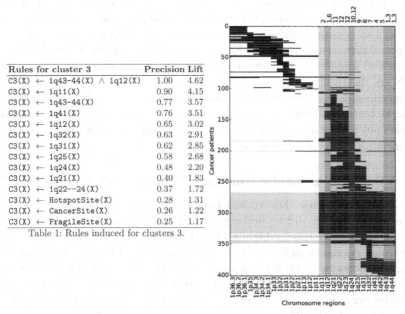

Rules for cluster 3	Precision	Lift
C3(X) ← 1q43-44(X) ∧ 1q12(X)	1.00	4.62
C3(X) ← 1q11(X)	0.90	4.15
C3(X) ← 1q43-44(X)	0.77	3.57
C3(X) ← 1q41(X)	0.76	3.51
C3(X) ← 1q12(X)	0.65	3.02
C3(X) ← 1q32(X)	0.63	2.91
C3(X) ← 1q31(X)	0.62	2.85
C3(X) ← 1q25(X)	0.58	2.68
C3(X) ← 1q24(X)	0.48	2.20
C3(X) ← 1q21(X)	0.40	1.83
C3(X) ← 1q22--24(X)	0.37	1.72
C3(X) ← HotspotSite(X)	0.28	1.31
C3(X) ← CancerSite(X)	0.26	1.22
C3(X) ← FragileSite(X)	0.25	1.17

Table 1: Rules induced for clusters 3.

Fig. 4. Rules induced for cluster 3 (left) and visualizations of rules and columns for cluster 3 (right) with relevant columns highlighted. A highlighted column denotes that an amplification in the corresponding region characterizes the instances of the particular cluster. A darker hue means that the region appears in more rules. The numbers on top right correspond to rule numbers. For example, the notations "1, 3" on top of rightmost column of cluster 3 indicates that the chromosome region appears in rules 1 and 3 tabulated in the left panel.

Table on the left panel of the Figure 4 show the rules induced for cluster 3, together with the relevant statistics. The induced rules quantify the clustering results obtained in Section 3.1 and confirmed by banded matrix visualization in Section 3.2. The banded matrix visualization depicted in Figure 3 shows that cluster 3 is marked by the amplifications in the regions 1q11–44. However, the rules obtained in Table on the left panel of the Figure 4 show that amplifications in all the regions 1q11–44 do not equally discriminate cluster 3. For example, rule

Rule 1: Cluster3(X) ← 1q43-44(X) ∧ 1q12(X)   characterizes cluster 3 best with a precision of 1. This means that amplifications in regions 1q43-44 and 1q12 characterizes cluster 3. It also covers 81 of the 88 samples in cluster 3. Nevertheless, amplifications in regions 1q11–44 shown in Figure 3 as discriminating regions, appear in at least one of the rules in the table on the left panel of the Figure 4 with varying degree of precision. Similarly, the second most discriminating rule for cluster 3 is: | Rule 2: `Cluster3(X)` ← `1q11(X)` | which covers 78 positive samples and 9 negative samples.

The rules listed in the table on the left panel of Figure 4 also capture the multiresolution phenomenon in the data. We input only one resolution of data to the algorithm but the hierarchy of different resolutions is used as background knowledge. For example, the literal 1q43–44 denotes a joint region in coarse resolution thus showing that the algorithm produces results at different resolutions. The results at different resolutions improve the understandability and interpretability of the rules [8]. Furthermore, other information added to the background knowledge are amplification hotspots, fragile sites, cancer genes, which are discriminating features of cancers but do not show to discriminate any specific clusters present in the data. Therefore, such additional information would be better utilized in situations where the dataset contains not only cancer samples but also control samples which is unfortunately not the situation here as our dataset has only cancer cases.

3.4 Visualizing Semantic Rules and Clusters with Banded Matrices

The second way to use the exposed banded structure of the data is to display columns that were found to be important due to appearing in rules from Section 3.3. We achieve this by highlighting the chromosomal regions which appear in the rules. As shown in Figure 4, the highlighted band for cluster 1 spans chromosome regions 1q32–44. For cluster 3, the entire q–arm of the chromosome is highlighted, as indeed the instances in cluster 3 have amplifications throughout the entire arm. The regions 1q11–12 and 1q43–44 appear in rules with higher lift, in contrast to the other regions showing that the amplifications on the edges of the region are more important for the characterization of the cluster.

In summary, Figures 3 and 4 together offer an improved view of the underlying data. Figure 3 shows all the clusters on the data while Figure 4 shows only specific cluster and its associated rules. We achieve this by reordering the matrix rows by placing similar items closer together to form a banded structure [5], which allows easier visualization of the clusters and rules. It is important to reorder the rows independently of the clustering process. Because the reordering selected does not depend on the cluster structure discovered, the resulting figures offer new insight into both the data and the clustering.

4   Summary and Conclusions

We have presented a three-part data analysis methodology: clustering, semantic subgroup discovery, and pattern visualization. Pattern visualization takes advantage of the structure—in our case the bandedness of the matrix. The proposed visualization allows us to explain the discovered patterns by combining different views of the data, which may be difficult to compare without a unifying visual display. In our experiments, we analyzed DNA copy number amplifications in the form of 0–1 data, where the clustering developed in previous work was augmented by explanatory rules derived from a semantic pattern mining approach combined by the facility to display the bandedness structure of the data.

The proposed semi–automated methodology provides complete analysis of a complex real-world multiresolution data. The results in the form of different clusters, rules, and visualizations are interpretable by the domain experts. Especially, the visualizations with banded matrix helps to understand the clusters and the rules generated by the semantic pattern mining algorithm. Furthermore, the use of the background knowledge enables us to analyze multiresolution data and garner results at different levels of multiresolution hierarchy. Similarly, the the obtained rules help to quantitatively prioritize chromosomal regions that are hallmarks of certain cancers among all the different chromosomal regions that are amplified in those cancer cases. In future work, we plan to extend the methodology and evaluate it using the wide variety of problems in comparison to some representative conventional methods.

Acknowledgement. This work was supported by Helsinki Doctoral Programme in Computer Science — Advanced Computing and Intelligent Systems (Hecse) and by the Slovenian Ministry of Higher Education, Science and Technology (grant number P-103), the Slovenian Research Agency (grant numbers PR-04431, PR-05540) and the SemDM project (Development and application of new semantic data mining methods in life sciences), (grant number J2–5478). Additionally, the work was supported by the Academy of Finland (grant number 258568), and European Commission through the Human Brain Project (Grant number 604102).

References

[1] Chen, C.-H., Hwu, H.-G., Jang, W.-J., Kao, C.-H., Tien, Y.-J., Tzeng, S., Wu, H.-M.: Matrix Visualization and Information Mining. In: Antoch, J. (ed.) COMP-STAT 2004 – Proceedings in Computational Statistics, pp. 85–100. Physica-Verlag HD (2004)

[2] Dempster, A.P., Laird, N.M., Rubin, D.B.: Maximum likelihood from incomplete data via the EM algorithm. Journal of the Royal Statistical Society, Series B (Methodological) 39(1), 1–38 (1977)

[3] Durkin, S.G., Glover, T.W.: Chromosome Fragile Sites. Annual Review of Genetics 41(1), 169–192 (2007)

[4] Futreal, P.A., Coin, L., Marshall, M., Down, T., Hubbard, T., Wooster, R., Rahman, N., Stratton, M.R.: A census of human cancer genes. Nature Reviews. Cancer 4(3), 177–183 (2004)

[5] Garriga, G.C., Junttila, E., Mannila, H.: Banded structure in binary matrices. Knowledge and Information Systems 28(1), 197–226 (2011)

[6] Hand, D., Mannila, H., Smyth, P.: Principles of Data Mining. Adaptive Computation and Machine Learning Series. MIT Press (2001)

[7] Hollmén, J., Seppänen, J.K., Mannila, H.: Mixture models and frequent sets: combining global and local methods for 0-1 data. In: Proceedings of the Third SIAM International Conference on Data Mining, pp. 289–293. Society of Industrial and Applied Mathematics (2003)

[8] Hollmén, J., Tikka, J.: Compact and understandable descriptions of mixtures of Bernoulli distributions. In: Berthold, M.R., Shawe-Taylor, J., Lavrač, N. (eds.) IDA 2007. LNCS, vol. 4723, pp. 1–12. Springer, Heidelberg (2007)

[9] Langohr, L., Podpecan, V., Petek, M., Mozetic, I., Gruden, K., Lavrač, N., Toivonen, H.: Contrasting Subgroup Discovery. The Computer Journal 56(3), 289–303 (2013)

[10] Lockwood, W.W., Chari, R., Coe, B.P., Girard, L., Macaulay, C., Lam, S., Gazdar, A.F., Minna, J.D., Lam, W.L.: DNA amplification is a ubiquitous mechanism of oncogene activation in lung and other cancers. Oncogene 27(33), 4615–4624 (2008)

[11] McLachlan, G.J., Peel, D.: Finite mixture models. Probability and Statistics – Applied Probability and Statistics, vol. 299. Wiley (2000)

[12] Myllykangas, S., Himberg, J., Böhling, T., Nagy, B., Hollmén, J., Knuutila, S.: DNA copy number amplification profiling of human neoplasms. Oncogene 25(55), 7324–7332 (2006)

[13] Myllykangas, S., Tikka, J., Böhling, T., Knuutila, S., Hollmén, J.: Classification of human cancers based on DNA copy number amplification modeling. BMC Medical Genomics 1(15) (May 2008)

[14] Novak, P., Lavrač, N., Webb, G.I.: Supervised Descriptive Rule Discovery: A Unifying Survey of Contrast Set, Emerging Pattern and Subgroup Mining. Journal of Machine Learning Research 10, 377–403 (2009)

[15] Shaffer, L.G., Tommerup, N.: ISCN 2005: An Intl. System for Human Cytogenetic Nomenclature (2005) Recommendations of the Intl. Standing Committee on Human Cytogenetic Nomenclature. Karger (2005)

[16] Tikka, J., Hollmén, J., Myllykangas, S.: Mixture Modeling of DNA copy number amplification patterns in cancer. In: Sandoval, F., Prieto, A.G., Cabestany, J., Graña, M. (eds.) IWANN 2007. LNCS, vol. 4507, pp. 972–979. Springer, Heidelberg (2007)

[17] Trajkovski, I., Železný, F., Lavrač, N., Tolar, J.: Learning Relational Descriptions of Differentially Expressed Gene Groups. IEEE Transactions on Systems, Man, and Cybernetics, Part C 38(1), 16–25 (2008)

[18] Vavpetič, A., Lavrač, N.: Semantic Subgroup Discovery Systems and Workflows in the SDM-Toolkit. The Comput. J. 56(3), 304–320 (2013)

[19] Vavpetič, A., Novak, P.K., Grčar, M., Mozetic, I., Lavrač, N.: Semantic Data Mining of Financial News Articles. In: Fürnkranz, J., Hüllermeier, E., Higuchi, T. (eds.) DS 2013. LNCS, vol. 8140, pp. 294–307. Springer, Heidelberg (2013)

[20] Vavpetič, A., Podpečan, V., Lavrač, N.: Semantic subgroup explanations. Journal of Intelligent Information Systems (2013) (in press)

[21] zur Hausen, H.: The search for infectious causes of human cancers: Where and why. Virology 392(1), 1–10 (2009)

Big Data Analysis of StockTwits to Predict Sentiments in the Stock Market

Alya Al Nasseri[1], Allan Tucker[2], and Sergio de Cesare[3]

[1] Brunel Business School, Brunel University, London, UK
[2] Department of Computer Science, Brunel University, London, UK
[3] Brunel Business School, Brunel University, London, UK

Abstract. Online stock forums have become a vital investing platform for publishing relevant and valuable user-generated content (UGC) data, such as investment recommendations that allow investors to view the opinions of a large number of users, and the sharing and exchanging of trading ideas. This paper combines text-mining, feature selection and Bayesian Networks to analyze and extract sentiments from stock-related micro-blogging messages called "StockTwits". Here, we investigate whether the power of the collective sentiments of StockTwits might be predicted and how these predicted sentiments might help investors and their peers to make profitable investment decisions in the stock market. Specifically, we build Bayesian Networks from terms identified in the tweets that are selected using wrapper feature selection. We then used textual visualization to provide a better understanding of the predicted relationships among sentiments and their related features.

Keywords: Wrapper feature selection, Bayesian Networks, Stock micro-blogging sentiment.

1 Introduction

Predicting the stock market is a very challenging task due to the fact that stock market data are noisy and time varying in nature [13]. In recent years, there has been increasing interest in stock market predictions using various statistical tools and machine learning techniques. Different methodologies have been developed with the aim of predicting the direction of securities' prices as accurately as possible [4]. The aim has been to create accurate models that have the ability to predict stock price behavioral movements in the stock market rather than predicting the investing decisions that derive from and cause the movement itself, such as the buying, selling and holding decisions. Investors' expectations and their psychological thinking from which the sentiments are derived are considered the main factors that affect stock price movements in capital markets [21]; it is therefore important to highlight the critical role played by trading decisions in the stock market. Trading decisions have a great effect on the profitability position of an investor in the capital market. Therefore, the ability to predict an intelligent trading support mechanism would help investors to make profitable investment decisions concerning a particular security in

S. Džeroski et al. (Eds.): DS 2014, LNAI 8777, pp. 13–24, 2014.

the capital market. In spite of researchers' continuing efforts to predict stock price movements, still little is known about the prediction of investing decisions (buy/sell/hold). This research study takes a different approach by integrating text-mining techniques with a Bayesian Network model to extract relevant features from StockTwits data to predict trading decisions (buy/hold/sell). The aim is to investigate the interactions between the selected features and their ability to predict investors' sentiments quarterly over different periods of the year. Previous research has proved the predictive ability of Bayesian Networks (BNs) in different domains both as classification and prediction tools [15]. The transparency and visibility of the connected relationships between nodes and parents in the Bayesian Networks model makes it better and more suitable approach for feature selection and prediction of sentiments in the stock market.

The investigation presented in this paper contributes to two different communities: the Financial Data Mining community and the Online Investing community. Feature selection is proposed to extract the most relevant words and terms to better predict investors' decision-making at different periods of the financial cycle. The paper also offers the potential to provide investors with an investment decision support mechanism by offering guidelines to help investors and traders determine the correct time to invest in the market, what type of stocks or sectors to invest in, and which ones yield maximum returns on their investments. Section 2 reviews the related literature on feature selection and Bayesian Networks. Section 3 presents the rationale of the model adopted for this research study. The experimental results are discussed in section 4. Section 5 presents and discusses the Bayesian Network Model for sentiment prediction. Section 6 explains textual visualization, which is utilized as a novel method when combined with both feature selection and the Bayesian Network model. We conclude our study in section 7.

2   Related Work

Many different soft computing methods and techniques have been utilized in the area of stock market predictions through the application of specific techniques for selecting important features [1]. Feature selection is one of the Data Mining techniques most commonly used to select a set of relevant features from datasets based on some predetermined criterion [18]. It is believed that the selected features subset provides a better representation of the original characteristics of the datasets. Feature selection is commonly used in the area of stock prediction. A considerable amount of literature has applied different feature selection methods to predict stock price movements. For example, [20] adopted the classification complexity of SVM as a feature selection criterion to predict the Shanghai Stock Exchange Composite Index (SSECI). [6] employed a wrapper approach to select the optimal feature subset and apply various classification algorithms to predict the trend in the Taiwan and Korea stock markets. [10] proposed a prediction model based on a hybrid feature selection method and SVM to predict the trend of the stock market. Although the above-mentioned studies have all demonstrated that feature selection methods might improve prediction accuracy, their methods lack the modeled interactions and causality between selected features.

3   Methods

3.1   Feature Selection

Feature selection is an essential pre-processing step in the text mining process. Removing features that have no discriminatory power [7] enables the classification performance to be obtained in a cost-effective and time-efficient manner, which often leads to more accurate classification results [5]. In general, two methods are associated with feature selection: the filter and wrapper [8]. The filter method evaluates the relevance of features and results in a subset of ranked features in accordance with relevancies. Here however, we exploit the wrapper method, which assesses the relevance of features by selecting the optimal features from the original subset using a specified classifier. This method has demonstrated much success in related areas [19,6].

3.2   Bayesian Networks

A Bayesian Network (BN) describes the joint distribution (a method of assigning probabilities to every possible outcome over a set of variables, $X_{1..}X_N$) by exploiting conditional independence relationships represented by a Directed Acyclic Graph (DAG) [16]. See Figure 1a for an example BN with 5 nodes. Each node in the DAG is characterised by a state, which can change depending on the state of other nodes and information about those states propagated through the DAG. This kind of inference facilitates the ability to ask 'what if?' questions of the data by entering evidence (changing a state or confronting the DAG with new data) into the network, applying inference and inspecting the posterior distribution (which represents the distributions of the variables given the observed evidence). For example, one could ask 'what is the probability of seeing a strong growth in the stock market if terms "bullish" and "confident" are commonly seen in tweets. There are numerous ways to infer both network structure and parameters from data. Search-and-score methods to infer BNs from data have been used often. These methods involve performing a search through the space of possible networks and scoring each structure. A variety of search strategies can be used. BNs are capable of performing many data analysis tasks including feature selection and classification (performed by treating one node as a class node and allowing the structure learning to select relevant features [3] (Figure 1b).

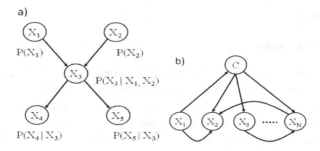

Fig. 1. a)A Simple Graphical representation of a Bayesian Network with 5 nodes and b) a Bayesian Classifier where C denotes the class node

3.3   Text Mining

The nature of the data to be collected (StockTwits posts) and the purpose of the data analysis (to extract sentiment from online financial text) inherently propose the need for text mining. The rationale of the model in this paper is that the models are trained from a corpus of manually labelled data to test the computational model, instead of using a sentiment lexicon, such as the SentiWordNet. Existing lexicons are not used in this paper mainly because this research is based on extracting sentiments from financial text, as the decision is to classify text into buy, sell or hold, not merely positive or negative. The vast majority of research papers in the sentiment analysis field focus mainly on domains, including emotional state [9], product review [22] and movie review [14], in which cases SentiWordNet is deemed a suitable lexicon. However, financial researchers have shown that dictionaries developed from other disciplines may not be effective for use in financial texts and may result in a misclassification of common words [11]. We use the tm package in R to preprocess the individual tweets [17]. Standard text mining procedures were employed for each paragraph type in order to remove stop-words, white spaces, punctuation and numbers, and to stem all necessary words. This results in an n (terms) by m (documents) matrix for each paragraph, where cells contain the number of times a term has appeared in the corresponding document.

4   Experiments and Analysis

In this paper, the experiment aims to predict investors' sentiments regarding a particular StockTwit post of DJIA companies on whether to buy, hold, or sell. The one-year training data are split into four subsets, each of which represents a quarter of the year's data. A prediction model is built for each subset using four different machine-learning algorithms: Bayes Net, Naïve Bayes, Random Forest and Sequential Minimum Optimal (SMO). The performance is used to evaluate the efficiency of each of these classifiers based on wrapper feature selection. A textual visualization tool called Wordle is used to visualize the posterior distribution of the selected terms based upon a Bayesian network model which is constructed for each quarter in order to investigate the causal relationships and interactions between the selected variables within each quarter's network.

4.1 Data Preparation and Pre-processing

Stock Tweets Data. The primary data for this study were obtained from Stocktwits.com (http://www.stocktwits.com). One year of StockTwits data are downloaded from the website's Application Programming Interface (API) for the period of April 2nd 2012-April 5th, 2013. StockTwits postings were pre-processed where those posts were without any ticker, or; those not in the DJIA index were removed, leaving 102,611 valid postings containing the dollar-tagged ticker symbol of the 30 stock tickers of the Dow 30.

General Inquirer's Harvard IV-4 Dictionary. General Inquirer is a well-known and widely-used program for text analysis. From the domain knowledge of Harvard-IV dictionary, more than 4,000 emotional words are tagged and classified as either positive or negative. Since a bull message indicates that an investor is optimistic and provides a "buy" signal to the market participants, it is therefore likely to associate positive emotions with the "buy" class. On the other hand, when an investor posts a bear message, this indicates that the investor is pessimistic and sends a "sell" signal to other market participants. The "hold" class is more likely to contain an equal balance of positive and negative emotions.

4.2 StockTwits Sentiment Manual Labelling

A random selection of a representative sample of 2,892 tweets on all 30 stocks on the Dow Jones Index are hand-labelled as either buy, hold or sell signals. These hand-labelled messages constitute the training set which is then used as an input for the model of different machine learning algorithms.

Performance Comparison. Four different machine learning classifiers (BN, NB, RandF and SMO) are applied in this paper; Table 1 presents the optimum feature attributes selected under the wrapper method for each classifier in each quarter independently, along with their average classification accuracy. Best first search was applied to Bayesian classifiers which we learnt using the K2 algorithm [12]. Applying a tenfold classification on the whole dataset, the last column in the Tables below represents predictions without feature selection (full feature set). The experimental results interestingly demonstrate that all classifiers perform well under the wrapper method in all quarters. This indicates that the wrapper approach, as a feature selection method, resulted in statistically significant improvements in classification performance over the use of the full feature set of all classifiers. Compared with the other machine learning algorithms, the Bayes net classifiers proved successful and can provide higher prediction accuracy.

Table 1. The experimental results of the feature selection andrelateda verage classification accuracy of (BN, NB, RandF and SMO) classifiers for all quarters (Qs)

(A) The Performance of four different classifiers on the first quarter (Q1)			
Classifiers	Attribute Selected	Classification Accuracy (%)	
		Selected Attributes	All Attributes
BN	24	69.96	66.05
NB	28	69.96	65.64
RandF	34	70.78	62.35
SMO	21	67.48	64.2

(C) The Performance of four different classifiers on the third quarter (Q3)			
Classifiers	Attribute Selected	Classification Accuracy (%)	
		Selected Attributes	All Attributes
BN	41	69.35	63.25
NB	27	68.13	63.49
RandF	23	68.38	62.03
SMO	15	64.59	64.84

(B) The Performance of four different classifiers on the second quarter (Q2)			
Classifiers	Attribute Selected	Classification Accuracy (%)	
		Selected Attributes	All Attributes
BN	22	67.72	61.96
NB	28	69.45	63.4
RandF	21	68.3	58.21
SMO	20	65.71	63.11

(d) The Performance of four different classifiers on the forth quarter (Q4)			
Classifiers	Attribute Selected	Classification Accuracy (%)	
		Selected Attributes	All Attributes
BN	39	72.74	70.16
NB	38	72.17	69.92
RandF	30	70.8	67.26
SMO	14	69.03	69.68

Feature Selection and Bayes Net Classifier. Since Bayes net classifiers proved effective in predicting sentiments of StockTwits data, it is worth pointing out at this stage the nature and type of the features selected in each quarter. Table 2 presents the wrapper-selected features using the Bayes net classifier of each quarter individually.

Table 2. Feature subset selected under Bayes Net classifier for individual quarters

Time	No of Selected Features	Selected Features Subset
Q1	24	aapl , **bearish**, bottom, bounc, bull, **cat,** flag, goog, head, **high**, jnj, lower, **nice**, sell, set, **short, sold,** stop, strong, support, weak, xom.
Q2	22	bac, **bearish**, bottom, bullish, **cat,** csco, current, cvx, **high**, jnj, jpm, look, low, move, **nice**, nke, **short, sold,** time, top, wmt.
Q3	41	**bearish**, bottom, bought, bounc, break, bull, bullish, buy, call, **cat**, channel, china, close, continu, cvx, dis, don, **high**, intc, jnj, jpm, level, lower, move, mrk, **nice**, nke, pfe, posit, quot, report, sell, **short, sold,** start, stop, strong, target, unh, xom,
Q4	39	**bearish**, bought, break, bull, bullish, call, **cat,** channel, close, csco, current, cvx, day, earn, entri, expect, gap, goog, **high**, hit, ibm, look, lower, mcd, **nice**, nke, posit, price, report, resist, **short, sold,** strong, support, trend, utx, via, volum.

As can be seen, a number of features appear in almost all quarters (see words in bold) while other features tend to appear in some of the quarters but not in others. An interesting observation from Table 2.is that some companies reappeared frequently in some quarters, such as Nike, Inc. "nke"and Chevron Corporation "cvx" and Johnson & Johnson "jnj", indicating that these companies were highly discussed in the StockTwits forum during that period. This suggests that new information about those discussed companies (e.g. earnings announcements) may be arriving in the market. [2] argues that, as messages are generally posted just before an event occurs, the forum may contain real-time information that is important for making investment decision.

5   Bayesian Network Model for Sentiment Prediction

In this paper, Bayesian Networks are built based on the selected features under the wrapper method for four datasets, one for each quarter. Each node in the network represents a term or word that exists in the tweet data whilst the class represents the sentiment. All term nodes in the networks are binary, i.e. having two possible states, which we will denote by T (True = feature appears in the tweet) and F (False = feature does not appear in the tweet) whilst the class can take on buy, hold or sell states. Figure 2 shows extracted versions of Bayesian Networks of each quarter, where the decision/sentiment "buy" is observed, giving a probability value of 1.

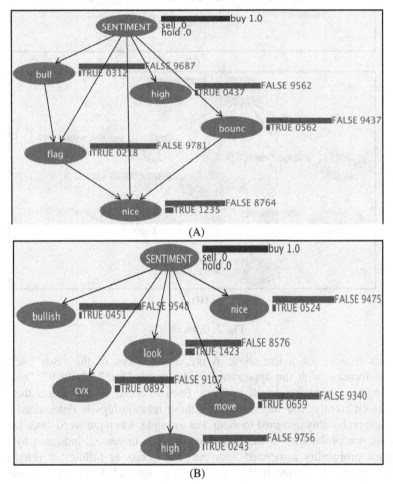

Fig. 2. Results of an extracted Bayesian Networks Model of Buy sentiment for (a) Q1, (b) Q2, (c) Q3 and (d) Q4 showing the most dominated words associated with Buy sentiment

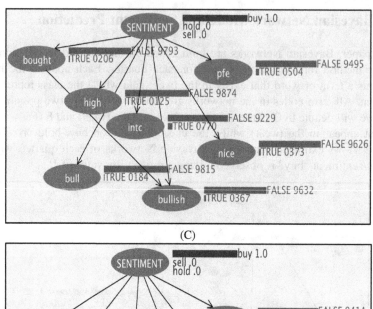

(C)

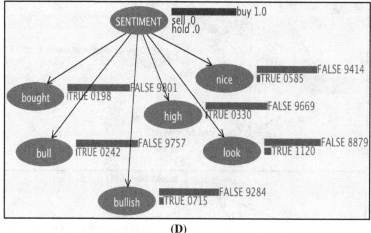

(D)

Fig. 2. (*continued*)

As can be seen from the above figure, the chances of the "Buy" sentiment occurring increase with the appearance of the words "bull", "bullish", "nice" and "high" in tweet messages in all quarters. Each of these words affects the "Buy" sentiment differently, and the strength of these relationships is determined by the conditional probability assigned to them. For example, when the word "nice" appears in a tweet, the probability of the "buy" sentiment is increased, indicated by a high conditional probability associated with the "buy" class as follows: P (buy| nice)= 0.1235, 0.0524, 0.0373 and 0.0585 for 1st, 2nd, 3rd and 4th quarter respectively. Similarly, there are a number of dominant words whose appearance increases the chance of the "Sell" sentiment. Those words are "short", "bearish", "lower" and "cat". For example, when the word "bearish" appears in a tweet, the probability of the "sell" sentiment is increased, indicated by a high conditional probability associated with the "sell" class as follows: P (sell| bearish)= 0.0445, 0.0929, 0.0454 and 0.0657for 1st, 2nd, 3rd and 4th quarters respectively. Since the sentiment event has three different states (buy, sell or hold), those words will affect each state differently based on the related weighted probability of their appearance. For example when the two child nodes such as "bearish" and "cat "are connected with a parent (sentiment), we can see that the probability of a sentiment occurring when both features appeared as P (Sell |Bearish, Cat) = 0.75, 0.056 and 0.5 for buy, sell and hold sentiments respectively, which means that when both words (bearish and cat) appeared together in a StockTwit message there is an excessive buy sentiment despite their individually prominent appearances in the sell sentiment. Therefore, a sentiment can sometimes be affected inversely depending on whether each word appears independently or in combination. For the "Hold" sentiment, we observe that some words are always likely to appear when the holding sentiment is "on", indicating either a company's ticker symbols (e.g. Chevron Corporation "cvx",Johnson & Johnson"jnj",Pfizer, Inc "pfe") or some neutral words (e.g. report, level). The change in conditional probability distributions of the most prominent words associated with the buy, sell and hold sentiment over time are shown in figure 3.

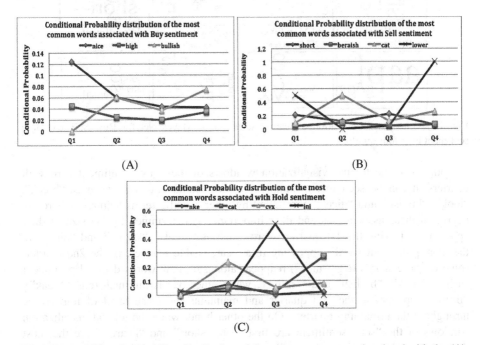

Fig. 3. The conditional probability distribution of the most common words related with the (A) buy, (B) sell and(C) hold sentiment

6   Textual Visualization of Features Selection Using Wordle

Wordle is a text analysis tool used to highlight the words that most commonly occur throughout StockTwits text (Wordle, http://www.wordle.net/creat). It creates an image that randomizes the words where the size of the words is determined according to the frequency with which they occur, highlighting their importance.In our case, the probability values of all features, which are obtained from a Bayesian network, are used to determine the prominence of those features. Figure 4 shows the visualized image of the selected features that are more associated with particular sentiments in all quarters.

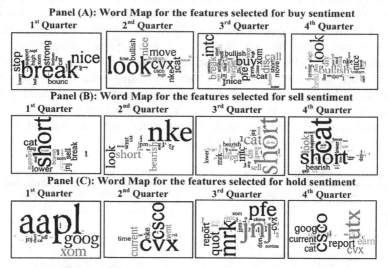

Fig. 4. The textual visualization of feature selections of Bayes Net for Buy, Sell and Hold sentiment over four quarters

Looking at the textual visualization windows of the "buy" sentiment across all quarters, it can be seen that a number of words stand out most, such as "break", "look", "bullish" and "nice", surrounded by other specific terms/features, where the degree of these associations and their importance is determined by the size of their appearance. During the 1st quarter, words such as "break", "strong" and "nice" are the most prominent words, appearing in a clear, visible manner. In the 2nd quarter, some of these words (e.g. "nice") reappear alongside new words deemed prominent such as "look", "bullish" and " move", while other words diminish (e.g. "break"). "Bullish" appears in the 2nd quarter and maintains the same level of importance throughout the remaining quarters. On the other hand, when the textual visualization windows of the "sell" sentiment are first seen, "short" and "bearish" are the most dominant words, standing out very clearly in every quarter. "cat","nke", "mrk", "jnj", "dis" and "intc" are companies clearly visible in sell sentiment that indicates a high bearishness where investors might tend to sell short their stocks of those companies. The textual visualization windows of the "hold" sentiment interestingly show that greater prominence is given to words that represent the company ticker symbols as well as some other words (e.g. "report", "qout", "don" and continue"), throughout the quarters. For example, the most dominant words associated with the "hold" sentiment are "aapl", "goog" and "xom", "csco","cvx" and "wmt", suggesting that these corporations are mostly being held during 1st and 2nd quarter. However some corporations reappear and demonstrate a holding position especially the largest corporations such as "csco" and "goog" which always seem to be associated with hold messages.

7   Conclusion

In this research paper, we proposed a novel approach by combining text mining, feature selection and Bayesian Networks models to predict investor sentiment from a stock micro-blogging forum (StockTwits) of DJIA companies. The experiments reported in this paper proved the predictive ability of Bayes Net classifiers in predicting StockTwist sentiments. In general, a look at the most prominent words per sentiment class indicate that Bayesian Network model and textual visualization using Wordle derived a plausible dictionary from our training set. Obviously, some features occur frequently in all sentiment classes (e.g. look). The positive emotions (e.g. nice and strong) are much more likely to be seen in the "buy" sentiment, while the "sell" sentiment contains many more negative emotions (e.g. "low" and "close"). The "buy" sentiment reflects the linguistic bullishness and is more likely to contain "bullish" words along with other technical words (e.g. "move" and "high") or trading words (e.g. "buy","bull" and "call"). On the other hand, the "sell" sentiment reflects bearishness and often combines bearish words with technical words (e.g. "support", "lower") or trading words (e.g. "sell" and "short"). The "hold" sentiment is more likely to contain neutral words (e.g. "report", "quote" and "time") or company names (e.g. the company ticker symbol; "cvx" and "jnj". An equal balance of negative and positive emotions is likely to be found in the "hold" sentiment. The findings of this research paper about the ability of Bayesian Networks to predict investors' sentiments in the stock market may yield promising insights into the potential provision of an investment support mechanism for analysts, investors and their peers. Practically, this could be used to determine the precise time when stocks are to be held, added (buy) or removed from a portfolio, thus yielding the maximum return on the investment for the investor.

References

1. Atsalakis, G.S., Valavanis, K.P.: Surveying stock market forecasting techniques – Part II: Softcomputing methods. Expert Systems with Applications 36(3), 5932–5941 (2009)
2. Claburn, T.: "Twitter growth surges 131% in March Information Week (2009), http://www.informationweek.com/news/internet/social_network/showArticle.jhtml?articleID=216500968 (retrieved October 25, 2010)
3. Friedman, N., Geiger, D., Goldszmidt, M.: Bayesian network classifiers. Machine Learning 29(2-3), 131–163 (1997)
4. Guresen, E., Kayakutlu, G., Daim, T.U.: Using artificial neural network models in stock market index prediction. Expert Systems with Applications 38(8), 10389–10397 (2011)
5. Guyon, I., Elisseeff, A.: An introduction to variable and feature selection. The Journal of Machine Learning Research 3, 1157–1182 (2003)

6. Huang, C.-J., Yang, D.-X., Chuang, Y.-T.: Application of wrapper approach and composite classifier to the stock trend prediction. Expert Systems with Applications 34(4), 2870–2878 (2008)
7. John, G.H., Kohavi, R., Pfleger, K.: Irrelevant Features and the Subset Selection Problem. In: ICML, vol. 94, pp. 121–129 (1994)
8. Kohavi, R., John, G.H.: Wrappers for feature subset selection. Artificial Intelligence 97(1), 273–324 (1997)
9. Kramer, A.D.: An unobtrusive behavioral model of gross national happiness. In: Proceedings of the 28th International Conference on Human Factors in Computing Systems, p. 287. ACM (2010)
10. Lee, M.-C.: Using support vector machine with a hybrid feature selection method to the stock trend prediction. Expert Systems with Applications 36(8) (2009)
11. Loughran, T., McDonald, B.: When is a liability not a liability? Textual analysis, dictionaries, and 10-Ks. The Journal of Finance 66(1), 35–65 (2011)
12. Hall, M., Frank, E., Holmes, G., Pfahringer, B., Reutemann, P., Witten, I.H.: The WEKA Data Mining Software: An Update. SIGKDD Explorations 11(1) (2009)
13. Ni, L.-P., Ni, Z.-W., Gao, Y.-Z.: Stock trend prediction based on fractal feature selection and support vector machine. Expert Systems with Applications 38(5), 5569–5576 (2011)
14. Pang, B., Lee, L., Vaithyanathan, S.: Thumbs up?: sentiment classification using machine learning techniques. In: Proceedings of the ACL 2002 Conference on Empirical Methods in Natural Language Processing, vol. 10, p. 79. Association for Computational Linguistics (2002)
15. Pazzani, M., Muramatsu, J., Billsus, D.: Syskill&Webert: Identifying Interesting Web Sites. In: Proceedings of the Thirteenth National Conference on Artificial Intelligence, pp. 54–61. AAAI Press, Portland (1996)
16. Pearl, J.: Probabilistic reasoning in intelligent systems: networks of plausible inference. Morgan Kaufmann (1988)
17. R Development Core Team: R: A language and environment for statistical computing. R Foundation for Statistical Computing, Vienna, Austria (2012), http://www.R-project.org/, ISBN 3-900051-07-0
18. Sima, C., Dougherty, E.R.: The peaking phenomenon in the presence of feature-selection. Pattern Recognition Letters 29(11), 1667–1674 (2008)
19. Stein, G., Chen, B., Wu, A.S.,Hua, K.A.: Decision tree classifier for network intrusion detection with GA-based feature selection. In: Proceedings of the 43rd Annual Southeast Regional Conference, vol. 2, pp. 136–141. ACM (March 2005)
20. Sui, X.-S., Qi, Z.-Y., Yu, D.-R., Hu, Q.-H., Zhao, H.: A novel feature selection approach using classification complexity for SVM of stock market trend prediction. In: International Conference on Management Science and Engineering, Harbin, China, pp. 1654–1659 (2007)
21. Tan, T.Z., Quek, C., Ng, G.S.: Biological brain-inspired genetic complementary learning for stock market and bank failure prediction. Computational Intelligence 23(2), 236–261 (2007)
22. Turney, P.D.: Thumbs up or thumbs down?: semantic orientation applied to unsupervised classification of reviews. In: Proceedings of the 40th Annual Meeting on Association for Computational Linguistics, p. 417. Association for Computational Linguistics (2002)
23. Wordle-Beautiful Word clouds (May 20, 2014), http://www.wordle.net/creat

Synthetic Sequence Generator for Recommender Systems – Memory Biased Random Walk on a Sequence Multilayer Network

Nino Antulov-Fantulin[1], Matko Bošnjak[2,*], Vinko Zlatić[3], Miha Grčar[4], and Tomislav Šmuc[1]

[1] Laboratory for Information Systems, Division of Electronics,
Rudjer Bošković Institute, Zagreb, Croatia
nino.antulov@irb.hr
[2] Department of Computer Science, University College London, London, UK
[3] Theoretical Physics Division, Rudjer Bošković Institute, Zagreb, Croatia
[4] Department of Knowledge Technologies - E8, Jožef Stefan Institute,
Ljubljana, Slovenia

Abstract. Personalized recommender systems rely on each user's personal usage data in the system, in order to assist in decision making. However, privacy policies protecting users' rights prevent these highly personal data from being publicly available to a wider researcher audience. In this work, we propose a memory biased random walk model on a multilayer sequence network, as a generator of synthetic sequential data for recommender systems. We demonstrate the applicability of the generated synthetic data in training recommender system models in cases when privacy policies restrict clickstream publishing.

Keywords: biased random walks, recommender systems, clickstreams, networks.

1 Introduction

Recommender systems provide a useful personal decision support in search through vast amounts of information on the subject of interest [1, 2] such as books, movies, research papers, and others. The operation and the performance of recommender systems based on collaborative data [3, 4] are necessarily tied to personal usage data, such as users' browsing and shopping history, and to other personal descriptive data such as demographical data. These data often conform to privacy protection policies, which usually prohibit their public usage and sharing, due to their personal nature. This, in turn, limits research and development of recommender systems to companies in possession of such vital data, and prevents performance comparisons of new systems between different research groups.

---

* Part of this research was done while the author was at the Rudjer Bošković Institute.

S. Džeroski et al. (Eds.): DS 2014, LNAI 8777, pp. 25–36, 2014.

In order to enable data sharing and usage, many published data sets were anonymized by removing all the explicit personal identification attributes like names and demographical data, among others. Nevertheless, various research groups managed to successfully identify personal records by linking different datasets over quasi-personal identifiers such as search logs, movie ratings, and other non-unique data, revealing as a composition of identifiers [5]. Due to successful privacy attacks, some of the most informative data for recommendation purposes, such as the personal browsing and shopping histories, are put out of the reach of the general research public. In their original form, usage histories are considered personal information, and their availability is heavily restricted. However, even with the personal information obfuscated, they remain a specific ordered sequence of page visits or orders, and as such can be uniquely tied to a single person through linkage attacks.

With usage histories often rendered unavailable for public research, recommender systems researchers have to manage on their own and often work on disparate datasets. Recently, a one million dollar worth Overstock.com recommender challenge released synthetic data, which shares certain statistical properties with the original dataset. The organizers noted that this dataset should have been used only for testing purposes, while the code itself had to be uploaded to RecLabs[1] for model building and evaluation against the real data. The challenge ended with no winner since no entry met the required effectiveness at generating lift. It would be useful both for contestants and the companies, if the synthetic data could be used for recommendation on real users.

We propose an approach to synthetic clickstream generation by constructing a memory biased random walk model (MBRW) on the graph of the clickstream sequences, which is a subclass of Markov chains [6, 7]. Random walks [8–10] have been used for constructing recommender systems on different types of graph structures originating from users' private data, but not to generate synthetic clickstreams. In this work we show that the synthetic clickstreams generated by the MBRW model share similar statistical properties to real clickstream. In addition, we use the MBRW model to generate synthetic clickstreams for the VideoLectures.NET[2] dataset from the ECML/PKDD 2011 Discovery Challenge [11] and publish it on-line. Finally, we demonstrate that synthetic data could be used to make recommendations to real users on the Yahoo! Music dataset released for the KDDCup challenge for the year 2011 [12] and the MovieLens dataset [3].

2   Methodology

The biased random walk on a graph [13, 14] is a stochastic process for modelling random paths on a general graph structure In our case, the graph we refer to is constructed from users' interaction history, i.e. clickstreams. Clickstream is a sequence of items (path on graph) $c^i = \{u_1^i, u_2^i, u_3^i, ..., u_n^i\}$, such as web pages,

---

[1] http://code.richrelevance.com/reclab-core/

[2] http://videolectures.net

[3] http://grouplens.org/datasets/movielens/

movies, books, etc., a user i interacted with. The set of all the clickstreams in a system is $C = \{c^1, c^2, ..., c^i, ..., c^m\}$. This set is usually used to generate an item history matrix, which is used by a recommender system algorithm for recommendation learning. In our work, we use two characteristic data generator matrices, obtained from the real clickstream data: the Direct Sequence matrix (DS) and the Common View Score matrix (CVS). The element $DS[m, n]$ of the matrix DS denotes the number of clickstreams in C in which the web page m immediately follows the web page n. The element $CVS[m, n]$ of the matrix CVS denotes the number of occurrences in which the web page m and the web page n belong to the same clickstream in C. In order to reconstruct synthetic clickstreams from these matrices, we introduce the memory component to the biased random walk [13], and obtain the memory biased random walk model.

The MBRW model is a discrete time Markov chain model, with a finite memory of m past states. Biases from the DS graph are the connecting probability of choosing the next item with respect to the current item, while biases from the CVS graph are the connecting probability of choosing the next item with respect to the the past m items in a clickstream. The initial vertex for the random walk can be chosen by either a stochastic or a deterministic rule.

Given an initial vertex u_1, the probability of choosing an adjacent vertex u_2 equals:

$$P(u_2|u_1) = \frac{DS_{u_2,u_1}}{\sum_k DS_{k,u_1}} \tag{1}$$

which, generates a clickstream $c^i = \{u_1, u_2\}$. The third vertex, u_3 in the clickstream is chosen with a probability of:

$$P(u_3|u_2, u_1) = \frac{DS_{u_3,u_2} CVS_{u_3,u_1}}{\sum_k DS_{k,u_1} CVS_{k,u_1}} \tag{2}$$

thus generating a clickstream $c^i = \{u_1, u_2, u_3\}$. Using a finite memory of size m, we choose the vertex u_n with the probability of:

$$P(u_n|u_{n-1}, ..., u_{n-m-1}) = \frac{DS_{u_n,u_{n-1}} \prod_{k=1}^{m} CVS_{u_n,u_{n-k-1}}}{\sum_j DS_{j,u_{n-1}} \prod_{k=1}^{m} CVS_{j,u_{n-k-1}}} \tag{3}$$

thus generating a clickstream $c^i = \{u_1, u_2, u_3, ..., u_n\}$ at the n-th step of the random walk.

The intuition behind (3) is that the probability of choosing the next item should be proportional to the product of direct sequence frequency DS and common view score frequencies CVS in the clickstream data. Direct sequence frequency DS measures the tendency of the current item preceding the next item in the clickstream data. The product of common view score frequencies measures the tendency of the next item appearing together with all the other items in a currently generated clickstream. The denominator of (3) is the normalization expression. In Figure 1, we demonstrate the transition probability calculation on a simple example.

We use the aforementioned MBRW model in a generative manner for constructing synthetic clickstreams. The procedure of generating a single clickstream

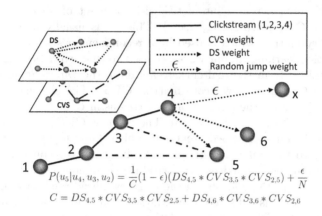

$$P(u_5|u_4, u_3, u_2) = \frac{1}{C}(1 - \epsilon)(DS_{4,5} * CVS_{3,5} * CVS_{2,5}) + \frac{\epsilon}{N}$$

$$C = DS_{4,5} * CVS_{3,5} * CVS_{2,5} + DS_{4,6} * CVS_{3,6} * CVS_{2,6}$$

Fig. 1. Simple example: at the current step the MBRW model ($m = 2$) has created a clicktream (u_1, u_2, u_3, u_4) and node u_4 has two neighbouring nodes u_5 and u_6 at the DS graph. The transition probability (see formula 3) to node u_5 is given, where the ϵ transition denotes the probability of a jump to some arbitrary node u_x, the C denotes the normalization, the factor $\frac{\epsilon}{N}$ denotes the probability of random jump back to node u_5 and N denotes total number of nodes in DS graph.

starts by randomly generating the first item. We then sample the length of the clickstream l from a discrete probability distribution L like Poisson, negative binomial, geometric, or from the real clickstream length distribution, if available. The next step is to iteratively choose the next $l - 1$ items with the MBRW model. In order to ensure additive smoothing over transition probabilities in the MBRW walk, we introduce a small ϵ probability of a random jump. At each step in the clickstream generation process a random walker produces a jump at some random item with the probability ϵ. This ϵ-smoothing technique turns all possible clickstreams to become non-forbidden in generation process. At the end of this process, the random walk path $c^i = \{u_1^i, u_2^i, ..., u_l^i\}$ presents one clickstream which is then appended to the synthetic clickstream set C^*. This whole clickstream generation process is iterated in K independent iterations to produce K synthetic clickstreams. The pseudo code for Memory Biased Random Walks with Random Jumps is provided in Algorithm 1. The code for the MBRW model is available on GitHub.[4]

3   Evaluation and Results

We analyse the statistical properties as well as the utility of the synthetic data in training recommender system models. In our experiments we used three datasets: (i) the Yahoo! Music dataset released for the KDDCup challenge for the year 2011 [12], (ii) the MovieLens 1M dataset and (iii) the VideoLectures.NET dataset

---
[4] http://github.com/ninoaf/MBRW

Algorithm 1. Memory Biased Random Walks with Random Jumps

Input: DS - Direct Sequence matrix, CVS - Common View Score matrix, K - number of synthetic clickstreams, ϵ - probability of random jump, m - memory length from prob. distr. M, L - clickstream length distribution
Output: $C^* = \{c^{*1}, c^{*2}, ..., c^{*K}\}$ synthetic clickstream set
$C^* = \emptyset$
for $i = 1 : \mathrm{K}$ **do**
  $c_i^* \sim \{u_1, u_2, ..., u_k\}$ // sample the initial item;
  $l \sim L$ // sample the clickstream length
  for $j = 2 : l$ **do**
    with $1 - \epsilon$ probability choose the next item u_j with MBRW walk on DS and CVS by using (3), otherwise with ϵ probability choose the next item u_j with a random jump;
    append new item: $c_i^* = c_i^* \cup u_j$
  end for
  append new synthetic clickstream: $C^* = C^* \cup c_i$
end for

---

from the ECML/PKDD 2011 Discovery Challenge [11]. As the privacy policies did not restrict publishing user preference data to particular items in both the KDDCup challenge 2011 and the MovieLens 1M dataset, we used them in our study as an experimental polygon to measure the performance of recommender systems models trained on synthetic data. Contrary, in the ECML/PKDD 2011 Discovery Challenge [11], only the content data and clickstrem statistics could be published but not the actual clickstreams. Therefore, we used our methodology on the VideoLectures.NET dataset to create and publish synthetic clickstream data. The first dataset used in our experiments is a subset of the Yahoo! Music dataset released for the KDDCup challenge for the year 2011 [12] which contains user preferences to particular musical items in a form of ratings, along with an appropriate time stamp. We extracted from this dataset a subset that represents a very good proxy for a set of sequential activities (clickstreams). For each user in our subset we retained a sequence of highly rated items in ascending order over time stamps (sequence activity or clickstream proxy). We limited the total number of items and users in our subset to 5000 and 10000 respectively, in order to be able to perform large set of computational experiments with resources on disposal. The reduced dataset, denoted with C, represents a set of clickstreams for 10000 users. This dataset reduction should not have any significant impact on the results and conclusions of the study. We will address this question later with the cross-validation technique. The second dataset contains approximately 10^6 anonymous ratings of approximately 3900 movies made by 6040 MovieLens users who joined MovieLens in 2000. Per each user, we extracted a sequence of highly rated items in the ascending order over time stamps from this dataset.

Our first hypothesis is that, given a sufficiently large synthetic dataset, basic statistical properties of DS^* and CVS^* matrices are preserved. We examined how statistical properties of the item preference matrix like DS and CVS are preserved in synthetic clickstream set, with respect to the original clickstream set. We calculated the DS and CVS matrices from the C dataset and created the synthetic clickstream set C^* by using the MBRW model. Memory parameter m was sampled from the Gaussian distribution $\mathcal{N}(3, 2^2)$, number of random walk hops parameter l was sampled from $\mathcal{N}(9, 2^2)$ and number of synthetic clickstreams parameter K varying from 10^4 to 10^6. Upon obtaining the synthetic clickstream set C^*, we calculated the DS^* and CVS^* matrices, and compared their statistical properties to the original matrices DS and CVS. We used the Spearman's rank correlation [15] measure between the corresponding rows in (DS, DS^*) and (CVS, CVS^*).

Table 1. Average rank correlation between (DS, DS^*) and (CVS, CVS^*) for different sizes (K) of generated synthetic clickstream set. Synthetic clickstream set is created using parameter m sampled from $\mathcal{N}(3, 2^2)$, parameter l sampled from $\mathcal{N}(9, 2^2)$.

Size	$AVG[r(DS, DS^*)]$	$STD[r(DS, DS^*)]$
$K = 10^4$	0.5700	0.3210
$K = 10^5$	0.8914	0.2224
$K = 10^6$	0.9294	0.0590
	$AVG[r(CVS, CVS^*)]$	$STD[r(CVS, CVS^*)]$
$K = 10^4$	0.4545	0.2677
$K = 10^5$	0.6050	0.2120
$K = 10^6$	0.7361	0.1784

Due to the fact that these matrices are sparse and that in the process of recommendation only top ranked items are relevant, we limited the rank correlation calculation to the first $z = 100$ elements. Rank correlation between complete rows would be misleadingly high due to row sparsity. Average rank correlation coefficient $AVG[r(DS, DS^*)] = 0.92$ and $AVG[r(CVS, CVS^*)] = 0.73$ over all corresponding rows was obtained for the first z most important elements, with the above parameters and $K = 10^6$. The rank correlation coefficients for different values of parameter K can be seen in Table 1. This shows highly correlated statistical properties (DS, DS^*) and (CVS, CVS^*).

Now, we analyse the ability to learn recommender system models from synthetic data and apply this model on real users. We measure and compare the recommender system models learned on real, synthetic and random data, and their corresponding performance on recommending items to real users. We take the standard Item-Knn [16] recommender system as a representative of similarity-based techniques and a state-of-the-art matrix factorization technique, namely Bayesian Personalized Ranking Matrix Factorization Technique [17]. We hypothesise that learning recommender systems models even from the synthetic data can help making predictions to real users.

In order to create proper training, query and test data for testing of our hypotheses, we create two splits: a vertical and horizontal split. The horizontal split of the clickstream dataset C randomly divides them to two disjoint,

Fig. 2. Three ways of splitting the original clickstream set used in computational experiments: A - Horizontal split, B - Vertical split and C - Horizontal and vertical split

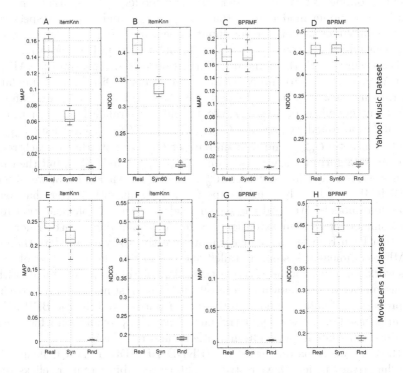

Fig. 3. Results for 10-folds cross-validation for MAP and NDCG measures for different datasets with Item-Knn [16] (plots A, B, E and F) and BPRMF [17] algorithm (plots C, D, G and H). Label "Real" represents performance on real dataset. Label: "Syn" represents synthetic data using MBRW with m sampled from $\mathcal{N}(3, 2^2)$, l sampled from real length distribution, $\epsilon = 0.0001$. Label "Rnd" represents random data generated by random jumps $\epsilon = 1.0$ on item graph. Plots: A, B, C and D are experiments on the Yahoo! Music dataset and plots: E, F, G and H are experiments on the Movielens 1M dataset. One can notice that results obtained with the Item-Knn and synthetic data are much lower than using real training data. This behaviour is more pronounced for the Yahoo! Music dataset, which is more sparse than the MovieLens 1M dataset. This confirms the hypothesis that Item-Knn is more sensitive to changes to the noise in local data distribution.

fixed-size clickstream sets C_{train} and C_{test}. Using the horizontal split on the Yahoo! Music dataset and the Movielens 1M dataset, we produced a training set C_{train} and then used the vertical split on the rest of the data to get the query set (C_{query}) and a test set (C_{test}). The vertical split, divides clickstream in C into two sets: first 50% of items are appended to first set C_{query}, whereas the rest of the clickstream items belong to a second set C_{test}. These splits are graphically represented in Figure 2. Experimental procedure is the following. We extract DS and CVS statistics from C_{train} and generate synthetic C^*_{train} with the MBRW model. The baseline random synthetic dataset C^*_{RND} is created by setting the parameter $\epsilon = 1$ (random jump model). Now, we create three different recommender system models: M (real model), M^* (synthetic model), and M_{RND} (random model) from the C_{train}, C^*_{train} and C^*_{RND}, respectively. Then recommender models for the input of real users C_{query} produce recommendations which are compared to C_{test} (ground truth). The performance on C_{test} is measured with the standard information retrieval measures: MAP [15] (Mean Average Precision) and NDCG [15] (Normalized Discounted Cumulative Gain: ranking measure). In order to estimate how performance results can generalize to independent datasets we use 10-fold cross-validation. Then in each cross-validation round we generate C^i_{train}, C^i_{test} and C^i_{query}. For each C^i_{train} we generate synthetic C^{i*}_{train} and random dataset C^{i*}_{RND}. Note that in each round recommender algorithms learn model on C^i_{train}, C^{i*}_{train} and C^{i*}_{RND} but their performance is measured for new users C^i_{query} on C^i_{test}. In Figure 3, we observe that BPRMF and Item-Knn models performed significantly better than baseline random models. We used the recommender system[5] implementation from the Recommender System extension [18–20] in RapidMiner. Furthermore, we notice that Item-Knn recommender is more sensitive to synthetic data than the BPRMF recommender system. Detailed analysis of this effects are out of scope of this work, but our hypothesis is that this behaviour of the Item-Knn algorithm is a consequence of well known high sensitivity of nearest neighbour approach to local properties of data and noise in the data. Contrary to this, the BPRMF algorithm is based on a low rank matrix factorization approximation which seems to produce same latent factors from synthetic and real data.

In the end, we focus on the ECML/PKDD 2011 Discovery Challenge [11], where the privacy policies have restricted public availability of users clickstream data on the VideoLectures.Net. Note, that here we did not have real clickstreams but only DS and CVS statistics. This challenge provided rich content data about items in a system and different statistics about users clickstream sequences. This motivated us to use the direct sequence statistics and common view statistics as generators of synthetic clickstreams with the proposed MBRW model. Direct sequence graph DS from this dataset consists of 7226 vertices in a single large, weakly connected component and common view score undirected graph CVS from this dataset consists of 7678 vertices in a large connected component.

---

[5] Item-Knn with $k = 20$ and BPRMF with num. factors: 10, user, item and negative Item regularization: 0.025, iterations: 30, learn rate: 0.05, initial mean: 0.0, initial std: 0.1 and fast sampling: 1024.

We produced and published[6] 20000 synthetic clickstreams for VideoLectures.net with the MBRW model with the memory parameter $m = 5$ and clickstream length L sampled from as a Geometric distribution with parameter 0.1 (expected length of clickstreams is 10).

4   Related Work and Discussion

The problems of privacy-preserving data publishing [21, 22] and privacy preserving data mining [23] are intensively researched within the database, the statistical disclosure, and the cryptography communities. Recently, a comprehensive survey [24] on the privacy challenges and solutions in privacy-preserving data mining has been published. Different privacy protection models already exists and here we will only mention the important ones.

Record linkage models like k-Anonymity model [25, 26] assure that the number of records with a quasi-identifier id is at least k and therefore assure the value of linkage probability of at most $1/k$. Attribute linkage models like L-diversity [27] are envisioned to overcome the problem of inferring sensitive values from k anonymity groups by decreasing the correlations between the quasi-identifiers and the sensitive values. Probabilistic models like ϵ-differential privacy model [28] ensure that individual's presence or absence in the database does not effect the query output significantly. Post-random perturbation (PRAM) methods [29, 30] change original values through probabilistic mechanisms and thus, by introducing uncertainty into data, reduce the risk of re-identification. Aggarwal et. al. [31] proposed an anonymization framework for string-like data. They used the condensation-based techniques to construct condensed groups and their aggregate statistics. From the aggregate statistics, they calculated the first and the second order information statistics of symbol distributions in strings, and generated synthetic, pseudo-string data. But still, many data-privacy researchers agree that high dimensional data poorly resist to de-anonymization [5] which poses privacy issues for companies, and prevent the usage of real-life datasets for research purposes.

Contrary to standard anonymization methods, synthetic data generation is an alternative approach to data protection in which the model generates synthetic dataset, while preserving the statistical properties of the original dataset. Several approaches for synthetic data generation have been proposed: (i) synthetic data generation by multiple imputation method [32], (ii) synthetic data by bootstrap method [33] (estimating multi-variate cumulative probability distribution, deriving similar c.d.f., and sampling a synthetic dataset), (iii) synthetic data by Latin Hypercube Sampling [34], (iv) and others such as a combination of partially synthetic attributes and real non-confidential attributes [35, 36]. These synthetic data generation strategies were mostly developed for database records with a fixed number of attributes but not for sequence data.

We proposed a novel approach for synthetic sequence generation by constructing the memory biased random walk (MBRW) model on the multilayer network

---

[6] http://lis.irb.hr/challenge/index.php/dataset/

of user sequences. Moreover, we demonstrated that this synthetic data can be used for learning recommender models which can be useful for applications on real users.

What are the potential privacy breach problems of our approach? Our method is based on the assumption that the sequence statistics: direct sequence DS and common view score CVS can be publicly available without breaking privacy of particular user. Why this is the case? We can view the clickstreams as a different way of writing the sequence statistics like finite state machines represent finite way of coding the infinite set of word from some regular language [37]. Note that the privacy breach can occur in a situation when the attacker can claim that individual unique synthetic subsequences could only be generated by using the unique transitions from particular user u. This is the reason why we need smoothing procedure (ϵ jumps) or k-anonymity filtering over the transition matrices DS and CVS. The ϵ random jumps in the generation process with small ϵ probability correspond to the additive smoothing of transition probabilities in MBRW model. Let us define the set of all possible combinatoric combinations of clickstreams with arbitrary length from set of items with Ω (infinite). Note that when $\epsilon = 0$ the MBRW model cannot create arbitrary clickstreams from the space of all clickstream combinations Ω due to the existence of zero values in DS and CVS matrices. As the additive smoothing technique turns all combinatoric clickstreams from Ω set possible, the attacker cannot claim that a certain unique user subsequence was used in the generation process. K-anonymity filtering can also be applied to CVS and DS directly by filtering all frequencies that are lower than k. This filtering enables that the presence or absence of individual transitions in DS or CVS cannot be detected. Therefore if the DS and CVS statistics can be publicly available without breaking privacy, our methodology can be applied.

5   Conclusion

The principle aim of our work was to construct a generator of real-like clickstream datasets, able to preserve the original user-item preference structure, while at the same time addressing privacy protection requirements. With respect to this aim, we investigated properties of the memory biased random walk model. We demonstrated that the basic statistical properties of data generators DS and CVS matrices are preserved in the synthetic dataset if we generate sufficiently large datasets. In addition, we demonstrated that the synthetic datasets created with it can be used to learn recommender system models applicable on real users.

Acknowledgments. This work was partially supported by the European Community 7th framework ICT-2007.4 (No 231519) "e-LICO: An e-Laboratory for Interdisciplinary Collaborative Research in Data Mining and Data-Intensive Science", partially by the EU-FET project MULTIPLEX (Foundational Research on MULTIlevel comPLEX networks and systems, grant no. 317532) and partially by the Croatian Science Foundation under the project number I-1701-2014.

References

1. Adomavicius, G., Tuzhilin, A.: Toward the next generation of recommender systems: A survey of the state-of-the-art and possible extensions. IEEE Transactions on Knowledge and Data Engineering 17(6), 734–749 (2005)
2. Rendle, S., Tso-Sutter, K., Huijsen, W., Freudenthaler, C., Gantner, Z., Wartena, C., Brussee, R., Wibbels, M.: Report on state of the art recommender algorithms (update). Technical report, MyMedia public deliverable D4.1.2 (2011)
3. Burke, R.: Hybrid recommender systems: Survey and experiments. User Modeling and User-Adapted Interaction 12(4), 331–370 (2002)
4. Resnick, P., Iacovou, N., Suchak, M., Bergstrom, P., Riedl, J.: Grouplens: an open architecture for collaborative filtering of netnews. In: Proceedings of the 1994 ACM Conference on Computer Supported Cooperative Work, CSCW 1994, pp. 175–186 (1994)
5. Narayanan, A., Shmatikov, V.: Robust de-anonymization of large sparse datasets. In: Proceedings of the 2008 IEEE Symposium on Security and Privacy, SP 2008, pp. 111–125 (2008)
6. Feller, W.: An introduction to probability theory and its applications, vol. 2. John Wiley & Sons (2008)
7. Kao, E.: An introduction to stochastic processes. Business Statistics Series. Duxbury Press (1997)
8. Bogers, T.: Movie recommendation using random walks over the contextual graph. In: Proceedings of the 2nd Intl. Workshop on Context-Aware Recommender Systems (2010)
9. Fouss, F., Faulkner, S., Kolp, M., Pirotte, A., Saerens, M.: Web recommendation system based on a markov-chain model. In: International Conference on Enterprise Information Systems, ICEIS 2005 (2005)
10. Gori, M., Pucci, A.: Research paper recommender systems: A random-walk based approach. In: Web Intelligence, pp. 778–781 (2006)
11. Antulov-Fantulin, N., Bošnjak, M., Žnidaršič, M., Grčar, M., Morzy, M., Šmuc, T.: ECML/PKDD 2011 Discovery Challenge overview. In: Proceedings of the ECML-PKDD 2011 Workshop on Discovery Challenge, pp. 7–20 (2011)
12. Dror, G., Koenigstein, N., Koren, Y., Weimer, M.: The Yahoo! music dataset and kdd-cup'11. In: Proceedings of KDD Cup 2011 (2011)
13. Zlatić, V., Gabrielli, A., Caldarelli, G.: Topologically biased random walk and community finding in networks. Phys. Rev. E 82, 066,109 (2010)
14. Newman, M.: Networks: An Introduction. Oxford University Press, Inc. (2010)
15. Manning, C.D., Raghavan, P., Schütze, H.: Introduction to Information Retrieval Cambridge University Press (2008)
16. Deshpande, M., Karypis, G.: Item-based top-n recommendation algorithms. ACM Transactions on Information Systems 22(1), 143–177 (2004)
17. Rendle, S., Freudenthaler, C., Gantner, Z., Schmidt-Thieme, L.: Bpr: Bayesian personalized ranking from implicit feedback. In: Proceedings of the 25th Conference on Uncertainty in Artificial Intelligence, UAI 2009, pp. 452–461 (2009)
18. Gantner, Z., Rendle, S., Freudenthaler, C., Schmidt-Thieme, L.: Mymedialite: A free recommender system library. In: Proceedings of the Fifth ACM Conference on Recommender Systems, pp. 305–308 (2011)
19. Mihelčić, M., Antulov-Fantulin, N., Bošnjak, M., Šmuc, T.: Extending rapidminer with recommender systems algorithms. In: Proceedings of the RapidMiner Community Meeting and Conference, pp. 63–75 (2012)

20. Bošnjak, M., Antulov-Fantulin, N., Šmuc, T., Gamberger, D.: Constructing recommender systems workflow templates in RapidMiner. In: Proc. of the 2nd Rapid-Miner Community Meeting and Conference, pp. 101–112 (2011)
21. Chen, B.C., Kifer, D., LeFevre, K., Machanavajjhala, A.: Privacy-preserving data publishing. Foundations and Trends in Databases 2(1-2), 1–167 (2009)
22. Fung, B.C., Wang, K., Fu, A.W.C., Yu, P.S.: Introduction to Privacy-Preserving Data Publishing: Concepts and Techniques, 1st edn. Chapman & Hall/CRC (2010)
23. Aggarwal, C.C., Yu, P.S. (eds.): Privacy-Preserving Data Mining. Models and Algorithms. Springer (2008)
24. Berendt, B.: More than modelling and hiding: towards a comprehensive view of web mining and privacy. Data Mining and Knowledge Discovery 24(3), 697–737 (2012)
25. Kenig, B., Tassa, T.: A practical approximation algorithm for optimal k-anonymity. Data Mining and Knowledge Discovery 25(1), 134–168 (2012)
26. Samarati, P.: Protecting respondents' identities in microdata release. IEEE Transactions on Knowledge and Data Engineering 13(6), 1010–1027 (2001)
27. Machanavajjhala, A., Kifer, D., Gehrke, J., Venkitasubramaniam, M.: L-diversity: Privacy beyond k-anonymity. ACM Transactions on Knowledge Discovery from Data 1(1) (2007)
28. Dwork, C.: Differential privacy. In: Bugliesi, M., Preneel, B., Sassone, V., Wegener, I. (eds.) ICALP 2006. LNCS, vol. 4052, pp. 1–12. Springer, Heidelberg (2006)
29. Wolf, P.P.D., Amsterdam, H.V., Design, C., Order, W.T.: An empirical evaluation of PRAM statistics. Netherlands Voorburg/Heerlen (2004)
30. Wolf, P.P.D., Gouweleeuw, J.M., Kooiman, P., Willenborg, L.: Reflections on PRAM. Statistical Data Protection, Luxembourg, pp. 337–349 (1999)
31. Aggarwal, C.C., Yu, P.S.: A framework for condensation-based anonymization of string data. Data Mining and Knowledge Discovery 16(3), 251–275 (2008)
32. Raghunathan, T., Reiter, J., Rubin, D.: Multiple imputation for statistical disclosure limitation. Journal of Official Statistics 19(1), 1–16 (2003)
33. Fienberg, S.: A radical proposal for the provision of micro-data samples and the preservation of confidentiality. Technical report, Department of Statistics, Carnegie-Mellon University (1994)
34. Dandekar, R.A., Cohen, M., Kirkendall, N.: Sensitive micro data protection using latin hypercube sampling technique. In: Domingo-Ferrer, J. (ed.) Inference Control in Statistical Databases. LNCS, vol. 2316, pp. 117–125. Springer, Heidelberg (2002)
35. Dandekar, R.A., Domingo-Ferrer, J., Sebé, F.: LHS-based hybrid microdata vs rank swapping and microaggregation for numeric microdata protection. In: Domingo-Ferrer, J. (ed.) Inference Control in Statistical Databases. LNCS, vol. 2316, pp. 153–162. Springer, Heidelberg (2002)
36. Reiter, J.: Inference for partially synthetic, public use microdata sets. Survey Methodology 29(2), 181–188 (2003)
37. Brookshear, J., Glenn, H.: Theory of Computation: Formal Languages, Automata, and Complexity. Benjamin/Cummings Publish Company, Redwood City (1989)

Predicting Sepsis Severity
from Limited Temporal Observations

Xi Hang Cao, Ivan Stojkovic, and Zoran Obradovic

Center for Data Analytic and Biomedical Informatics
Temple University, Philadelphia, PA, USA
{xi.hang.cao,ivan.stojkovic,zoran.obradovic}@temple.edu
http://www.dabi.temple.edu/dabi

Abstract. Sepsis, an acute systemic inflammatory response syndrome caused by severe infection, is one of the leading causes of in-hospital mortality. Our recent work provides evidence that mortality rate in sepsis patients can be significantly reduced by Hemoadsorption (HA) therapy with duration determined by a data-driven approach. The therapy optimization process requires predicting high-mobility group protein B-1 concentration 24 hours in the future. However, measuring sepsis biomarkers is very costly, and also blood volume is limited such that the number of available temporal observations for training a regression model is small. The challenge addressed in this study is how to balance the trade-off of prediction accuracy versus the limited number of temporal observations by selecting a sampling protocol (biomarker selection and frequency of measurements) appropriately for the prediction model and measurement noise level. In particular, to predict HMGB1 concentration 24 hours ahead when limiting the number of blood drawings before therapy to three, we found that the accuracy of observing HMGB1 and three other cytokines (Lsel, TNF-alpha, and IL10) was comparable to observing eight cytokines that are commonly used sepsis biomarkers. We found that blood drawings 1-hour apart are preferred when measurements are noise free, but in presence of noise, blood drawings 3 hours apart are preferred. Comparing to the data-driven approaches, the sampling protocol obtained by using domain knowledge has a similar accuracy with the same cost, but half of the number of blood drawings.

Keywords: health informatics, acute inflammation, therapy optimization, limited temporal data, model predictive control.

1 Introduction and Motivation

Sepsis is a serious condition resulting from uncontrolled systematic inflammatory response to some pathogen infections. This condition is characterized by fast progression, severe symptoms and high mortality rate. In fact this is the number one cause of in hospital death in the USA [1]. Despite the high importance of the problem and substantial amount of researchers effort, not much progress has been achieved in resolving it. The vast heterogeneity of clinical manifestations

S. Džeroski et al. (Eds.): DS 2014, LNAI 8777, pp. 37–48, 2014.

makes identification of sepsis severity challenging. Another difficulty lies in rapid progression of the condition, where a patient goes from mild symptoms of infection to life threatening systematic inflammation condition in just several hours. Treatment consists of administering cocktails of various antibiotics in order to cover the spectrum of possible pathogens (usually bacteria) as much as possible. Even this aggressive treatment often is not enough since mortally in severe sepsis is as high as 30% and up to 70% when septic shock occurs [2].

Two main challenges arise in the problem of reducing the lethality of the sepsis. First it is very important to devise accurate diagnostic techniques that are also able to classify condition as early as possible. After correct diagnosis, it may be even more important that therapy is applied timely and appropriately. Since, in order to be effective, sepsis therapy should be aggressive, treating a person that is healthy is almost as undesirable as not treating an ill patient. The problems of early and accurate diagnostics have been addressed in a number of articles, such as [4], [5] and [8]. Recently, a form of blood purification called Hemoadsorption (HA) was proposed as a complement to antibiotic therapy. It was shown that HA is beneficial when used in animal models of sepsis [6]. It is based on removing certain cytokines from the blood, which are involved in mechanisms of systemic inflammation. Systemic inflammation takes place when these biomarkers enter a positive feedback loop with immune cells resulting in uncontrollable increase in inflammation. This process is known as a cytokine storm and it plays major role in number of conditions including sepsis. By cytokine reduction, HA therapy attempts to regain control over the inflammation process and return it to normal mode.

Given their roll in development of sepsis, observing cytokines over time is beneficial in both diagnostic and therapeutic purposes. Recently, they were used in the task of early classification of septic patients [8]. It is shown that applying HA therapy can be guided according to the predicted future values of cytokines in the Model Predictive Control framework [3]. On the other hand, there are also some constraints on cytokines use in the task of predicting sepsis progression. Constraints are mainly posed by limits on various resources. With current technologies fairly large volume of blood is needed to measure a particular (single) cytokine. However, there are at least 150 different cytokines, and many of them are involved in the inflammation process. Instead of measuring all cytokines, in clinical applications just a few of the most informative should be identified. Even when just a few cytokines are measured, measurement needs to be done on a number of different chronological occasions in order to catch the temporal dynamics of their change, which also increases demands on blood that needs to be drawn. In reality the amount of blood that can be drawn from a subject is a limiting factor in temporal observations of cytokines. The total amount of blood drawn over some period of time is limited by cost, data extraction time and even medical protocols. An additional constraint in small animals experiments (e.g mice and rats) is that drawing too much blood can interfere with the state of the subject since volume of the subject's bodily fluids is small.

In this article we are therefore addressing the problem of predicting progression of cytokines from a limited number of temporal observations. Here, we propose an approach for learning from limited temporal observations by utilizing prior knowledge of the interconnections of biomarkers and important internal states of sepsis progression. Using this approach, we discovered a blood drawing and biomarker measuring protocol which balances the constraints, cost, and accuracy.

The rest of the paper is organized as follows. In the second section the dynamical model of sepsis progression along with the process of virtual patient generation is presented. The third section is comprised of several subsections: in section 3.1, a detailed problem formulation of this study is provided; in section 3.2, a domain knowledge based approach is proposed; in section 3.3 and 3.4, alternative data-driven approaches are introduced. In section 4, experiments corresponding to different approaches are described in details, and results are analyzed. Summary and conclusion of this study is provided in section 5.

2   Sepsis Model and Data Generation

2.1   Model of Sepsis Progression

A set of Ordinary Differential Equations (ODE) describing the evolution of severe sepsis in rats is introduced in [7]. The network of interactions included in the model consists of 19 variables and 57 parameters. Out of 19 states 11 are unobservable: CLP protocol (CLP), Bacteria (B), Anti-Inflammatory state (AI), Pro-Inflammatory state (PI), Tissue Damage (D) and five types of Neutrophiles (Nr, Np, Na, Ns, Nt and Nl). Unobservable states are interconnected through equations with each-other and with eight cytokines, which represents variables that can really be measured. These eight observable states are the following plasma cytokines: tumor necrosis factor (TNF), three kinds of interleukins (IL-1b, IL-6 and IL-10), Lselectin (Lsel), high mobility group box1 (HMGB1), creatinine (CRT) and alanine aminotransferase (ALT). Domain knowledge was utilized to relate particular biomarkers that serves as proxies for particular unobservable states. Three cytokines TNF, IL-1b and IL-6 are well known as major Pro-Inflamatory mediators. Similarly IL-10 favors Anti-Inflamaton, while Lsel is related to Neutrophiles. The remaining three cytokines HMGB1, CRT and ALT are indicators of tissue damage and ODEs are devised accordingly. Most of the parameters in the model were fitted from real experimental data, while only a few were adopted from the literature. Experimental data were collected from a set of 23 rats where sepsis was induced by the CLP protocol. Eight longitudinal measurements of eight cytokines were collected at 18, 22, 48, 72, 120, 144, and 168 hours after sepsis induction.

The devised model, although coarse, serves to allow insight into plausible mechanism that drives the progression of sepsis. Moreover it provides a tool for performing experiments on in silico patients, which in turn can lead to new promising hypotheses that could later be evaluated in real experiments.

2.2 Generation of Virtual Patients

For the purpose of conducting experiments on the prediction of sepsis biomarkers, we used the ODE model for the generation of in silico or virtual patients. Every virtual patient behaves according to the mentioned dynamical equation, but each of the patients has a unique set of parameters and therefore unique response to the CLP induction of sepsis. Sets of parameters characterizing each patient were obtained using the following 3-step protocol: First, the valid ranges of parameter values are adopted from [7], and parameters are randomly sampled from those intervals. Next the 19 states model is simulated over time for chosen set of parameters. Finally, the likelihood that the evolution of 8 observable states follows the evolution of the real data from [7] is calculated, and if the likelihood is high enough then the virtual patient has been accepted as valid, or rejected otherwise. In that way, a number of virtual patients is generated for the purpose of training, validating or testing in the conducted experiments, for which setups and results are reported in following section.

3 Biomarkers Selection for Prediction of Sepsis Severity from Temporal Observations

3.1 Problem Definition

To determine the proper duration of HA therapy, the severity progression of sepsis is assessed, based on temporal observations of relevant variables, i.e. extending duration if sepsis severity is predicted to increase. In this paper, a cytokine called high-mobility group protein B-1 (HMGB1) is used as the biomarker indicating severity of sepsis. Recently, it has been shown that using HMGB1 in the objective function for model predictive control, the rescue rate was significantly improved [3]. Therefore, our objective is to estimate the value of HMGB1 in the future (typically 24 hours ahead) before applying therapy. Simulations have shown that 18% of septic patients could be rescued with a 4-hour duration HA therapy from the 18th hour since sepsis induction [7]. Therefore, in this problem setup, we would like to predict the value of HMGB1 at the 42nd hour since sepsis induction, while the start of therapy is scheduled at the 18th hour. One may think that this is a typical time series prediction problem, because once we measure HMGB1 for 18 hours, we can deploy any regression model to make predictions. However, in practice we cannot make observations at all historic time points. In our application the minimum time interval between two consecutive blood drawings is 1 hour. Because of this, no more than 18 blood drawings corresponding to hourly observation are possible before the start of therapy. In practice, the number of blood drawings is also limited by blood volume and medical regulations. However, at each blood sample we could measure multiple biomarkers, including HMGB1. That brings up another problem; each individual measurement of a biomarker is very costly, e.g. the cost of measuring 10 biomakers in a blood sample is 10 times as costly as measuring 1 biomarker. Sepsis biomarkers are correlated, and so measurement costs can be reduced as we can predict from less measurements by utilizing their relationships. In summary, the problem addressed in this study is to balance the number of blood drawings, the number of biomarker measurements, and the prediction accuracy.

In the following 3 subsections we describe 3 methods for determining when to do blood drawings and what biomarkers to measure. A brute-force approach would consist of measuring 8 biomarkers hourly. Following such a protocol, in 18-hours, the number of measurement would be 144 ($8 \times 18=144$). Considering all 2^{144} combinations of biomarker measurements is infeasible. Therefore, in this paper, we propose using a domain knowledge based approach to select biomarkers. This method is compared to two data-driven approaches based on feature selection and L1 regularization.

3.2   Domain Knowledge Based Sepsis Biomarkers Selection

Numerous sepsis-related studies resulted in understanding of the basic mechanism of sepsis. We propose using existing domain knowledge as clues about biomarkers that are closely related to sepsis progression. In particular, we use domain knowledge to identify biomarkers related to the prediction target HMGB1.

The prediction target HMBG1 is related to tissue damage (D). From the model description in Section 2.1, other cytokines related to D are creatinine (CRT) and alanine aminotrasferase (ALT). Thus, we assume that measuring CRT and ALT could provide information about changes of HMGB1 in the future. However, since CRT, ALT and HMGB1 are proxies for the same internal state, the changes of these three tissue-damage-related biomarkers over time should be similar. Therefore, in the proposed approach we decided not to measure them all, but just select one of them to measure. In order to increase the information gain, we propose that we should select biomarkers that are proxies to different internal states. In other words, for each blood sample, in addition to measuring HMGB1 which is the observable biomarker that estimates tissue damage and is also the prediction target of our interest, we propose measuring L-selectin (Lsel), which is a proxy for peritoneal neutrophil, tumor necrosis factor-α (TNFα), which is a proxy for systemic pro-inflammatory response, and interleukin-10 (IL10), which is a proxy for systemic anti-inflammatory response.

A reduction from 8 cytokines to measuring 4 (Lsel, HMGB1, TNFα, and Il10) in each blood sample is not sufficient, as the problem of when to draw blood remains. Intuitively, measurement at the 18th hour (start of the therapy) gives us the latest status information of the patient. Therefore, we will always draw blood and take measurements at the 18th hour. In the proposed approach we assume that the time between two consecutive blood drawings is the same. If a blood drawing is definite at the 18th hour, the problem becomes finding the most suitable sampling interval of blood drawings. We restrict the number of blood drawings/samples to three, as three is a reasonable number of blood drawings, and it allows us to investigate a variety of different choices of blood sampling intervals in the initial 18-hour period from infection time to the beginning of therapy. We expect that suitable sampling intervals would vary under different situations, such as the level of noise in the measurement data. Thus, we conduct experiments to see how the preferred sampling interval changes under various conditions.

3.3   Forward Feature Selection Based Biomarkers Identification

We can also treat the problem described in Section 3.1 as a traditional feature selection problem in machine learning. If measurements of 8 biomarkers are available at every hour in the 18-hour history we have 144 features. To reduce measurements we can apply a greedy forward feature selection technique. The forward selection algorithm will try to add features to the candidate set. If the criterion function decreases after adding a feature to the candidate set, that feature will be included to the candidate set (Algorithm 1). In this case, the criterion function is the average root mean squared error (RMSE) in the training set using 5-fold cross validation, while a Linear Regression (LR) model is used as the predictor.

---

input : X, feature set; $f(.)$, criterion function
output: X_c, candidate set
initialization: $S = \infty$; $gain = true$; X_c is $empty$; x_a is $empty$;
while $gain = true$ **do**
   gain = false;
   foreach $feature\ x\ in\ X$ **do**
      add x to X_c;
      if $f(X_c) < S$ **then**
         $S = f(X_c)$; gain = true; $x_a = x$;
      end
      remove x from X_c;
   end
   if $gain = true$ **then**
      add x_a to X_c; remove x_a from X;
   end
end

Algorithm 1. Biomarkers Identification by Forward Selection

3.4   Lasso Regression Based Biomarkers Selection

The Lasso Regression Model is a Linear Regression model that includes an L_1-norm regulation term to enhance the sparsity of the coefficients (β). The values of the coefficients are found by solving the optimization function (1). Thus, features with non-zero coefficients are relevant to the prediction task. So, Lasso Regression has a built-in functionality of feature selection.

$$\min_{\beta} ||X\beta - y||_2 + \lambda||\beta||_1 \qquad (1)$$

where X is the augmented feature matrix, y is the target vector, β is the coefficient vector, and λ is the regulation coefficient.

4 Experiments and Results

The data used in our experiments were generated by using the system of equations described in Section 2.1. The value of each biomarker measurement is between 0 and 1. In order to simulate real-life conditions, we will add various levels of uniform noise to the generated data.

4.1 Using Domain Knowledge to Select Biomarkers

In the proposed approach based on using prior knowledge, three blood drawings would be made; one of the three blood drawings would be always at the 18th hour. This experiment was designed to answer the following questions: 1. What is the most suitable time interval between two consecutive blood drawings? 2. How do different choices of biomarkers used in the model would affect the prediction accuracy? 3. How does the number of virtual patients in training affect the accuracy? 4. How does the noise level in the data affect the accuracy. For purposes of comparison, a linear model and a nonlinear model were used for prediction. The linear model was Linear Regression (LR), and the nonlinear model was Support Vector Regression (SVR) [10] with radial basis kernel. SVR was implement by using the LIBSVM package [11].

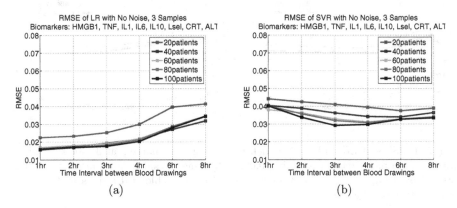

Fig. 1. RMSE's of LR and SVR model, measuring 8 biomarkers at 3 blood drawings noise free. Results shown from models trained on 1, 2, 3, 4, 6, 8 hours interval between blood drawings and on data from 20 to 100 subjects.

As a baseline, we compare predictions with measurements of all eight biomarkers in the blood drawings. The number of virtual patients in training varied from 20 to 100, with increments of 20, and the time intervals between blood drawings were 1 hour, 2 hours, 3 hours, 4 hours, 6 hours, and 8 hours. The prediction error is measured by root mean squared error (RMSE) on 2,000 virtual patients. Figure 1a and Figure 1b show the RMSE's using different numbers of virtual patients in training, and using different time intervals between blood drawings when number of blood drawings was 3, and with no noise present in measurements. The RMSE's of LR are lower than SVR when the time interval is small. From the figures, we learn that if measurements are noise-free, a short time interval between measuremetns (1 hour) will provide lower errors. We also learn that the model trained on observations from 40 virtual patients performs much better than the model trained on 20 virtual patients. However, training with more than 40 virtual patients has not further reduce prediction error. With uniform noise in range of [-0.02, 0.02] present in the measurements, on Figure 2a and Figure 2b, we learn that the RMSE's of LR and SVR here are very similar. In presence of noise, drawing blood with short time intervals was less accurate, and the effect of noise on larger time intervals was less significant. Larger time intervals between blood drawings are more robust to additive noise. In the case of LR, large time intervals between blood drawings result in lower error.

The obtained results provide evidence that including more virtual patients in training would not reduce errors. Therefore, in the following experiments, the number of virtual patients in training is fixed to 100.

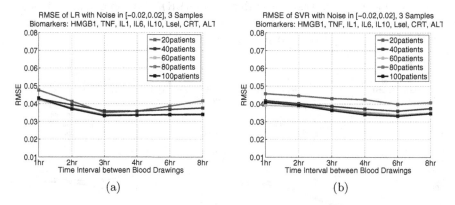

Fig. 2. RMSE's of LR and SVR model, measuring 8 biomarkers at 3 blood drawings with uniform measurement noise in [-0.02, 0.02] range. Results shown are from models trained by using 1, 2, 3, 4, 6, 8 hours interval between blood drawings and on data from 20 to 100 subjects.

Measuring all 8 biomarkers in each blood drawing is not desirable, as the total number of measurements is 24 when taking 3 blood drawings. We would like to obtain similar accuracy by measuring fewer biomarkers. The domain knowledge based approach described in Section 3.2 enabled us to do so. Figure 3a shows the RMSE's of the LR model trained by 3 blood drawings with noise-free, and noisy measurements of HMGB1, TNFα, IL10, and Lsel biomarkers which are related to different internal states that reflect severity of sepsis. In the obtained results, RMSE's in noise-free condition are smaller than the ones in noisy conditions; as the noise level increases, the errors increase. For uniform noise in the [-0.02, 0.02] range errors using these 4 biomarkers are similar to the ones based on all 8 biomarkers (black line in Figure 2a). So, since the number of blood drawings is the same, we could use half the number of measurements to achieve a very similar error. The prediction error when using HMGB1, CRT, and ALT biomarkers is shown in Figure 3b. For noise-free measurements, using these three biomarkers can achieve low error with 1-hour time intervals between blood drawings. For additive uniform noise in [-0.02, 0.02] the errors increases significantly, especially when the time interval between blood drawings is 1 hour. When noise is present, the overall errors using these three biomarkers are significantly higher than the ones when predicting based on 4 biomarkers.

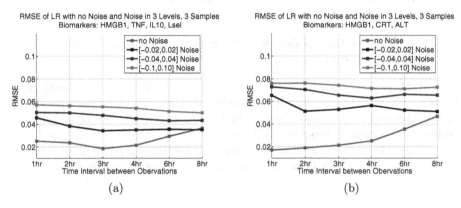

(a)   (b)

Fig. 3. RMSE's in LR model using 4 and 3 biomarkers measured at 3 blood drawings on (a) and (b) respectively. Results are based on models trained on 100 subjects under noise-free, and different noisy conditions, with sampling interval of 1, 2, 3, 4, 6, and 8 hours between blood drawings.

4.2   Using Forward Selection for Biomarkers Identification

The training set consisted of 100 virtual patients and measurements had [-0.02, 0.02] additive uniform noise. The criterion function of the selection procedure was the average RMSE using 5-fold cross validation on the training set. The selection procedure was repeated 20 times. Selected biomarkers are shown in Figure 4. In all the trails, number of biomarker measurements ranges from 7 to 14, number of required blood drawings ranges from 5 to 8. After testing the model on 2000 virtual patients in each trial, the range of RMSE is from 0.0356 to 0.0623. The minimum RMSE is achieved by 12 biomarker measurements from 7 blood drawings. The minimum RMSE is similar to the one achieved by the domain knowledge based approach, but the number of blood drawings is more than twice large (7 v.s 3). We found that about 21% of the biomarkers were selected from the 18th hour; this result consistent with our intuition that recent measurements are very informative for prediction. Other than the 18th hour, selected biomarkers uniformly span the whole 18-hour period.

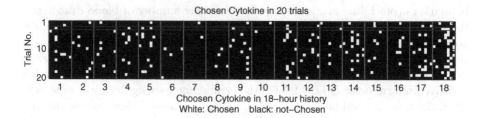

Fig. 4. Biomarker selection using sequential forward feature selection method. A matrix shows the biomarkers selected in 20 trials. The matrix dimension is 20 by 144, where 20 indicates 20 trials and 144 indicates 8 biomakers in 18 hour period (8×18=144).

4.3   Using Lasso Regression for Biomarker Selection and Sepsis Severity Prediction

100 virtual patients with [-0.02,0.02] uniform noise were used for training. 100 different values of the regularization coefficient λ were used to generate models with different numbers of non-zero coefficients. We tested 100 trained linear models (with different non-zero coefficients) on 2,000 virtual patients, and obtained the RMSE of each model. We found that the RMSE's remain low (about 0.035) when models with 12 or more non-zero coefficients (see Figure 5).

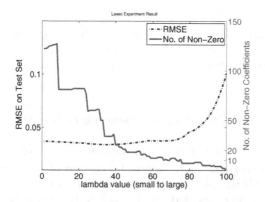

Fig. 5. Lasso regulation for feature selection

Models with 12 non-zero coefficients have RMSE's in range from 0.0347 to 0.0359. In the model achieves minimum RMSE, number of required blood drawings is 6, and number of required biomarker measurements is 12. Although, the minimum RMSE is similar to the one in the domain knowledge based approach, number of blood drawings is twice larger (6 v.s 3).

4.4 Overall Comparison of Different Approaches

The sampling protocol design objectives were low prediction error as well as small number of blood drawings and biomarker measurements. The prediction error (RMSE) of different approaches, their required number of blood drawings, and number of biomarkers measurements are shown at Table 1. Uniform noise in [-0.02, 0.02] range was added to the signal to simulate reality. The smallest RMSE was achieved by measuring all eight observable biomarkers. However, the error was just slightly larger when using only half of measurements selected based on knowledge of sepsis mechanism.

Table 1. Comparison of Different Approaches under Uniform Noise in [-0.02,0.02]

Approach	Best RMSE in test	No. of blood drawings	No. of biomarker measurements
Data-Driven: Forward Selection	0.0356	7	12
Data-Driven:Lasso Regression	0.0347	6	12
Domain Knowledge 8 biomarkers	0.0338	3	24
Domain Knowledge: 4 biomarkers	0.0341	3	12
Domain Knowledge: 3 biomarkers	0.0511	3	9

5 Summary and Conclusion

In this study, we used different approaches to characterize options for obtaining temporal observations of biomarkers in an 18-hour period to predict the value of HMGB1 in the future 24th hour. From the data-driven approaches, we learned that with blood drawings at proper times, 12 biomarker measurements were sufficient to make good predictions. Additional biomarker measurements would not improve the prediction accuracy. Inspired by the data-driven results, we came up with an approach that utilized domain knowledge of the interconnections of biomarkers and important internal states of sepsis progression. Using this approach, we discovered a blood drawing and biomarker measuring protocol which balances the constraints, cost, and accuracy.

Acknowledgment. This work was funded, in part, by DARPA grant [DARPAN66001-11-1-4183] negotiated by SSC Pacific grant.

References

1. Liu, V., Escobar, G.J., Greene, J.D., Soule, J., Whippy, A., Angus, D.C., Iwashyna, T.J.: Hospital Deaths in Patients With Sepsis From 2 Independent Cohorts. Journal of American Medicine Association (May 18, 2014)
2. Russel, J.A.: The current management of septic shock. Minerva Med. 99(5), 431–458 (2008)
3. Ghalwash, M., Obradovic, Z.: A Data-Driven Model for Optimizing Therapy Duration for Septic Patients. In: Proc. 14th SIAM Intl. Conf. Data Mining, 3rd Workshop on Data Mining for Medicine and Healthcare, Philadelphia, PA, USA (April 2014)
4. Ghalwash, M., Radosavljevic, V., Obradovic, Z.: Utilizing Temporal Patterns for Estimating Uncertainty in Interpretable Early Decision Making. In: Proc. 20th ACM SIGKDD Conf. on Knowledge Discovery and Data Mining, New York, NY, USA (August 2014)
5. Ghalwash, M., Ramljak, D., Obradovic, Z.: Patient-Specific Early Classification of Multivariate Observations. International Journal of Data Mining and Bioinformatics (in press)
6. Peng, Z.Y., Wang, H.Z., Carter, M., Dileo, M., Bishop, J.V., et al.: Acute removal of common sepsis mediators does not explain the effects of extracorporeal blood purification in experimental sepsis. Kidney Int. 81, 363–369 (2011)
7. Song, S.O.K., Hogg, J., Peng, Z.-Y., Parker, R.S., Kellum, J.A., Clermont, G.: Ensemble models of neutrophil tracking in severe sepsis. PLoS Computational Biology 8(3) (2012)
8. Ghalwash, M.F., Radosavljevic, V., Obradovic, Z.: Early diagnosis and its benefits in sepsis blood purification treatment. In: IEEE International Conference on Healthcare Informatics (ICHI), International Workshop on Data Mining for Healthcare, Philadelphia, PA, USA (September 2013)
9. Hogg, J.S., Clermont, G., Parker, R.S.: Real-time optimization of cytokine-selective hemoadsorption devices for treatment of acute inflammation. Journal of Critical Care 26(2), e14 (2011)
10. Vapnik, V., Golowich, S.E., Smola, A.: Support vector method for function approximation, regression estimation, and signal processing. In: Advances in Neural Information Processing Systems, pp. 281–287. Morgan Kaufmann Publishers (1997)
11. Chang, C.-C., Lin, C.-J.: LIBSVM: a library for support vector machines. ACM Transactions on Intelligent Systems and Technology (TIST) 2(3), 27 (2011)

Completion Time and Next Activity Prediction of Processes Using Sequential Pattern Mining

Michelangelo Ceci, Pasqua Fabiana Lanotte, Fabio Fumarola,
Dario Pietro Cavallo, and Donato Malerba

Dipartimento di Informatica, University of Bari "Aldo Moro", Bari, Italy
{michelangelo.ceci,pasquafabiana.lanotte,fabio.fumarola,
donato.malerba}@uniba.it, dario.pt.cavallo@gmail.com

Abstract. Process mining is a research discipline that aims to discover, monitor and improve real processing using event logs. In this paper we describe a novel approach that (i) identifies partial process models by exploiting sequential pattern mining and (ii) uses the additional information about the activities matching a partial process model to train nested prediction models from event logs. Models can be used to predict the next activity and completion time of a new (running) process instance. We compare our approach with a model based on Transition Systems implemented in the ProM5 Suite and show that the attributes in the event log can improve the accuracy of the model without decreasing performances. The experimental results show how our algorithm improves of a large margin ProM5 in predicting the completion time of a process, while it presents competitive results for next activity prediction.

1 Introduction

Today, many organizations store event data from their enterprise information system in structured forms such as *event logs*. Examples of such logs are audit trails of workflow management systems, transaction logs from enterprise resource planning systems, electronic patient records, etc.. Here, the goal is not to just collect as much data as possible, but to extract valuable knowledge that can be used to compete with other organizations in terms of efficiency, speed and services. These issues are taken into account in *Process Mining*, whose goal is to discover, monitor and improve processes by providing techniques and tools to extract knowledge from event logs. In typical application scenarios, it is assumed that events are available and each event: (i) refers to an *activity* (i.e., a well-defined step in some process), (ii) is related to a *case* (i.e., a process instance), (iii) can have a *performer* (the actor executing or initiating the activity), and (iv) is executed at a given *timestamp*. Moreover, an event can carry additional process-specific attributes (e.g. the cost associated to the event, the place where the event is performed).

Event logs such as the one shown in Table 1 are used as the starting point for mining. As described in [12], we distinguish four different analyses: (1) *Discovery*, (2) *Conformance*, (3) *Enhancement* and (4) *Operational Support*. In *Discovery*, a process model is discovered based on event logs [7,14]. For example, the α-algorithm [14] mines a process model represented as a *Petri-Net* [4,2] from event

S. Džeroski et al. (Eds.): DS 2014, LNAI 8777, pp. 49–61, 2014.
© Springer International Publishing Switzerland 2014

logs. In *Conformance analysis*, an existing process model is compared with event logs to check and analyze discrepancies between the model and the log. Viceversa, the idea behind *Enhancement* is to extend or improve an existing process model using information about the actual process recorded in some event logs. Types of enhancement can be *extension*, i.e., adding new perspectives to a process model by cross-correlating it with a log, or *repair*, i.e., modify an discovered model to better reflect reality. It is noteworthy that in all the analysis considered above, it is assumed that process mining is done *offline*. Processes are analyzed thereafter to evaluate how they can be improved or extended. On the contrary, *Operational Support* techniques are used in *online* settings. Given a process model built over some event logs and a partial trace, operational support techniques can be used for detecting deviation at runtime (**Detect**), predicting the remaining processing time (**Predict**) and recommending the next activity (**Recommend**).

At the best of our knowledge, classical algorithms presented in the literature for operational support, (1) build a process model in form of Transition Systems [11] or Petri-Nets [4,2], (2) re-analyze the log to extend the model with temporal information and aggregated statistics [13], and finally, (3) learn a regression or a classification model to support prediction and recommendation activities. However, as noted in [5], these operational support methods naturally fit cases where processes are very well-structured (i.e. perfectly matching some predefined schema), for real-life logs they suffer of problems related to "incompleteness" (i.e. the model represents only a small fraction of the possible behavior due to the large number of alternatives), "noise" (i.e., logs containing exceptional/infrequent activities that should not be incorporated in the model), "overfitting" and "under-fitting", thereby resulting in a spaghetti-like model, which is rather useless in practice. Other approaches, such as computational intelligence systems [7], which overcome these problems, tend to be inefficient and, thus, have problems to scale in case of a huge amount of activities which are correlated each other (by means ofprecedence/causality dependencies). In any case, existing solutions do not take into account additional process-specific attribute values which change in running processes.

In this paper we present a novel approach for operational support which deals with the problems presented before, that is, "incompleteness", "robustness to noise" and "overfitting". The solution we propose aims at identifying partial process models to be used for training predictive models. In our approach, two types of predictive models are inferred: for the prediction of the next activity and for the estimation of the completion time. In details, we identify frequent partial processes in form of frequent activity sequences. These sequences are extracted by adapting an efficient frequent pattern mining algorithm and are represented in form of sequence trees. Afterwards, we associate at each node of the tree a specific prediction model that takes into account, in addition to classical attributes (such as the performer of each activity), also additional attributes such as the cost associated to the event or the place where the event is performed. We call this last prediction model "nested". While the sequence mining algorithm allows us to deal with incompleteness, robustness to noise and overfitting by removing

Table 1. An example of event log

CID	Act	Time	Perf	X	Y	CID	Act	Time	Perf	X	Y
1	A	0	p1	x1	y1	4	C	22	p1	-	-
1	B	6	p1	-	-	4	D	28	p1	-	-
1	C	12	p1	-	-	5	A	18	p1	x1	y2
1	D	18	p1	-	-	5	C	22	p2	-	-
2	A	10	p2	x1	y1	5	B	26	p1	-	-
2	C	14	p2	-	-	5	D	32	p2	-	-
2	B	26	p2	-	-	6	A	19	p1	x2	y2
2	D	36	p2	-	-	6	E	28	p3	-	-
3	A	12	p3	x2	y2	6	D	59	p2	-	-
3	E	22	p3	-	-	7	A	20	p1	x2	y1
3	D	56	p3	-	-	7	C	25	p3	-	-
4	A	15	p1	x1	y1	7	B	36	p3	-	-
4	B	19	p1	-	-	7	D	44	p2	-	-

unfrequent behaviors, the nested models guarantee some flexibility. In fact, it is possible to *i)* plug-in any classification/regression learning algorithm and *ii)* enable a different representation of the data, one for each node of the trees. Our solution has its inspiration in works which face with the associative classification task [3], where descriptive data mining techniques are exploited for predictive purposes using a hybrid data mining approach.

The paper is organized as follows: in the next section we describe the proposed approach. Section 3 is devoted to present the empirical evaluation of the proposed solution. Finally, Section 4 concludes the paper and draws some future work.

2 Methodology

This section describes our two-stepped online operational support approach.

First Phase: Process Discovery

In this phase, we look at a (partial) process as a sequence of activities and we apply a sequential pattern mining algorithm in order to generate a partial process model. This model allows us to represent both complete and partial traces which are found frequent by the algorithm. The algorithm we adopt in this phase is FAST [9] which guarantees low computational costs and allows us to represent frequent sequences in a compact way by means of *sequence trees*. FAST, by focusing only on frequent sequences, leads to predictive models (see Section 16) which are robust to noise and do not suffer from overfitting problems. Moreover, since FAST is able to extract frequent non-contiguous (and partial) sequences of activities, we are also able to deal with incompleteness problems.

In this work we extend FAST by allowing it to also extract: *1)* only contiguous (partial) sequences and *2)* only contiguous (partial) sequences that represent only processes since their beginning. Obviously, these extensions generate smaller sequence trees, improve robustness to noise, reduce overfitting, but increase problems related to incompleteness. All these aspects are due to a smaller number of sequence patterns and, thus, to less specialized models. It is noteworthy that the idea of reducing the size of the process model is not new and in [13] the authors convert sequences into sets and multi-sets. However, such approach results in loosing the exact order of events and the number of the occurrences.

In our approach, an event log is represented as a sequence database *(SDB)*, that is, a set of tuples $\langle CID, S \rangle$, where CID is a case id and S is a sequence of ordered set of events (i.e. set of activities). Figure 1a shows the sequence database extracted from the event log reported in Table 1. For instance, in this figure, the cases 1 and 4 are composed by the sequence of activities A, B, C, D.

To formally describe the task solved in this phase, we give some definitions.

Definition 1 (Sub/Super-Sequence). *A sequence $\alpha = \langle a_1, a_2, \ldots, a_k \rangle$ is called a sub-sequence of another sequence $\beta = \langle b_1, b_2, \ldots, b_m \rangle$, denoted as $\alpha \sqsubseteq \beta$, if there exist integers $1 \leq j_1 \leq j_2 \leq \ldots \leq j_k \leq m$ such that $a_1 = b_{j_1}, a_2 = b_{j_2}, \ldots, a_k = b_{j_k}$. For example, if $\alpha = \langle A, B \rangle$ and $\beta = \langle A, C, B \rangle$, where A, C and B are events, then α is sub-sequence of β and β is a super-sequence of α.*

Definition 2 (Frequent sequences). *Let $\alpha = \langle a_1, a_2, \ldots, a_k \rangle$ be a sequence of activities, SDB be a sequence database and minsup a user-defined threshold. α is frequent if its support $\sigma(\alpha, SDB)$ (i.e. the number of sequences in SDB which are super-sequences of α) is greater than minsup.*

Definition 3 (Contiguous frequent sequences). *Let $\alpha = \langle a_1, a_2, \ldots, a_k \rangle$ be a frequent sequence in a sequence database SDB, according to a user-defined threshold minsup. We define the contiguous support of α (denoted as $\sigma_{cs}(\alpha, SDB)$) as the number of sequences in SDB containing α such that, for each $i = 1 \ldots k - 1$, the activity a_{i+1} is observed immediately after the activity a_i. If $\sigma_{cs}(\alpha, SDB) \geq minsup$, then α is a contiguous sequence.*

Definition 4 (Opening frequent sequence). *Let $\alpha = \langle a_1, a_2, \ldots, a_k \rangle$ be a contiguous frequent sequence in a sequence database SDB and minsup a user-defined threshold. We define the opening support of α (denoted $\sigma_{os}(\alpha, SDB)$) as the number of sequences in SDB containing α, and having a_1 in the first position. If $\sigma_{os}(\alpha, SDB) \geq minsup$, then α is an opening sequence.*

Example 1. Given the sequence database SDB in Figure 1a and $minsup = 1$, we analyze the sequence $\alpha = \langle A, B \rangle$. α is frequent because it is present in the tuples with $CID = \{1, 2, 4, 5, 7\}$. Now we check if α can be marked as *contiguous frequent sequence*. The activities A, B appear contiguously only in the tuples with $CID = \{1, 4\}$ (i.e. $\sigma_{cs}(\alpha, SDB) = 2$). Since $\sigma_{cs}(\alpha, SDB) \geq minsup$, α is marked as contiguous frequent sequence. Finally, we check if α can be marked as *opening frequent sequence*. In this case, we have to count how many times the first activity of α (i.e. A) is observed in first position in the SDB and its next activity is B. Since this happens for $CID = \{1, 4\}$ (i.e. $\sigma_{os}(\alpha, SDB) = 2$), α is an opening frequent sequence.

The above definitions allow us to push constraints in the patterns used to build partial process models. Methodologically, we face the following task: given a sequence database SDB, where each sequence represents a sequence of events and given a user-specified minimum support threshold *minsup*, the task of process

CID	Sequence
1	<A,B,C,D>
2	<A,C,B,D>
3	<A,E,D>
4	<A,B,C,D>
5	<A,C,B,D>
6	<A,E,D>
7	<A,C,B,D>

[2]
[3]
NULL
[2]
[3]
NULL
[3]

(a)   (b)

Fig. 1. (a) Sequence Database extracted from Table 1 and (b) VIL for $\langle A, B \rangle$

discovery is to find either: *i)* the frequent sequences in SDB, *ii)* the contiguous frequent sequences in SDB or *iii)* the opening sequences in SDB. In all the cases, sequences are expressed in the form of sequence trees. The original version of FAST can generate a sequence tree of activities in two phases. In the first phase all frequent activities are selected then, in the second phase, these activities are used to populate the first level of the sequence tree and to generate sequences with size greater than one. The mined sequence tree is characterized by the following properties: 1) each node in the tree corresponds to a sequence and the root corresponds to the null sequence ($<>$); 2) if a node corresponds to a sequence s, its children are generated by adding to s the last activity of its siblings. Only frequent children are stored in the tree. Figure 2 shows the sequence tree extracted by FAST from the database in Table 1a ($minsup = 1$).

To represent in optimized way the sequence dataset and to perform efficient support counting of activities and sequences, FAST uses a data structure called *vertical id-list* (VIL). In the following we give a brief definition of a VIL.

Definition 5 (Vertical Id-list). *Let SDB be a sequence database of size* n *(i.e.* $|SDB| = n$*),* $S_j \in SDB$ *its j-th sequence (*$j \in \{1, 2, \ldots, n\}$*), and* α *a sequence associated to a node of the tree, its vertical id-list, denoted as* VIL_α*, is a vector of size* n*, such that for each* $j = 1, \ldots, n$

$$VIL_\alpha[j] = \begin{cases} [posAct_{\alpha,1}, posAct_{\alpha,2}, \ldots, posAct_{\alpha,m}] & \text{if } S_j \text{ contains } \alpha \\ null & \text{otherwise} \end{cases}$$

where $posAct_{\alpha,i}$ *is the end position of the i-th occurrence (*$i \leq m$*) of* α *in* S_j*.*

Example 2. Figure 1b shows the VIL of the sequence $\alpha = \langle A, B \rangle$. Values in VIL_α represent the end position of the occurrences of the sequence α in the sequences of Figure 1a. In particular, the first element (list with only value 2) represents the position of the first occurrence of activity B, after the activity A (i.e. B is the last activity in α), in the first sequence S_1. The second element is (list with only value 3) the position of the first occurrence of B (after A) in the sequence S_2. The third element is null since α is not present in S_3. The other values are respectively list with only value 2 (for sequence S_4), list with only value 3 (for S_5), *null* (for S_6) and list with only value 3 (for S_7).

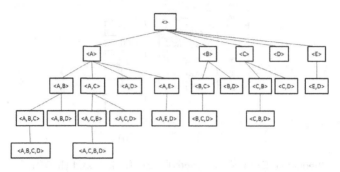

Fig. 2. Sequence tree learned with FAST

Given the sequence α and its sibling β FAST builds a new node γ and its VIL, using VIL_α and VIL_β. In particular, for each $j = 1, \ldots, n$, given: •$VIL_\alpha[j] = [posAct_{\alpha,1}, \ldots, posAct_{\alpha,m_{\alpha,j}}]$,

- an index i (initialized to 1) on $VIL_\alpha[j]$,
- $VIL_\beta[j] = [posAct_{\beta,1}, \ldots, posAct_{\beta,m_{\beta,j}}]$,
- an index z (initialized to 1) on $VIL_\beta[j]$.

FAST checks whether $posAct_{\alpha,i} < posAct_{\beta,z}$. That is, the last activity of the first occurrence of α precedes the last activity of the first occurrence of β. If the condition is not satisfied, FAST increments z. This process is applied until either $posAct_{\alpha,i} < posAct_{\beta,z}$ is satisfied or the *null* value is found. In the first case FAST sets $VIL_\gamma[j] = [posAct_{\beta,z}, \ldots, posAct_{\beta,m_{\beta,j}}]$, otherwise $VIL_\gamma[j] = null$. The support of γ is then computed as: $\sigma(\gamma, SDB) = |\{j \mid VIL_\gamma[j] \neq null\}|$.

In our extension of FAST, we employ the VILs not only in the extraction of frequent sequences, but also in the extraction of contiguous and opening frequent sequences. In particular, the VIL structure is used to generate three different types of trees, representing frequent sequences (see Figure 2), contiguous frequent sequences (see Figure 3a) and opening sequences (see Figure 3b), respectively. It is interesting to note that this last type of sequence tree corresponds to a transition system model [12] obtained from the same event log, after removing unfrequent activities. These three different types of sequence trees are hereafter denoted as (proper) *sequence tree*, *contiguous tree* and *opening tree*. The contiguous tree is obtained by substituting, in the original implementation of FAST, the common definition of support with $\sigma_{cs}(n, SDB) = |\{j|vil = contiguous(n, T) \land vil[j] \neq NULL\}|$. In this formula, $contiguous(n, T)$ is the VIL returned by the application of the function described in Algorithm 1. This algorithm iteratively looks for possible holes in the sequences by bottom-up climbing the sequence tree. A hole is found when the condition at line 10 is not satisfied. Similarly, the opening tree is obtained by substituting, in FAST, the common definition of support with $\sigma_{os}(n, SDB) = |\{j|vil = contiguous(n, T) \land vil[j][1] = 1\}|$.

Second Phase: Nested-Model Learning

Once the partial process model is learned, it is used to build a nested model for operational support. In particular, for each node α of the partial process model, the description of the processes, which contribute to the support of the

(a) (b)

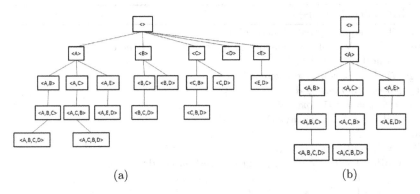

Fig. 3. (a) Contiguous and (b) Opening trees associated to the tree in Figure 2

$CID_{<A>}$	$Perf_{<A>}$	$X_{<A>}$	$Y_{<A>}$	$Completion_{<A>}$	$Next_{<A>}$
1	p1	x1	y1	18	B
2	p2	x1	y1	26	C
3	p3	x2	y2	44	E
4	p1	x1	y1	13	B
5	p1	x1	y2	14	C
6	p1	x2	y2	40	E
7	p1	x2	y1	24	C

$CID_{<A,B>}$	$Perf_{<A,B>}$	$Completion_{<A,B>}$	$Next_{<A,B>}$	$Perf_{<A>}$	$X_{<A>}$	$Y_{<A>}$
1	p1	12	C	p1	x1	y1
4	P1	9	C	p2	x1	y1

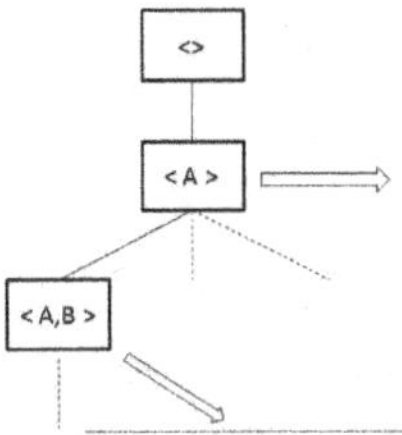

Fig. 4. Datasets associated with nodes $\langle A \rangle$ and $\langle A, B \rangle$

pattern expressed at that node, is used as a training set for a predictive learning algorithm. Learned prediction models are then used to predict the completion time and the next activity of a running process.

In the generation of the training set associated with each node α, all the informative attributes associated with each event in α, and available in an event log, can be used. Uninformative attributes, such as the CID, are removed, while two additional attributes, that is, *Completion time* and *Next Activity* can be created and populated for predictive purposes. Figure 4 shows the datasets associated with the nodes representing $\langle A \rangle$ and $\langle A, B \rangle$ of the opening tree reported in Figure 3b. Obviously, the attribute *Completion time* is only available when training a regression model for *Completion time*, while the attribute *Next Activity* is only available when training a classification model for *Next Activity*. In the construction of the nested prediction models, any traditional machine learning algorithm for regression/classification can be used.

In the prediction phase, the algorithm traverses the (proper, contiguous or opening) frequent sequence tree in order to identify the regression/classification model to be used. The search starts from the root and proceeds towards the leaves of the sequence tree. For each new activity of a running process, the algorithm moves to the corresponding next level of the tree. If there is no corresponding

---

Algorithm 1. contiguous(T,n)

---

    input : T: a sequence tree; n: node of T
    output: vil: the VIL of the sequence at node n such that the contiguous
             condition is satisfied.

1   $vil =$ getVIL(n); $parent =$ getParent(n);
2   **while** $parent \mathrel{!=} root(T)$ **do**
3       $parentVil =$ getVIL($parent$);
4   **foreach** $j = 1 \ldots$ length(vil)$;$
5   **do**
6           $i = 0$; $contiguous =$ FALSE;
7   **while** $++i < len(vil[j])$ *and not contiguous* **do**
8              $z = 0$;
9   **while** $++z \le i$ *and not contiguous* **do**
10   **if** $vil[j][i] = parentVil[j][z] + 1$ **then**
11                    $vil[j] = parentVil[j][z..(len(parentVil[j]) - 1)]$;
12                    $contiguous = TRUE$;

13   **if** *not contiguous* **then**
14                 $vil[j] = NULL$;

15     $n = parent$; $parent =$ getParent(n);
16 **return** vil;

---

node, that is, the complete sequence was not found frequent during the partial
process model construction, the algorithm does not move to the next level and
remains in the current node until a new activity of the running process allows us
to move to the next level. At each point of the running process, the prediction
model associated to the current node is used for predictive purposes.

Example 3. Let $p = \langle A, B, E, C \rangle$ be the running process for which we intend to
predict either the next activity or the completion time. Let the sequence tree in
Figure 3b be the learned partial model. Starting from the root of the tree, when
the first activity of the process p arrives, the algorithm moves first in the node
associated with the sequence $\langle A \rangle$, then, when the second activity of the process
p arrives, it moves in the child node associated with $\langle A, B \rangle$. Since no child node
of $\langle A, B \rangle$ associated with the sequence $\langle A, B, E \rangle$ exists, when the third activity
of the process p arrives, the algorithm remains in the current node. When the
next activity C of p arrives, the algorithm moves in the node associated with
the pattern $\langle A, B, C \rangle$ and uses the nested models in this node for prediction.

3   Experiments

In this section we present the empirical evaluation of the proposed algorithm. For
evaluation purposes, we used a 10-fold cross-validation schema and, we collected
the average classification rate (C-RATE, i.e. the number of processes for which

Table 2. Number of sequences extracted during the sequential pattern mining phase for ProM and THINK3

minsup	40%			30%			25%		
tree type	1	2	3	1	2	3	1	2	3
Prom	607	31	6	735	53	11	4671	66	13
Think3	104	26	7	208	40	9	855	68	21

we were able to obtain predictions), the average predictive accuracy of the next activity and the error in the completion time estimation. As for this last error, we use the symmetric mean absolute percentage error (SMAPE) defined in Equation (1), since it is more robust to effect of zero or near-zero values than traditional error measures [6] and the classical root relative mean squared error (RRMSE) defined in Equation (2) [10]. In both equations y_i is the actual completion time, $\hat{y}_i$ is the estimated completion time and $\overline{y}_i$ is the average completion time.

$$SMAPE = \left(\sum_{i=1}^{n} |\hat{y}_i - y_i|\right) / \left(\sum_{i=1}^{n} (\hat{y}_i + y_i)\right) \tag{1}$$

$$RRMSE = \sqrt{\left(\sum_{i=1}^{n} (\hat{y}_i - y_i)^2\right) / \left(\sum_{i=1}^{n} (\overline{y}_i - y_i)^2\right)} \tag{2}$$

In our implementation, we use as nested learning algorithms C4.5 [8] (to predict the next activity) and M5' [15] (to estimate the completion time). Results are collected by varying the minimum support threshold and the type of the sequence tree. We denote proper sequence trees with "tree type 1", contiguous trees with "tree type 2" and opening trees with "tree type 3". Completion time results obtained with our approach are compared with results obtained from the transition systems implemented in ProM5 Suite, where the prediction is made based on the average time to completion for process instances in a similar state [13]. Unfortunately, the ProM5 Suite does not include tools for next activity prediction. Since Tree type 3 is, as stated before, the model more similar to a transition system, we consider this setting as baseline for our comparisons.

The evaluation is performed on two real datasets that is, ProM and THINK3. ProM concerns repairing telephones of a communication company. The event log contains 11,855 activities and 1,104 cases, while the number of distinct performers is 29. Activities are classified as complete (1,343), schedule (6,673), resume (178), start (809), suspend (166) and unknown (remaining). Additionally, ProM stores several properties like name, timestamp, resources (in term of roles). The second dataset, THINK3 [1] is an event log presenting 353,490 cases in a company, for a total of 1,035,119 events executed by 103 performers. Activities are classified as administrator tools (131), workflow (919,052), namemaker (106,839), delete (2,767), deleteEnt (2,354), prpDelete (471), prpSmartDelete (53), prpModify (34) and cast (1,430).

Results on ProM are extracted by using three minimum support thresholds: 0.4, 0.3 and 0.25. Table 3 shows results for both the considered predictive tasks. As it is possible to see, even if we consider a partial process model, we are

Table 3. Averaged cross-validation results for ProM with different tree type (tree type). Column "gain" indicates the RRMSE gain over the ProM5 Suite for completion time prediction.

tree type	minsup	Completion time prediction			Next activity prediction	C-RATE
		RRMSE	SMAPE	gain(%)	ACCURACY	
1	40%	0.71	0.19	29%	0.72	1.00
	30%	0.70	0.20	30%	0.74	1.00
	25%	0.69	0.19	31%	0.78	1.00
2	40%	0.83	0.24	17%	0.60	1.00
	30%	0.69	0.19	31%	0.64	1.00
	25%	0.69	0.20	31%	0.68	1.00
3	40%	0.83	0.24	17%	0.60	1.00
	30%	0.69	0.19	31%	0.64	1.00
	25%	0.69	0.20	31%	0.68	1.00

Table 4. Average running times for ProM (sec.)

tree type	minsup	seq. pattern discovery	datasets generation	construction nested model	total
1	40%	0.448	7.783	2.898	11.129
	30%	0.498	8.514	2.965	11.977
	25%	0.863	162.313	2.892	166.068
2	40%	0.448	5.478	3.155	9.081
	30%	0.498	5.892	3.707	10.097
	25%	0.863	6.666	3.434	10.963
3	40%	0.448	6.008	3.245	9.701
	30%	0.498	5.853	3.497	9.848
	25%	0.863	6.352	3.452	10.667

able to provide a prediction for (almost) all the sequences (C-RATE). Moreover, we can observe that trees of type 2 and 3 are more robust to noise and to incompleteness with respect to trees of type 3. For the next activity prediction task, by reducing minsup, predictive accuracy increases. Moreover, the partial model based on propers sequence trees (tree type 1) leads to the best results. By comparing these results with those reported in Table 2, we can see that best results are obtained with the most complete trees. This means that $minsup=0.25$ is still enough to do not suffer from overfitting problems. As for completion time prediction, we show that results do not change significantly varying the tree type. Best results are obtained with proper sequence trees (i.e tree type 1) and $minsup=0.25$; contiguous trees (i.e. tree type 2) and $minsup=0.3$; opening trees (i.e. tree type 3) and $minsup=0.3$. This means that, in this case, the abstraction introduced in trees of type 2 and 3 is beneficial. Moreover, the comparison with the ProM5 Suite shows that our approach leads to reduce the error of a great margin (between 17% and 31% of the RRMSE). By analyzing results reported in Table 4, we see that the generation of trees of types 2 and 3 is significantly more efficient than that of trees of type 1. This means that, while for next activity prediction, high running times of tree type 1 are justified by effectiveness, this is not true for completion time prediction, where tree types 2 and 3 are the best solutions in terms of efficiency and effectiveness.

Table 5. Averaged cross-validation results for THINK3 with different configurations (conf.). Column "gain" indicates the RRMSE gain over the ProM5 Suite for completion time prediction.

tree type	minsup	Completion time prediction			Next activity prediction	C-RATE
		RRMSE	SMAPE	gain(%)	ACCURACY	
	15%	0.91	0.49	9%	0.51	1.00
1	10%	0.91	0.49	9%	0.54	1.00
	5%	0.97	0.73	3%	0.62	1.00
	15%	0.94	0.41	6%	0.49	1.00
2	10%	0.89	0.39	11%	0.49	1.00
	5%	0.92	0.47	8%	0.54	1.00
	15%	0.95	0.49	5%	0.49	0.99
3	10%	0.94	0.44	6%	0.49	0.99
	5%	0.94	0.41	6%	0.54	1.00

Table 6. Average running times for THINK3 (sec.)

tree type	minsup	seq. pattern discovery	datasets generation	construction nested model	total
	15%	0.848	413.355	67.067	481.270
1	10%	2.081	443.652	69.628	515.361
	5%	3.573	572.952	69.970	646.495
	15%	0.848	343.444	74.323	418.615
2	10%	2.081	351.362	72.796	426.239
	5%	3.573	369.396	74.447	447.416
	15%	0.848	393.726	73.169	467.743
3	10%	2.081	461.615	92.444	556.140
	5%	3.573	567.422	127.741	698.736

Results on THINK3 are obtained with three minimum support thresholds: 0.05, 0.1 and 0.15. In Table 5, we show results obtained for both the considered predictive tasks. Also in this case our approach is able to provide a prediction for (almost) all the sequences (C-RATE). Similarly to what observed for the ProM dataset, best results for the next activity prediction are obtained with proper sequence trees (i.e. tree type 1). As for the prediction of the completion time, the best results are obtained with the contiguous trees (i.e. tree type 2, *minsup=0.1*). This setting is also one of the best settings in terms of running times (see Table 6). Moreover, the comparison with the ProM5 Suite shows that our approach leads, as in the case of ProM data, to reduce the RRMSE in all cases (up to 11%). This is an interesting result if we consider that the THINK3 dataset has less attributes than the ProM dataset.

4   Conclusions and Future Works

This paper faces the problem of operational support in process mining and, in particular, the prediction of the next activity and of the completion time. The proposed approach is two-stepped and combines descriptive data mining for partial model mining and predictive data mining for mining nested classification/regression models. This solution provides incompleteness-robust and non-overfitted prediction models thanks to the first phase, where a tailored sequential pattern mining algorithm is adopted. Moreover, with this method, we can apply any traditional classification/regression techniques thanks to the construction of the nested model. As can be seen, using this approach, completion time predictions can be significantly improved over state-of-the-art applications (approximately 30% with ProM Data and 11% with THINK3 Data).

For future work, we intend to extend the experiments with additional (noisy) cases, to check the effectiveness of the proposed approach to noise, and to exploit "closed" sequential pattern mining instead of frequent sequential pattern mining to further reduce the number of nested models to learn. Moreover, we intend to consider the use of other algorithms for sequential pattern mining with constraints, in addition to FAST as well as give more importance to recent activities in the model construction, as typically done in data stream mining.

Acknowledgements. This work fulfils the research objectives of the UE FP7 project MAESTRA (Grant number ICT-2013-612944). This work is also partially supported by the Italian Ministry of Economic Development (MISE) through the project LOGIN.

References

1. Appice, A., Ceci, M., Turi, A., Malerba, D.: A parallel, distributed algorithm for relational frequent pattern discovery from very large data sets. Intell. Data Anal. 15(1), 69–88 (2011)
2. Carmona, J., Cortadella, J., Kishinevsky, M.: A Region-Based Algorithm for Discovering Petri Nets from Event Logs. In: Dumas, M., Reichert, M., Shan, M.-C. (eds.) BPM 2008. LNCS, vol. 5240, pp. 358–373. Springer, Heidelberg (2008)
3. Ceci, M., Appice, A.: Spatial associative classification: propositional vs structural approach. J. Intell. Inf. Syst. 27(3), 191–213 (2006)
4. Dongen, B., Busi, N., Pinna, G., Aalst, W.: An Iterative Algorithm for Applying the Theory of Regions in Process Mining. In: Proceedings of the Workshop on Formal Approaches to Business Processes and Web Services, pp. 36–55 (2007)
5. Folino, F., Greco, G., Guzzo, A., Pontieri, L.: Mining usage scenarios in business processes: Outlier-aware discovery and run-time prediction. Data Knowl. Eng. 70(12), 1005–1029 (2011)
6. Hyndman, R.J., Koehler, A.B.: Another look at measures of forecast accuracy. International Journal of Forecasting, 679–688 (2006)
7. Medeiros, A.K., Weijters, A.J., Aalst, W.M.: Genetic process mining: An experimental evaluation. Data Min. Knowl. Discov. 14(2), 245–304 (2007)
8. Quinlan, J.R.: C4.5: Programs for Machine Learning. Morgan Kaufmann Publishers Inc., San Francisco (1993)
9. Salvemini, E., Fumarola, F., Malerba, D., Han, J.: FAST sequence mining based on sparse id-lists. In: Kryszkiewicz, M., Rybinski, H., Skowron, A., Raś, Z.W. (eds.) ISMIS 2011. LNCS, vol. 6804, pp. 316–325. Springer, Heidelberg (2011)
10. Stojanova, D., Ceci, M., Appice, A., Malerba, D., Džeroski, S.: Global and local spatial autocorrelation in predictive clustering trees. In: Elomaa, T., Hollmén, J., Mannila, H. (eds.) DS 2011. LNCS, vol. 6926, pp. 307–322. Springer, Heidelberg (2011)

11. van der Aalst, W.M.P.: Process Mining: Discovery, Conformance and Enhancement of Business Processes, 1st edn. Springer Publishing Company, Incorporated (2011)
12. van der Aalst, W.M.P., Pesic, M., Song, M.: Beyond process mining: From the past to present and future. In: Pernici, B. (ed.) CAiSE 2010. LNCS, vol. 6051, pp. 38–52. Springer, Heidelberg (2010)
13. van der Aalst, W.M.P., Schonenberg, M.H., Song, M.: Time prediction based on process mining. Inf. Syst. 36(2), 450–475 (2011)
14. van der Aalst, W.M.P., Weijter, A., Maruster, L.: Workflow mining: Discovering process models from event logs. IEEE Transactions on Knowledge and Data Engineering 16, 2004 (2003)
15. Wang, Y., Witten, I.H.: Induction of model trees for predicting continuous classes (1996)

Antipattern Discovery
in Ethiopian Bagana Songs

Darrell Conklin[1,2] and Stéphanie Weisser[3]

[1] Department of Computer Science and Artificial Intelligence
University of the Basque Country UPV/EHU, San Sebastián, Spain
[2] IKERBASQUE, Basque Foundation for Science, Bilbao, Spain
[3] Université Libre de Bruxelles, Brussels, Belgium

Abstract. This paper develops and applies sequential pattern mining to a corpus of songs for the bagana, a large lyre played in Ethiopia. An important aspect of this repertoire is the unique availability of rare motifs that have been used by a master bagana teacher in Ethiopia. The method is applied to find antipatterns: patterns that are surprisingly rare in a corpus of bagana songs. In contrast to previous work, this is performed without an explicit set of background pieces. The results of this study show that data mining methods can reveal with high significance these antipatterns of interest based on the computational analysis of a small corpus of bagana songs.

1   Introduction

Sequences are a special form of data that require specific attention with respect to alternative representations and data mining techniques. Sequential pattern mining methods [2,1,19] can be used to find frequent and significant patterns in datasets of sequences, and also sequential patterns that contrast one data group against another [16,23]. In music, sequential pattern discovery methods have been used for the analysis of single pieces [9], for the analysis of a corpus of pieces [8], and also to find short patterns that can be used to classify melodies [21,22,18,7].

Further to standard pattern discovery methods, which search for frequent patterns satisfying minimum support thresholds [24], another area of interest is the discovery of rare patterns. This area includes work on rare itemset mining [14] and negative association rules [4]. For sequence data, rare patterns have not seen as much attention, but are related to *unwords* [15] in genome research (i.e. absent words that are not subsequences of any other absent word), and *antipatterns* [10] in music (patterns that are surprisingly rare in a corpus of music pieces). Antipatterns may represent structural constraints of a music style and can therefore be useful for classification and generation of new pieces.

The Ethiopian lyre *bagana* is played by the Amhara, inhabitants of the Central and Northern part of the country. The bagana is a large lyre, equipped with ten gut strings, most of them plucked with the fingers. According to Amhara tradition, the bagana is the biblical instrument played by King David and was

S. Džeroski et al. (Eds.): DS 2014, LNAI 8777, pp. 62–72, 2014.
© Springer International Publishing Switzerland 2014

brought to Ethiopia, together with the Ark of Covenant, by the legendary Emperor Menelik, mythical son of King Solomon and Queen of Sheba. The bagana belongs to the spiritual sphere of Amhara music, even though it is not played during liturgical ceremonies. Because of its mythical origin and connection to the divine, the bagana is highly respected, as the instrument of kings and nobles, played by pious men and women of letters [26].

The analysis of the learning process used by the most revered player, Alemu Aga, has shown that the first phase of this process is based on exercises composed of short motifs [25]. These exercises correspond, according to Alemu Aga, to motifs that are either frequently or rarely encountered in his real bagana songs. They are meant to familiarize the student with the playing technique, the numbered notational system (see below) as well as with the sound colour of the instrument, which is, due to its buzzing quality, unique in the Amhara musical systems.

The study of a bagana corpus provides a unique opportunity for evaluation of pattern discovery techniques, because there exist known rare motifs that also have functional significance. The aim of this paper is to explore whether sequential pattern discovery methods, specifically methods for the discovery of rare or absent patterns in music [10], can reveal the known rare patterns and possibly other rare patterns in a corpus of bagana songs.

Fig. 1. The Ethiopian lyre bagana

2   Bagana Background

This section provides some background on the Ethiopian bagana, presenting how the fingers are assigned to strings, the tuning and scales of the bagana, and finally the encoding of a corpus of bagana songs for computational analysis.

2.1   Bagana Notation

The bagana has 10 strings which are plucked by the left hand, with the fingers numbered from 1 (thumb) to finger 5 as described in Table 1.

Table 1. Fingering of the bagana, with finger numbers assigned to string numbers

string	1	2	3	4	5	6	7	8	9	10
fingering	1	r	2'	2	r	3	r	4	r	5

In Table 1, "r" (for "rest") indicates a string that is not played, but rather is used as a rest for the finger after it plucks the string immediately next to it. Strings 3 and 4 are both played by finger number 2 (string 3 being therefore notated as finger 2'), otherwise the assignment of finger number to string number is fixed.

Table 2. Tuning of the bagana, in two different scales, and the nearest Western tempered note corresponding to the degrees of the scales

finger	1	2' or 2	3	4	5
string	1	3 or 4	6	8	10
scale tezeta	E/F	C	D	A	G
scale anchihoye	F	C	D♭	G♭	A
scale degree	$\hat{3}$	$\hat{1}$	$\hat{2}$	$\hat{5}$	$\hat{4}$

As with the other Amhara instruments, the bagana is tuned to a traditional pentatonic scale. Usually, the player chooses between the tezeta scale and the anchihoye scale. Tezeta is anhemitonic (without semitones) and is relatively close to Western tempered degrees (see Table 2). Anchihoye, however, is more complex and comprises two intervals smaller than the Western tempered tone.

To illustrate the notation in Table 2, for example, in the tezeta scale the ascending pentatonic scale (C, D, F, G, A) would be notated by the sequence (2, 3, 1, 5, 4). The scale degree notation is also useful to measure the diatonic interval between two strings, as will be used in Section 4.

Figure 2 shows the placement of the left hand on the 10 bagana strings, along with the information from Table 2: the finger numbering used, and the notes played by the fingers.

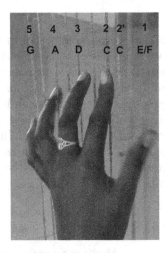

Fig. 2. Placement of left hand on the strings of the bagana

2.2   Bagana Corpus

Bagana songs, also called *yebagana mezmurotch* in Amharic, are based on a relatively short melody, repeated several times with different lyrics, except for the refrain (*azmatch*) for which the lyrics do not vary. Bagana songs are usually preceded by instrumental preludes, called *derdera* (pl. *derderotch*). The analyzed corpus comprises 29 melodies of bagana songs performed by 7 players (5 men, 2 women), and 8 derderotch. These 37 pieces were recorded by Weisser [25] between 2002 and 2005 in Ethiopia (except for 2 of them recorded in Washington DC). In this paper, no differentiation will be made between derdera and bagana songs. A total of 1903 events (finger numbers) are encoded within the 37 pieces (events per song: $\mu = 51$, $\sigma = 30$, min $= 13$, max $= 121$). Figure 3 shows an example of a fragment of a bagana song, encoded as a sequence of finger numbers, corresponding to the fingering of the song.

Fig. 3. A short fragment encoded in score and finger notation, from the beginning of the song Abatachen Hoy ("Our Father"), one of the most important bagana songs, as performed by Alemu Aga (voice not shown). Transcribed in 2006 (see [27]).

2.3   Rare Motifs

Table 3 shows four motifs that correspond, according to the bagana master Alemu Aga, to motifs that are rarely encountered in his real bagana songs and are used during practice to strengthen the fingers with unusual finger configurations [25]. The first two motifs in Table 3 are short bigram patterns. In Section 4 it will be explored whether these two rare patterns can be discovered from corpus analysis. The third and fourth motifs of Table 3 (bottom) correspond to longer pentagram patterns that form ascending and descending pentatonic scales and are also used for didactic purposes. Since most pentagram patterns will be rare in a small corpus, an additional question that will be explored in Section 4 is whether these two pentagram patterns are surprisingly rare.

Table 3. Rare motifs, from [25], page 50

	Motifs in numeric notation
First exercise	$(1, 4)$
Second exercise	$(1, 2)$
Third exercise	$(2, 3, 1, 5, 4)$
Fourth exercise	$(4, 5, 1, 3, 2)$

3   Antipattern Mining

In this work we apply data mining to discover antipatterns in the bagana corpus. The task is an instance of *supervised descriptive rule discovery* [20], a relatively new paradigm for data mining which unifies the areas of subgroup discovery [17], contrast set mining [6,12], and emerging pattern mining [13].

Referring to Figure 4, in the supervised descriptive mining paradigm, data may be partitioned into two sets, an analysis class $\oplus$ with $n^{\oplus}$ objects, and a background set $\ominus$ with $n^{\ominus}$ objects. The partitioning is flexible and the background set may contain instances labelled with multiple different classes. A *pattern* is a predicate that is satisfied by certain data objects. The number of occurrences of a pattern P in the set $\oplus$ is given by $c_P^{\oplus}$, and in the set $\ominus$ by $c_P^{\ominus}$. The goal is to discover patterns predictive of the $\oplus$ class, covering as few of the $\ominus$ objects as possible. If under- rather than over-represented patterns are desired, the reversal of the roles of the analysis and background classes $\ominus$ and $\oplus$ can naturally lead to the discovery of patterns frequent in $\ominus$ and rare or absent from $\oplus$ [10]. In this case the inner box of Figure 4 would be shifted downwards into the $\ominus$ region.

In the original studies of subgroup discovery and contrast data mining [17,6,13] objects and subgroups are described using attribute-value representations. Later work has shown that contrast data mining can be applied to sequence data: Ji et al. [16] consider *minimal distinguishing subsequence patterns* and Deng and Zaïane [11] consider *emerging sequences* (sequential patterns frequent in one group but infrequent in another).

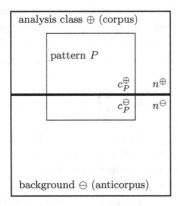

Fig. 4. The schema for contrast set mining, showing the major regions of objects involved. The top part of the outer box encloses data labelled with the class of interest, below this the background. The inner box contains the objects described by a pattern, and the top part of the inner box the contrast set described by a discovered pattern.

Music can be represented as sequences of events for the purposes of supervised descriptive data mining. In music the analysis and background set are called the *corpus* and *anticorpus* (Figure 4). In pattern discovery in music, the counting of pattern occurrences can be done in two ways [9]: either by considering *piece count* (the number of pieces containing the pattern one or more times, i.e. analogous to the standard definition of pattern support in sequential pattern mining) or by considering *total count* (the total number of positions matched by the pattern, also counting multiple occurrences within the same piece). The latter is used when a single music piece is the target of analysis. For the bagana, even though several pieces are available, total count is used, because we consider that a motif is frequent (or rare) if it is frequently (or rarely) encountered within any succession of events. Therefore in this study the set $\oplus$ (resp., $\ominus$) comprises all *suffixes* in the corpus (anticorpus), and practically $n^{\oplus}$ ($n^{\ominus}$) is therefore the total number of suffixes in the corpus (anticorpus), and $c_P^{\oplus}$ ($c_P^{\ominus}$) the total number of sequences in $\oplus$ ($\ominus$) for which the pattern P is a prefix. In this study overlapping pattern occurrences are excluded from the total count of a pattern.

For antipattern mining of bagana songs, unlike for previous antipattern mining studies with Basque folk songs [10], an interesting and important feature is that there is no naturally available anticorpus to contrast with the corpus. Therefore a different method was needed to reveal those patterns whose count within the corpus is significantly low, evaluated here using a binomial distribution of pattern counts along with a zero-order model of finger numbers to compute the pattern probabilities.

3.1 Patterns and Expectation

In this study of the bagana corpus, a *pattern* is a contiguous sequence of events, represented by finger numbers. To contrast the occurrences of a pattern between a set $\oplus$ and a set $\ominus$, the *empirical background probability* of a pattern $P = (e_1, \ldots, e_\ell)$ may be computed simply as $c_P^\ominus / n^\ominus$, if a large background set $\ominus$ is available [13]. Without a large background corpus, the background probability of the pattern must be estimated analytically, for example using a zero-order model of the corpus:

$$b_P = \prod_{i=1}^{\ell} c_{e_i}^\oplus / n^\oplus$$

where $c_{e_i}^\oplus$ is the total count of event e_i, and $n^\oplus$ is the total number of events in the corpus. The background probability b_P therefore gives the probability of finding the pattern in exactly ℓ contiguous events.

A useful quantity derived from the background probability is the *expected total count*. Letting X be the random variable that models the total count of a pattern P, the expected total count is:

$$\mathbb{E}(X) = b_P \times t_P$$

where t_P is the maximum number of non-overlapping positions that can be possibly matched by the pattern, approximated here simply by $n^\oplus / \ell$.

In this study a zero-order analytic model of the corpus is used to permit the detection of over- or under-representation in bigram or longer patterns. A first- or higher-order analytic model would not be able to detect bigram patterns because the expected total count of a pattern would be equivalent to its actual count.

3.2   Antipatterns and Statistics

An *antipattern* is a pattern that is rare, or even absent, in a corpus. For data mining, this definition is not operational because almost any sequence of events is an antipattern, that is, most possible event sequences will never occur in a corpus, with their count rapidly falling to zero with increasing length. Most of these patterns are not interesting because it is expected that their total count is zero. Therefore we want to know which are the *significant* antipatterns: those that are *surprisingly* rare or absent from a corpus.

Antipatterns are evaluated according to a *p*-value, which gives the probability of finding an equal or fewer number of occurrences than the number observed. Low *p*-values are desired, because it means such patterns are surprisingly rare in the corpus. The *p*-value of finding $c_P^\oplus$ or fewer occurrences in the corpus is modelled using the binomial distribution:

$$\mathbb{P}(X \le c_P^\oplus) = \sum_{i=0}^{c_P^\oplus} \mathbb{B}(i, t_P, b_P) \tag{1}$$

where $\mathbb{B}(i, t_P, b_P)$ is the binomial probability of finding exactly i occurrences of the pattern P, in t_P possible placements of the pattern, and b_P is the background probability of the pattern. Low *p*-values indicate patterns that are statistically surprising and therefore potentially interesting.

3.3   Discovery Algorithm

The antipattern discovery task is stated simply as: given a corpus, and a significance level α, find all patterns P with a p-value (Equation 1) of at most α:

$$\mathbb{P}\big(X \leq c_P^{\oplus}\big) \leq \alpha \tag{2}$$

Furthermore, for presentation we consider only those significant antipatterns that are *minimal*, that is, those that are not contained within any other significant antipattern [10].

The discovery of minimal significant antipatterns can be efficiently solved by a refinement search of pattern space [8,10], using a method similar to the SPAM algorithm [5]. A depth-first search starts at the most general (empty) pattern, and the search at a particular node of a search tree is continued only while the pattern is not significant. In this work only the S-step refinement operator [5], which extends a sequential pattern on the right hand side by one element, is used: an I-step is not necessary because events here have only one feature (the finger number).

The complexity of the antipattern discovery algorithm is determined by the significance level α, because with low α the search space must be more deeply explored before a significant pattern is reached. Nevertheless, similar to the statistical significance pruning method of Bay and Pazzani [6] who evaluate contrast sets using a χ^2 statistic, it is possible to compute the minimal p-value (Equation 1) achievable on a search path. This can lead to the pruning of entire paths that will not visit a pattern meeting the significance level of α.

4   Results and Discussion

The pattern discovery method described in Section 3 was used to find all minimal antipatterns at the significance level of $\alpha = 0.01$ (Equation 2). The method revealed exactly ten significant antipatterns (Table 4): five patterns and their retrogrades (reversal). The third column shows the total count of the pattern, and in brackets their piece count (number of pieces containing the pattern one or more times). Interestingly, all minimal antipatterns are bigrams. The two patterns (4,1) and (2,1) presented at the top part of Table 4 (with their retrogrades, which are also significant) are the *most significant* antipatterns discovered, and correspond to the retrogrades of two of the didactic rare motifs (Table 3).

The second column of Table 4 presents the undirected diatonic interval formed by the pattern, in the tezeta scale (Table 2). Interestingly, all of the discovered antipatterns form a melodic interval of a major third or greater (P4, M3, and P5).

It is worth noticing that these results prove the rarity of the interval of fifths (perfect in tezeta, diminished and augmented in anchihoye). According to authoritative writings in ethnomusicology [3], the perfect fifth and the cycle of fifths play a founding role in anhemitonic pentatonic scales such as tezeta. The rarity of the actual fifths in the songs is therefore significant, and it can be speculated

Table 4. Bagana antipatterns discovered at $\alpha = 0.01$. Top: known rare bigram patterns; middle: novel rare patterns; bottom: pentagram patterns of Table 3

P	diatonic interval	$c_P^{\oplus}$	$\mathbb{E}(X)$	p-value
(4,1)	P4	2 (2)	48	6.3e-19
(1,4)		21 (14)	48	7.5e-06
(2,1)	M3	6 (5)	50	1.8e-15
(1,2)		13 (7)	50	2.6e-10
(4,3)	P5	2 (2)	30	3.8e-11
(3,4)		3 (3)	30	4.0e-10
(5,2)	P5	17 (10)	38	6.6e-05
(2,5)		16 (9)	38	2.7e-05
(3,5)	P4	5 (4)	26	5.7e-07
(5,3)		11 (8)	26	0.00077
(2,3,1,5,4)	ascending scale	8 (7)	0.11	1
(4,5,1,3,2)	descending scale	4 (3)	0.11	1

that this interval is a mental reference that is never (and does not necessarily need to be) performed. Similarly, the (3,5), a perfect fourth, is the inversion, i.e. the interval to be added to another one to constitute an octave. Intervals and their inversions are usually connected in several musical cultures, including Western art music.

For completeness with the results of Weisser [25], at the bottom of Table 4 are the two pentagram patterns from Table 3. As expected, these patterns are not frequent, but surprisingly they are not significant according to their p-value (Equation 1) (therefore they are not found by the pattern discovery method). In fact, they occur in the corpus many times more than their expected total count.

From a small corpus of bagana songs, antipattern discovery is able to find the two published rare bigram motifs. The results suggest several directions for future studies. The novel antipatterns (3,4), (2,5), and (3,5) (with their retrogrades) found by the method (Table 4) may have new implications for the study of didactic and distinctive motifs of the bagana. The study of these patterns is left for future work. Further, in this study only melodic aspects have been considered, and not rhythmic aspects. Though good results have been obtained with melodic information only, rhythmic patterns could also be of interest, especially when linked with melodic aspects, although data sparsity for this corpus may become an issue. A transcription and encoding of rhythmic information for the corpus from bagana recordings is in progress. It is also planned to explore positive pattern as well as antipattern discovery, partitioning the corpus in different interesting ways, for example according to performer and the scale employed in a performance. Finally, the use of antipatterns as structural constraints during the process of generating new bagana song instances will be explored.

Acknowledgements. This research is supported by the project Lrn2Cre8 which is funded by the Future and Emerging Technologies (FET) programme within the Seventh Framework Programme for Research of the European Commission, under FET grant number 610859. Special thanks to Kerstin Neubarth, Dorien Herremans, and Louis Bigo for valuable comments on the manuscript.

References

1. Adamo, J.M.: Data Mining for Association Rules and Sequential Patterns. Springer (2001)
2. Agrawal, R., Srikant, R.: Mining sequential patterns. In: Proceedings of the Eleventh International Conference on Data Engineering, Washington, DC, pp. 3–14 (1995)
3. Arom, S.: Le "syndrome" du pentatonisme Africain. Musicae Scientiae 1(2), 139–163 (1997)
4. Artamonova, I., Frishman, G., Frishman, D.: Applying negative rule mining to improve genome annotation. BMC Bioinformatics 8, 261 (2007)
5. Ayres, J., Gehrke, J., Yiu, T., Flannick, J.: Sequential pattern mining using a bitmap representation. In: Proceedings of the International Conference on Knowledge Discovery and Data Mining, Edmonton, Canada, pp. 429–435 (2002)
6. Bay, S., Pazzani, M.: Detecting group differences: Mining contrast sets. Data Mining and Knowledge Discovery 5(3), 213–246 (2001)
7. Conklin, D.: Melody classification using patterns. In: MML 2009: International Workshop on Machine Learning and Music, Bled, Slovenia, pp. 37–41 (2009)
8. Conklin, D.: Discovery of distinctive patterns in music. Intelligent Data Analysis 14(5), 547–554 (2010)
9. Conklin, D.: Distinctive patterns in the first movement of Brahms' String Quartet in C Minor. Journal of Mathematics and Music 4(2), 85–92 (2010)
10. Conklin, D.: Antipattern discovery in folk tunes. Journal of New Music Research 42(2), 161–169 (2013)
11. Deng, K., Zaïane, O.R.: Contrasting sequence groups by emerging sequences. In: Gama, J., Costa, V.S., Jorge, A.M., Brazdil, P.B. (eds.) DS 2009. LNCS, vol. 5808, pp. 377–384. Springer, Heidelberg (2009)
12. Dong, G., Bailey, J. (eds.): Contrast Data Mining: Concepts, Algorithms, and Applications. Chapman & Hall/CRC Data Mining and Knowledge Discovery Series. Chapman and Hall/CRC (2012)
13. Dong, G., Li, J.: Efficient mining of emerging patterns: discovering trends and differences. In: Proceedings of the Fifth ACM SIGKDD International Conference on Knowledge Discovery and Data Mining, KDD 1999, San Diego, pp. 43–52 (1999)
14. Haglin, D.J., Manning, A.M.: On minimal infrequent itemset mining. In: Stahlbock, R., Crone, S.F., Lessmann, S. (eds.) Proceedings of the 2007 International Conference on Data Mining, Las Vegas, Nevada, USA, pp. 141–147. CSREA Press (2007)
15. Herold, J., Kurtz, S., Giegerich, R.: Efficient computation of absent words in genomic sequences. BMC Bioinformatics 9, 167 (2008)
16. Ji, X., Bailey, J., Dong, G.: Mining minimal distinguishing subsequence patterns with gap constraints. Knowledge and Information Systems 11(3), 259–296 (2007)
17. Klösgen, W.: Explora: A Multipattern and Multistrategy Discovery Assistant. In: Fayyad, U., Piatetsky-Shapiro, G., Smyth, P. (eds.) Advances in Knowledge Discovery and Data Mining, pp. 249–271. MIT Press, Cambridge (1996)

18. Lin, C.-R., Liu, N.-H., Wu, Y.-H., Chen, A.L.P.: Music classification using significant repeating patterns. In: Lee, Y., Li, J., Whang, K.-Y., Lee, D. (eds.) DASFAA 2004. LNCS, vol. 2973, pp. 506–518. Springer, Heidelberg (2004)
19. Mooney, C.H., Roddick, J.F.: Sequential pattern mining – approaches and algorithms. ACM Comput. Surv. 45(2), 19:1–19:39 (2013)
20. Novak, P.K., Lavrač, N., Webb, G.I.: Supervised descriptive rule discovery: A unifying survey of contrast set, emerging pattern and subgroup mining. Journal of Machine Learning Research 10, 377–403 (2009)
21. Sawada, T., Satoh, K.: Composer classification based on patterns of short note sequences. In: Proceedings of the AAAI 2000 Workshop on AI and Music, Austin, Texas, pp. 24–27 (2000)
22. Shan, M.K., Kuo, F.F.: Music style mining and classification by melody. IEICE Transactions on Information and Systems E88D(3), 655–659 (2003)
23. Wang, J., Zhang, Y., Zhou, L., Karypis, G., Aggarwal, C.C.: CONTOUR: an efficient algorithm for discovering discriminating subsequences. Data Min. Knowl. Disc. 18, 1–29 (2009)
24. Webb, G.I.: Discovering significant patterns. Machine Learning 71(1), 131 (2008)
25. Weisser, S.: Etude ethnomusicologique du bagana, lyre d'Ethiopie. Ph.D. thesis, Université Libre de Bruxelles (2005)
26. Weisser, S.: Ethiopia. Bagana Songs. Archives Internationales de Musique Populaire (AIMP)-VDE Gallo, Liner notes of the CD (2006)
27. Weisser, S.: Transcrire pour vérifier: le rythme des chants de bagana d'Éthiopie. Musurgia XIII(2), 51–61 (2006)

Categorize, Cluster, and Classify: A 3-C Strategy for Scientific Discovery in the Medical Informatics Platform of the Human Brain Project

Tal Galili[1,*], Alexis Mitelpunkt[1,*], Netta Shachar[1], Mira Marcus-Kalish[2], and Yoav Benjamini[1,2,**]

[1] Department of Statistics and Operations Research,
[2] The Sagol School for Neurosciences
Tel Aviv University, Israel
ybenja@tau.ac.il

Abstract. One of the goals of the European Flagship Human Brain Project is to create a platform that will enable scientists to search for new biologically and clinically meaningful discoveries by making use of a large database of neurological data enlisted from many hospitals. While the patients whose data will be available have been diagnosed, there is a widespread concern that their diagnosis, which relies on current medical classification, may be too wide and ambiguous and thus hides important scientific information.

We therefore offer a strategy for a search, which combines supervised and unsupervised learning in three steps: Categorization, Clustering and Classification. This 3-C strategy runs as follows: using external medical knowledge, we categories the available set of features into three types: the patients' assigned disease diagnosis, clinical measurements and potential biological markers, where the latter may include genomic and brain imaging information. In order to reduce the number of clinical measurements a supervised learning algorithm (Random Forest) is applied and only the best predicting features are kept. We then use unsupervised learning in order to create new clinical manifestation classes that are based on clustering the selected clinical measurement. Profiles of these clusters of clinical manifestation classes are visually described using profile plots and analytically described using decision trees in order to facilitate their clinical interpretation. Finally, we classify the new clinical manifestation classes by relying on the potential biological markers. Our strategy strives to connect between potential biomarkers, and classes of clinical and functional manifestation, both expressed by meaningful features. We demonstrate this strategy using data from the Alzheimer's Disease Neuroimaging Initiative cohort (ADNI).

Keywords: medical informatics, bioinformatics, disease profiling, categorization, clustering, classification.

---

* The first two authors contributed equally to this work.
** Corresponding author.

S. Džeroski et al. (Eds.): DS 2014, LNAI 8777, pp. 73–86, 2014.
© Springer International Publishing Switzerland 2014

1   Introduction

One of the goals of the European Flagship Human Brain Project is to create a platform that will enable scientists to search for new biologically and clinically meaningful discoveries by making use of a large database of neurological data enlisted from many hospitals. While the patients whose data will be available have been diagnosed, there is a widespread concern that their diagnosis, which relies on current medical classification, may be too wide and ambiguous and thus hide important scientific information. This is also the case with Alzheimer's disease (AD).

Alzheimer's disease is the most common form of dementia. The disease is characterized by the accumulation of b-amyloid (Ab) plaques and neurofibrillary tangles composed of tau amyloid fibrils associated with brain cells damage and neurodegeneration. The degeneration leads to progressive cognitive impairment. There is currently no known treatment, nor one that slows the progression of this disorder. According to the 2010 World Alzheimer report, about 35.6 million people worldwide are living with dementia, at a total cost of more than US$600 billion in 2010. The incidence of AD throughout the world is expected to double in the next 20 years. There is a pressing need to find markers to both predict future clinical decline and for use as outcome measures in clinical trials of disease-modifying agents and foster the development of innovative drugs[1].

The diagnosis of Alzheimer's disease requires histopathologic examination, which in most cases, can be found only by post-mortem pathological examination (if such is done). Therefore, the diagnosis of AD is often based on clinical criteria. In 2013, an updated criteria was published by the American Psychiatric Association in the Diagnostic and Statistical Manual of Mental Disorders 5-th edition (DSM - 5)[2]. The clinical criteria are based on a history of insidious onset and progressive deterioration, exclusion of other etiologies, and documentation of cognitive impairments that interfere with independence in everyday activities, in one or more of the following domains: Learning and memory, Language, Executive function, Complex attention, Perceptual-motor, Social cognition. A detailed cognitive and general neurologic examination is essential for the clinical decision. The DSM-5 also mentions that "Evidence of a causative Alzheimer disease genetic mutation from family history or genetic testing" can be used as part of the diagnostic criteria[2]. In contrast, neuropsychological testing is not specific to AD, even though it may provide confirmatory information on cognitive impairment and can aid in patient management. The role of laboratory and imaging investigations is mainly to exclude other diagnoses. Some studies suggest that certain biomarkers including increased levels of tau protein[3] and decreased levels of beta-amyloid protein ending at amino acid 42 in cerebrospinal fluid (CSF) or plasma [4],[5] elevated ApoE and ApoE4 plasma levels[6], and others may have predictive value for AD in healthy and in patients with minimal cognitive impairment (MCI). These may also aid in distinguishing AD from other forms of dementia, and may identify subsets of patients with AD at risk for a rapidly progressive course. However, the role for these measurements in clinical practice has not been established.

Brain imaging using magnetic resonance imaging (MRI) is part of the diagnostic process for dementia. It is mainly used to exclude other possible diagnosis rather than AD for the condition. In some studies it has been postulated that a decreased volume of certain brain areas is related to AD but contradicting studies found a general process of volume reduction with aging. Functional brain imaging with [18F] fluoro-deoxyglucose positron emission tomography (FDG-PET), functional MRI (fMRI), perfusion MRI, or perfusion single photon emission computed tomography (SPECT) reveals distinct regions of low metabolism and hypoperfusion in AD. These areas include the hippocampus, the precuneus (mesial parietal lobes) and the lateral parieto-temporal cortex. Clinical studies suggest that FDG-PET may be useful in distinguishing AD from frontotemporal dementia, but this result have not become a standard for diagnosis.

The Alzheimer's disease Neuroimaging Initiative (ADNI) was conceived at the beginning of the millennium as a North American multicenter collaborative effort funded by public and private bodies [1], in order to facilitate a progression in the understanding, assessing and treating AD. The initiative obtains data on patients of normal cognitive state, early and late mild cognitive impairment (MCI) and AD. Clinical, neuropsychological, biological markers, imaging and genetic data is collected on the patients. Many articles have been published in vast aspects of AD research from correlations between different measures through prediction of disease course and classification of patients. Much of the current research focuses on classification methods proposed to (i) help diagnose AD patients, (ii) distinguish them from MCI patients as early as possible, or (iii) identify MCI patients with high risk of converting to AD. Some of the proposed methods focus on one or two specific and promising biomarkers such as Magnetic Resonance Imaging results[7], FDG-PET imaging [8] or CSF biomarkers [9], others are trying to combine a number of biomarkers [10].

The combined-biomarkers methods are often based on various machine learning algorithms: Kohannim et al. [11] implemented the support vector machines (SVM) tool in order to classify AD and MCI patients. The authors considered age, sex, BMI, MRI summaries, ApoE, FDG-PET and CSF as possible biomarkers for classification and found that while MRI measures contributed most to the classification of AD the FDG-PET and CSF biomarkers were more useful in classifying MCI. Hinrichs et al. [12] tried to predict conversion from MCI to AD using a multi kernel learning framework on a dataset containing MRI, FDG-PET, CSF assays, ApoE genotype and scores from NPSE exam, reported better results than an SVM (but only by 3%-4%). Zhang et al. [13] presented a three steps methodology in which we first select a relevant subset of features using multi-task feature learning methods, then a kernel-based multimodal-data-fusion method is applied in order to effectively fuse data from the different modalities and finally a support vector regression is trained. Whalhovd et al [14] used a repeated measures general linear model regression and logistic regressions in order to evaluate the sensitivity of CSF, MR and FDG-PET to the diagnosis and to the longitudinal changes of scores for MMSE exam and CDR-SB.

Our approach deviates from these lines of research in two major ways. First, we avoid using the available diagnosis as the ultimate true status of the patients. We do acknowledge that the available diagnosis are somewhat informative, and thus can guide us towards the goal, but do not treat them as targets to be predicted. Thus the available features are *categorized* to clinical variables that reflect the functionality of the patient, and to biological variables that can serve as potential biological markers. The available diagnosis is used to select via classification methods a relevant and important set of features among the clinical variables. This subset is then *clustered* in an unsupervised way to create a set of disease categories that take the role of the available diagnosis.

Second, these categories are also predicted from the biological markers in two stages: informative subset of potential biological markers is selected by random forest model that is difficult to interpret, and then a simpler decision tree *classification* model is built, which caters to the medical decision-making process. (See Figure 1)

In this paper we demonstrate the proposed approach on a limited part of ADNI data, and on a limited part of the available information about each subject. This is but a first step in a longer effort that will include evaluation and further adaptations, before returning to the original problem of using hospital data on a grand scale.

2   Data and Methods

Data

We used the ADNIMERGE table, extracted from the ADNIMERGE R package (version 0.0.1), in order to combine data from different domains of ADNI data. Variables were chosen to reflect both clinical and potential biomarkers. We only used baseline data of ADNI stage II + ADNI Go, out of which 796 subjects had no missing values on the clinical measurements.

Pre-processing

We dropped the "ADAS11" CM since it had near one correlation with the CM "ADAS13". In Addition, we removed from the analysis the CM EcogSPTotal and the CM EcogPtTotal as they are both they are derived from some of the other sub measurments. In order to reduce skewness of some of the CMs, log transformations (for ADAS13, EcogPtMem), logit transformations (for EcogPtDivatt, EcogPtVisspat, EcogSpDivatt, EcogSpVisspat, MMSE, MOCA) and inverse transformations (for CDRSB, EcogPtLang, EcogPtOrgan, EcogPtPlan, EcogSpLang, EcogSpOrgan, EcogSpPlan, FAQ, Ravlt.prec.forgetting) were utilized. The following CMs needed no transformation: RAVLT.forgetting, RAVLT.immediate and RAVLT.learning. Six new CM were defined to be the difference between the transformed patients' report and the partner's report on certain everyday cognition (Ecog) variable. Finally, all the variables were scaled to have mean 0 and a variance of 1.

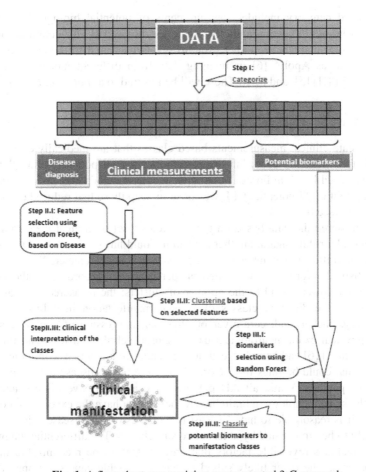

Fig. 1. A flow chart summarizing our suggested 3-C approach

Stage I: Categorization of Variables

Categorization was done using expert medical knowledge.

(1) The first category is the disease diagnosis variable as assigned in the ADNI database. This *assigned diagnosis* has five levels: Cognitively Normal (CN), Significant Memory Concern (SMC), Early Mild Cognitive Impairment (EMCI), Late Mild Cognitive Impairment (LMCI), Alzheimr Disease (AD).

(2) The second category is of *clinical measurements* (CM) that reflect the functionality of the patient. They encompass scores of different cognitive and psycho-neurological tests and ratings, according to clinical assessment and patient's or partners' report. This battery of cognitive and functional assessment scores include: Clinical Dementia Rating Sum of Boxes (CDR-SB), Alzheimer's disease assessment scale (ADAS), mini–mental state examination (MMSE), Rey Auditory Verbal Learning Test (RAVLT), Functional assessment questionnaire (FAQ) Montreal Cognitive Assessment (MoCA), Everyday Cognition (Ecog).

(3) The third category includes measurements of potential biological markers, which were proposed to have a predictive value for disease risk, for deterioration, or for severity. These markers are either proteins levels measured in the cerebrospinal fluid (CSF) such as ApoE4 [6] or imaging data from different modalities: FDG-PET[14], AV45 PET[15], and MRI. These will be referred to as *potential biomarkers*.

Stage II: Feature Selection and Clustering

In order to create clinical measurements based classes that are medically easy to interpret, a feature selection procedure was performed on all potential clinical measurements. We used Random Forest, but of course other methods may prove as useful or even more. Out of 27 potential CM, we chose to keep those that reduced error-rate by 15% or more (see figure 2)

We then clustered the data based on the selected subset of clinical measurements using k-means algorithm (again, another algorithm could have been used). Of course, in any such algorithm the number of clusters is a crucial parameter. We chose to combine statistical information with medical perspectives. According to the latter, there is a natural lower bound to the number of clusters: the measured clinical data should represent the different classes of clinical manifestation including patients' medical history, background, care-giver or physician impression of cognitive state, symptoms, physical exam, neurological exam, neuropsychological tests and ratings. From the literature [16] and knowledge about dementia we know that within the clinical spectrum that could be lines from Normal to Alzheimer's disease there are some sub-classes of patients. While all AD patients have a progressive disease and we would find the same pathology in brain biopsy they do not have the exact same course of illness, so it is reasonable to have at least two classes of AD disease. The normal group is likely to be represented by at least two sub-classes. The differential diagnosis of dementia includes reversible factors that are assumed to be recognizable in the screening process. The other likely sub-class has irreversible causes, among which Alzheimer's disease accounts for about half the cases. Other considerable causes are vascular dementia, dementia with lewy bodies (DLB) and frontotemporal dementia[17]. We therefor assess that at least three subclasses should be represented encompassing these phenomena. The last reasonable subclass would be for elderly with minimal cognitive impairment that will not progress to one of the mentioned diseases. In summary, it appears that at least 8 subclasses should be considered.

This medical insight into the potential number of classes was combined with the statistical point of view, utilizing the gap statistic plot (see figure 3). A cutoff of 15% error-reduction was set to choose the number of clusters.

Stage III: Classification Using Potential Biomarkers

At this stage we classify the clinical measurements based classes using the set of potential biomarkers. In principle this stage also consists of two parts. First, using importance analysis by, say, random forests, where a promising subset of the biomarkers is selected.   The final classification step is done using hierarchical decision trees, or other rule based analysis, utilizing the selected subset. This is essential in order to give easy interpretation to the diagnosis process. In the envisioned application to hospital data the number of potential biomarkers may increase to thousands, before incorporating genomic information. Thus, the subset selection stage may be essential. In the current analysis we skip this stage as the number of potential biomarkers is small.

Algorithms and Software

Analyzes were performed using R [18] . For assessing the importance of the clinical measurements (as a preparation to the Clustering stage) we used the classification method of the {randomForest} R package [19]. Importance was measured as the marginal loss of classification accuracy for each variable by randomly permuting it on the test (out of bag) validation set. A "junk" variable was added, taking the form of an independent random uniform [0,1] variable, in order to signal variables above the noise level. For clustering we used the R package {cluster} [20], using the gap statistic [21] to choose the number of clusters. The gap statistic was based on 100 bootstarps and calculated for up to 20 clusters. Clustering was done using k-means with 10 iterations at most, based on the Hartigan and Wong algorithm. Classification and regressions tree (CART) was constructed using the {rpart} R package, the tree was constructed with the minimal possible number of observations for a split set to zero, minimal number of observation in a leaf set to zero, and with a 10-fold cross validations for tuning the complexity parameter. Scatter plot matrix was produces using the {psych} R package [22].

3   Results

Stage IIa: Out of 27 potential CM, we chose to keep the 7 CM that reduced error-rate of predicting the assigned diagnosis by 15% or more. (See figure 2)

Stage IIb: These variables were clustered using k-means with varying number of means and the gap statistics plot for aid in the choice of the number of clusters. The first local maxima of the gap statistics above the clinically determined 8 was chosen to indicate that 10 clinically determined classes are needed. In order to discuss the meaning of the newly created classes we present their cross-classification with the assigned diagnosis in Table 1 and a profiles plot in Figure 4.

Classes 1 and 3 contain nearly all the participants with an assigned diagnosis of AD. Class 3 might be a class of more severe AD cases (see minimal average level on all coordinates of the profile plot). From this plots we also see that Class 1 members score higher on EcogPtLang and EcogPtMem than those classified to 3. Classes 4,5 and 10 hold the majority of patients whose assigned diagnosis is CN. It is interesting that while these classes have a very small amount of patients with different diagnoses they were still separated to three classes based on their clinical manifestation. Class 4 has the highest "MMSE" and a low "CDRSB" scores which points to a group of clinically normal participants, but the score in "ECogPtLang" is lower than other classes which means that these participants are more concerned of their personal observation

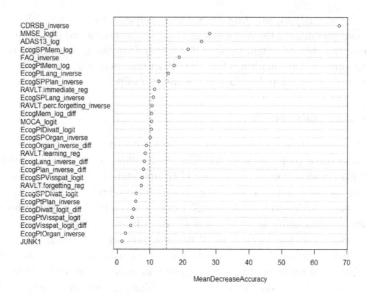

Fig. 2. Feature Selection for Clinical measurements. Only CM with mean decrease of accuracy of 15% or more were kept.

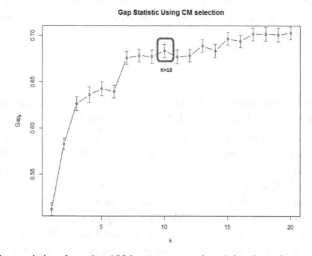

Fig. 3. The Gap statistics plot using 100 bootstrap samples. A local maximum at k=10 suggest 10 clusters are appropriate to the data at hand.

of language difficulties. Classes 6 and 7 include normal and mildly affected participants (sharing the same branch in the decision tree), but differ from each other especially in their patients' "ADAS13" scores. Another group of classes is 2,8,9 in which patients are distributed almost uniformly but their disease manifestation differ from one another (though they all seem to have a progressive disease but not to a level which qualifies as AD).

Table 1. Cross-classification table of originally assigned diagnosis **vs** clinical classes

	1	2	3	4	5	6	7	8	9	10
CN	0	5	0	37	59	11	15	2	3	44
SMC	1	5	0	26	16	2	31	1	6	6
EMCI	23	67	5	1	3	0	19	62	79	4
LMCI	35	19	32	0	0	0	1	28	31	0
AD	44	0	71	0	0	0	0	2	0	0

Inspecting the decision tree representing the classes (figure 5), we can see that "FAQ" feature had much influence on the clustering: the classes with low "FAQ" are 1,3,9 meaning those patients would be likely to have a progressive disease. Further down on the right branch of the tree, class 3 has a low "MMSE" score and class 1 has low score "ECogPtLang" representing the interference of language impairment of the patient's life in his own perception. Walking down the left branch of the tree the first split sends down the right branch all patients with a "CDRSB" score of over 0.33, this by definition of the inclusion criteria will not allow normal participants in that branch. Clusters 2,8,9 occupy that branch of the CART decision tree.

This decision tree gives the possibility to determine rules and to explain to the physician the way the classes where created from the data. The first two branches divide the participants into "Normal" and "Not Normal". This division of the data is done using 2 variables: "FAQ" and "CDRSB", that are related to disease state definitions. Then, both branches use the level of "MMSE" to create a separation within each branch between normal and AD affected. In a lower and fine distinction the next junction divides them to a class of participants according to ECogPtLang value meaning that the participant feels at least occasionally that his language ability is worse than it was 10 years earlier.

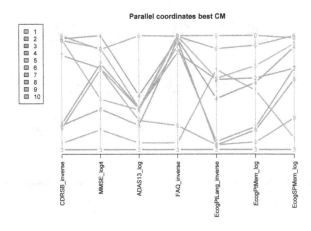

Fig. 4. Profiles plot (parallel coordinates) demonstrating the values of the CM across classes. Note that some of the CM were reversed so that high level of all CMs will represent normal status.

Stage III: The potential biomarkers used to classify subjects to the ten clinically relevant classes consist of "ApoE", "AV45", "FDG" and some gross imaging volume measurements: "Entorhinal", "Fusiform", "Hippocampus", "ICV", "Midtemp", "Ventricles", "Whole-brain". The results of the use of the Random Forest algorithm are presented in table 2 in a Confusion matrix of classification from the Potential Biomarkers (PB) to the 10 clusters created in step II. Even though error rates are quite high, if we take into consideration the interpretation of the prior step, other explanation could be considered. For example, as shown above, classes 1,3 contain most of severely ill patients. If we look at the confusion matrix most of the cases of class 3 are either classified as class 1 or 3. Other classes are rarely assigned. If we were join these two classes error rate might drop, but since two clusters were formed it might be interesting to further investigate the clinical difference between them and look for new biomarkers that are able to capture these difference. The PB decision tree (figure 6) distinguishes primarily between the patients designated to class 3 concurrent with severely symptomatic disease. This branches according to the value of "FDG". This coincides with that FDG is a known marker of AD.

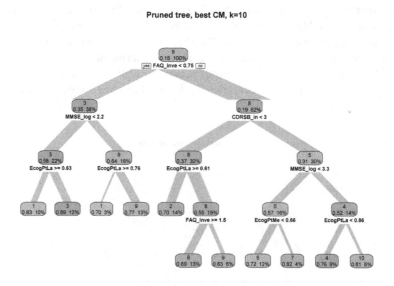

Fig. 5. Decision tree representing the classes resulted from the clustering step

Table 2. Confusion matrix resulting from class prediction based on potential biomarkers

	1	2	3	4	5	6	7	8	9	10	class error
1	13	6	26	0	3	0	0	6	8	0	79%
2	4	26	2	4	6	0	2	16	18	4	68%
3	15	2	40	0	1	0	0	4	9	1	44%
4	0	9	1	1	4	0	3	7	7	1	97%
5	2	16	0	1	5	0	2	4	19	6	91%
6	0	6	0	0	0	0	0	2	1	0	100%
7	1	8	1	0	6	0	3	5	6	1	90%
8	5	19	6	2	3	0	3	10	28	2	87%
9	8	21	16	2	4	0	1	12	27	3	71%
10	0	15	1	1	5	0	0	6	10	5	88%

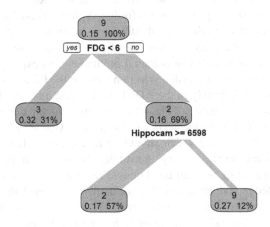

Decision tree, PB classification

Fig. 6. Decision tree representing the prediction of classes from PB

4   Discussion

The criteria used today according to DSM-5[2] for diagnosis of AD relies on the clinical and functional ability of the patient. Most biological exams, such as imaging results, are mainly used to rule out other possible diagnoses. In our strategy we therefore differentiate between variables which are descriptive of the patient's functional conditions and variables which are collected in order to try and find possible disease causes, the latter being targets for drug development or surrogate markers for disease stage or trajectory.

A diagnosis depends on the time it was made and the available knowledge at that moment. DSM V presented ten etiological subtypes which did not appear in prior editions. Other than the explicit link to specific known etiologies, most of these sub-types' criteria are largely similar to one another. However, there are important and often subtle differences between these disorders [2]. We present an approach separating patient to groups according to their clinical data. Interestingly, our data also identifies 10 classes that might represent a more accurate distinction of the patient rather than 5 diagnosis criteria given by the ADNI protocol.

We do not claim that our findings present our best current views on the problem. We are very aware that this was but a sketch of strategy that happened to offer some new insights. Further exploration is needed on a few fronts: The use of the raw exams data instead of combined scores, adding potentially important measurements, enlightenment of the data by expert knowledge such as differing questions to the different cognitive function domain measured could all help in creating more subtle and fine clusters of patient's disease presentation. From a statistical point of view, different clustering procedures and/or different selection procedures may yield better results under different settings, an issue we have not started to address at all.

We believe that the attempt to predict the assigned diagnosis from very specific potential biomarker is futile. The route we have taken is to predict more subtle disease manifestation classes. Such a process needs further exploration but has the potential to fit a small biomarker arrow to the clinical bull's eye. In our data, only a few biomarkers were available and therefore we used all of which in the attempt to predict the classes. Had the data been richer in potential biomarkers, as expected from hospital data, we may had to perform a variable selection step for the PB as well.

In many studies and definitely in the ADNI study a vast amount of measurable information is collected. Is it enough? The tacit knowing held and applied by proficient practitioners represents a valuable form of clinical knowledge, which has been acquired through experience, and which should be investigated, shared, and contested [23]. In clinical work, tacit knowing constitutes an important part of diagnostic reasoning and judgment of medical conditions. Practitioners apply a broad range of experiential knowledge and strategies that are hardly mentioned in the textbooks or implicated in the analysis of research results. If we could use this known, yet sometimes ignored, information and quantify it - a valuable aspect of analysis and interpretation of the results could be added. In this work we implemented clinical categorization of features to better model the diagnosis process. Further exploration is needed of both the data nuances and methods, before trying to scale to the much harder problem associated with regular hospital data. We do believe that the strategy we have outlined in this work is capable of achieving that.

References

1. Weiner, M.W., Veitch, D.P., Aisen, P.S., Beckett, L.A., Cairns, N.J., Green, R.C., Harvey, D., Jack, C.R., Jagust, W., Liu, E., Morris, J.C., Petersen, R.C., Saykin, A.J., Schmidt, M.E., Shaw, L., Shen, L., Siuciak, J.A., Soares, H., Toga, A.W., Trojanowski, J.Q.: The Alzheimer's Disease Neuroimaging Initiative: A review of papers published since its inception. Alzheimer's Dement 9(5), e111–e194 (2013)

2. American Psychiatric Association, DSM-5 criteria for major neurocognitive disorder due to AD, 5th edn. Arlington, VA (2013)
3. Sonnen, J.A., Montine, K.S., Quinn, J.F., Kaye, J.A., Breitner, J.C.S., Montine, T.J.: Biomarkers for cognitive impairment and dementia in elderly people. Lancet Neurol. 7(8), 704–714 (2008)
4. Sunderland, T., Linker, G., Mirza, N., Putnam, K.T., Friedman, D.L., Kimmel, L.H., Bergeson, J., Manetti, G.J., Zimmermann, M., Tang, B., Bartko, J.J., Cohen, R.M.: Decreased beta-amyloid1-42 and increased tau levels in cerebrospinal fluid of patients with Alzheimer disease. JAMA 289(16), 2094–2103 (2094)
5. Yaffe, K., Weston, A., Graff-Radford, N.R., Satterfield, S., Simonsick, E.M., Younkin, S.G., Younkin, L.H., Kuller, L., Ayonayon, H.N., Ding, J., Harris, T.B.: Association of plasma beta-amyloid level and cognitive reserve with subsequent cognitive decline. JAMA 305(3), 261–266 (2011)
6. Gupta, V.B., Laws, S.M., Villemagne, V.L., Ames, D., Bush, A.I., Ellis, K.A., Lui, J.K., Masters, C., Rowe, C.C., Szoeke, C., Taddei, K., Martins, R.N.: Plasma apolipoprotein e and Alzheimer disease risk: The AIBL study of aging. Neurology 76(12), 1091–1098 (2011)
7. Evans, M.C., Barnes, J., Nielsen, C., Kim, L.G., Clegg, S.L., Blair, M., Leung, K.K., Douiri, A., Boyes, R.G., Ourselin, S., Fox, N.C.: Volume changes in Alzheimer's disease and mild cognitive impairment: cognitive associations. Eur. Radiol. 20(3), 674–682 (2010)
8. Langbaum, J.B.S., Chen, K., Lee, W., Reschke, C., Fleisher, A.S., Alexander, G.E., Foster, N.L., Michael, W., Koeppe, R.A., Jagust, W.J., Reiman, E.M.: Categorical and Correlational Analyses of Baseline Fluorodeoxyglucose Positron Emission Tomography Images from the Alzheimer's Disease. Neuroimage 45(4), 1107–1116 (2010)
9. Tosun, D., Schuff, N., Truran-Sacrey, D., Shaw, L.M., Trojanowski, J.Q., Aisen, P., Peterson, R., Weiner, M.W.: Relations between brain tissue loss, CSF biomarkers and the ApoE genetic profile: A longitudinal MRI study. Neurobiol. Aging 31(8), 1340–1354 (2011)
10. Cui, Y., Liu, B., Luo, S., Zhen, X., Fan, M., Liu, T., Zhu, W., Park, M., Jiang, T., Jin, J.S.: Identification of conversion from mild cognitive impairment to Alzheimer's disease using multivariate predictors. PLoS One 6(7), e21896 (2011)
11. Kohannim, O., Hua, X., Hibar, D.P., Lee, S., Chou, Y.-Y., Toga, A.W., Jack, C.R., Weiner, M.W., Thompson, P.M.: Boosting power for clinical trials using classifiers based on multiple biomarkers. Neurobiol. Aging 31(8), 1429–1442 (2010)
12. Hinrichs, C., Singh, V., Xu, G., Johnson, S.C.: Predictive markers for AD in a multimodality framework: An analysis of MCI progression in the ADNI population. Neuroimage 55(2), 574–589 (2011)
13. Zhang, D., Shen, D.: Multi modal multi task learning for joint prediction of multiple regression and classification variables in Alzheimer's disease. Neuroimage 59(2), 895–907 (2013)
14. Walhovd, K.B., Fjell, M., Brewer, J., McEvoy, L.K., Fennema-Notestine, C., Hagler, D.J., Jennings, R.G., Karow, D., Dale, M.: Combining MR imaging, positron-emission tomography, and CSF biomarkers in the diagnosis and prognosis of Alzheimer disease. AJNR. Am. J. Neuroradiol. 31(2), 347–354 (2010)
15. Johnson, K.A., Sperling, R.A., Gidicsin, C., Carmasin, J., Maye, J., Coleman, R.E., Reiman, E.M., Sabbagh, M.N., Sadowsky, C.H., Fleisher, A.S., Doraiswamy, P.M., Carpenter, A.P., Clark, C.M., Joshi, A.D., Lu, M., Grundman, M., Mintun, M.A., Pontecorvo, M.J., Skovronsky, D.: Florbetapir (F18-AV-450) PET to assess amyloid burden in Alzheimer's disease dementia, mild cognitive impairment, and normal aging. Alzheimer's Dement 9 (2013)

16. Shadlen, M.-F., Larson, E.B.: UpToDate: Evaluation of cognitive impairment and dementia
17. Longo, D., Fauci, A., Kasper, D., Hauser, S., Jameson, J., Loscalzo, J.: Harrison's Principles of Internal Medicine, 18th edn., National Institute of Health, Bethesda, MD, National Institute of Allergy and Infectious Diseases, Brigham and Women's Hospital (2011)
18. R Core Team, R: A language and environment for statistical computing
19. Liaw, A., Wiener, M.: Classification and Regression by randomForest. R News 2, 18–22 (2002)
20. Maechler, M., Rousseeuw, P., Struyf, A., Hubert, M., Hornik, K.: Cluster Analysis Basics and Extensions. R package version 1.14.4. CRAN (2013)
21. Tibshirani, R., Walther, G., Hastie, T.: Estimating the number of clusters in data set via the gap statistic. Journal of the Royal Statistical Society: Series B, Part 2, 411–423 (2001)
22. Revelle, W.: psych: Procedures for psychological, psychometric, and personality research, pp. 0–90. Northwest. Univ. Evanston, Illinois (2010)
23. Malterud, K.: The art and science of clinical knowledge: evidence beyond measures and numbers. Lancet 358(9279), 397–400 (2001)

Multilayer Clustering: A Discovery Experiment on Country Level Trading Data

Dragan Gamberger[1], Matej Mihelčić[1], and Nada Lavrač[2,3]

[1] Rudjer Bošković Institute, Bijenička 54, 10000 Zagreb, Croatia
`dragan.gamberger@irb.hr`
[2] Jožef Stefan International Postgraduate School, Jamova 39,
1000 Ljubljana, Slovenia
[3] University of Nova Gorica, Vipavska 13, 5000 Nova Gorica, Slovenia

Abstract. The topic of this work is the presentation of a novel clustering methodology based on instance similarity in two or more attribute layers. The work is motivated by multi-view clustering and redescription mining algorithms. In our approach we do not construct descriptions of subsets of instances and we do not use conditional independence assumption of different views. We do bottom up merging of clusters only if it enables reduction of an example variability score for *all* layers. The score is defined as a two component sum of squared deviates of example similarity values. For a given set of instances, the similarity values are computed by execution of an artificially constructed supervised classification problem. As a final result we identify a small but coherent clusters. The methodology is illustrated on a real life discovery task aimed at identification of relevant subgroups of countries with similar trading characteristics in respect of the type of commodities they export.

1 Introduction

Clustering is an optimisation task which tries to construct subpopulations of instances so that distances between instances within each subpopulation are small while distances between instances in different subpopulations are as large as possible [1]. The main problem of clustering algorithms is to define an appropriate measure of distance between instances. It is well-known that different measures may result in identification of different clusters [2,3]. The most common measure is Euclidean distance that is well defined for numerical attributes [1]. Nominal attributes can be handled only after some transformations. When dealing with numerical attributes it is necessary to normalize the data in the preprocessing step in order to ensure equal relevancy of all attributes regardless of their absolute values [1]. Results obtained by clustering are unreliable in the sense of the number of constructed clusters and in the sense of instances included in clusters.

A multi-view learning uses more than one set of attributes in order to improve quality of both supervised and unsupervised techniques [4,5]. Redescription mining can be interpreted as a clustering approach in which the quality of the results is ensured by the condition that resulting clusters must have meaningful

S. Džeroski et al. (Eds.): DS 2014, LNAI 8777, pp. 87–98, 2014.
© Springer International Publishing Switzerland 2014

interpretations in independent attribute layers [6,7]. In this work we present an approach to reliable clustering that reuses the basic ideas of multi-view clustering and redescription mining in a novel setting. We call it multilayer clustering because it has been originally developed for analysis of network data available in more than one layer [8]. In contrast to redescription mining, we do not construct descriptions of subsets of instances and in contrast to multi-view clustering we do not assume conditional independence of layers.

The first step is to determine the similarity of instances by executing a supervised machine learning task on an artificial problem in which the target set of instances are positive examples and negative examples are obtained by random shuffling of positive examples. We compute similarity tables for each attribute layer independently and then search for clusters that satisfy similarity conditions in *all* available layers. The main characteristic of the approach is that the resulting clusters are small but very coherent. Additionally, the methodology can be directly implemented on original attribute values without any transformation and normalization. When compared to redescription mining, results are less sensitive in respect of noise. The novel methodology is presented in Section 2. Its application is illustrated on a real world problem of recognizing groups of countries with similar economical profile based on export data for 106 different commodity types. It is a good example of clustering in a domain with a lot of noisy and imprecise data. Besides export data, we have a separate set of 105 attributes describing socio-economic characteristics of the countries. In this way we have a typical multi-view setting with two independent attribute layers for a fixed set of examples consisting of 155 countries. The obtained results are presented in Section 3.

2   Clustering Related Variability Reduction Algorithm

In machine learning we have a set of examples E that are described by a set of attributes A. Specifically in redescription mining, it is assumed that the set of attributes may be partitioned in at least two disjoint parts (layers). The partitioning is not random but a consequence of the meaning of the attributes or the way the data have been collected. For example, in a medical domain the first layer may contain anamnestic data (medical history of patients) while the second layer may contain laboratory measurements. In some other domain, different layers may contain the same attributes but collected in various time periods. The goal is to construct coherent clusters, that are as large as possible, in the complete attribute space.

2.1   Single Layer Clustering

Let us assume a basic clustering task in which we have only one layer of attributes. The approach consists of two steps. In the first step we compute the so called example similarity table. It is an N times N symmetric matrix, where N is the number of examples. All its values are in the range 0.0 - 1.0. A large value at a position i, j $(i \neq j)$ denotes large similarity between examples i and j. In the second step we use the table in order to construct clusters.

Example Similarity Table (EST) Computation

We start from the original set of N examples represented by nominal and numerical attributes that may contain unknown values. The next step is to define an artificial classification problem so that the examples from the original set make positive examples while we artificially construct negative examples by shuffling values of the positive examples. Shuffling is done at the level of attributes so that we randomly mix values among examples. The values remain within the same attribute as in the original example. As a result, we have the same values in positive and negative examples but in negative examples we have randomized connections between attributes. Typically we construct 4 times more negative examples than positive examples.

Next, we use a supervised machine learning to build a predictive model for the discrimination between positive cases (original examples) and negative cases (examples with shuffled attribute values). The goal of learning is not the predictive model itself but information on similarity of examples. Machine learning approaches in which we can determine if some examples are classified in the same way are appropriate for this task. For example, in decision tree learning it means that examples end in the same leaf node while in covering rule set induction it means that examples are covered by the same rule. In order to estimate similarity between examples it is necessary to do a statistics over a potentially large set of classifiers. Additionally, a necessary condition for a good result is that classifiers are as diverse as possible and that each of them is better than random. All these conditions are satisfied by Random Forest [9] and Random Rules algorithms [10]. We use the latter approach in which we typically construct about 1500 rules for each EST computation.

Similarity of examples is determined so that for each pair of examples we count how many rules are true for both examples. The example similarity table presents the statistics for positive examples (original set of examples). A pair of similar examples will be covered by many rules while no rules or a very small number of rules will cover pairs that are very different in respect of their attribute values. Final EST values are obtained by the normalization of the determined counts by the largest detected value.

Table 1 presents an example of the similarity table for a set of 6 examples extracted from a real case with 155 examples. On the left side is the table with number of rules covering pairs of examples. Diagonal elements represent total number of rules covering each example. By the normalization of this table we obtain EST that is presented on the right side. It can be noticed that we have two very similar examples (examples 2 and 5), three similar (examples 1,3, and 4), and one very different example 6. The maximal value in the table on the left side is 97 and EST values (the table on the right side) are obtained by normalization with this value.

Table 1. Example of an EST

	ex1	ex2	ex3	ex4	ex5	ex6
ex1	38	0	27	28	0	7
ex2	0	97	3	1	97	3
ex3	27	3	47	16	3	1
ex4	28	1	16	45	1	4
ex5	0	97	3	1	97	3
ex6	7	3	1	4	3	39

	ex1	ex2	ex3	ex4	ex5	ex6
ex1	0.39	0.0	0.28	0.29	0.0	0.07
ex2	0.0	1.0	0.03	0.01	1.0	0.03
ex3	0.28	0.03	0.48	0.16	0.03	0.01
ex4	0.29	0.01	0.16	0.46	0.01	0.04
ex5	0.0	1.0	0.03	0.01	1.0	0.03
ex6	0.07	0.03	0.01	0.04	0.03	0.40

Clustering Related Variability (CRV) Score

The second step in the process of clustering starts from the EST. The goal is to identify subsets of examples that can reduce variability of values in the EST. For this purpose we define a so called Clustering Related Variability (CRV) score. It is the basic measure which guides the search for iterative bottom up clustering. CRV score is not the other name for some type of example similarity measure. It is defined for a single example but so that the value depends on the examples it is clustered with. A cluster may consist of a single example.

Clustering related variability for an element i contained in a cluster C is denoted by CRV_i. It is the sum of squared deviates of EST values in row i ($X_i = \{x_{i,j}, j \in \{1, \ldots, N\}\}$) computed separately for examples that are within and outside cluster C. $CRV_i = CRV_{i,wc} + CRV_{i,oc}$.

Within cluster value $CRV_{i,wc} = \sum_{j \in C} (x_{i,j} - x_{mean,wc})^2$ is computed as a summation over columns j of row i corresponding to examples included in the same cluster with example i. In this expression $x_{mean,wc}$ is the mean value of all $x_{i,j}$ in the cluster. When example i is the only example in cluster C then $CRV_{i,wc} = 0$ because we compute the sum only for value $x_{i,i}$ and $x_{mean,wc} = x_{i,i}$.

Outside cluster value $CRV_{i,oc}$ is defined in the same way as $CRV_{i,wc}$ but for $x_{i,j}$ values of row i not included in cluster C. The used $x_{mean,oc}$ is the mean value of the EST element values not included in the cluster and it is different from the $x_{mean,wc}$ used to compute $CRV_{i,wc}$. When example i is the only example in a cluster then $CRV_{i,oc}$ is the sum of squared deviates for all values in row i except $x_{i,i}$.

The final CRV value of a cluster C is the average sum of all the CRV values for the elements contained in the cluster. That is, $CRV_C = \frac{\sum_{i \in C} CRV_i}{|C|}$

Example of CRV Computation

We will use the data from the EST, presented in Table 1, to compute the CRV value for the example (ex1) contained in the cluster C. In this demonstration we will concentrate on three main cases: when a cluster contains only example ex1, when ex1 is clustered with ex3, and finally when it is clustered both with ex3 and ex4. By visual inspection of EST we can immediately notice some similarity among examples $\{ex1, ex3, ex4\}$. The goal is to demonstrate the CRV value computation and to show how its value decreases when clusters contain similar examples.

If example $ex1$ is the only example in a cluster: $C = \{ex1\}$ then:

$CRV_{ex1,wc} = (0.39 - 0.39)^2 = 0$

$CRV_{ex1,oc} = (0.0-0.13)^2 + (0.28-0.13)^2 + (0.29-0.13)^2 + (0.0-0.13)^2 + (0.07 - 0.13)^2 = 0.08$

$CRV_{ex1} = 0.08$

When we add a new element (ex3) to this cluster: $C = \{ex1, ex3\}$

$CRV_{ex1,wc} = (0.39 - 0.34)^2 + (0.28 - 0.34)^2 = 0.01$

$CRV_{ex1,oc} = (0.0 - 0.09)^2 + (0.29 - 0.09)^2 + (0.0 - 0.09)^2 + (0.07 - 0.09)^2 = 0.06$

$CRV_{ex1} = 0.07$

Finally, when we have: $C = \{ex1, ex3, ex4\}$

$CRV_{ex1,wc} = (0.39 - 0.32)^2 + (0.28 - 0.32)^2 + (0.29 - 0.32)^2 = 0.01$

$CRV_{ex1,oc} = (0.0 - 0.02)^2 + (0.0 - 0.02)^2 + (0.07 - 0.02)^2 = 0.00$

$CRV_{ex1} = 0.01$

Single Layer Algorithm

It is possible to define the following bottom up clustering algorithm that is based on the CRV score.

CRV score based single layer clustering

1) Each example is in its own cluster
2) Iteratively repeat steps 3-6
3) For each pair of clusters x,y compute
   $CRVx$ (mean CRV_i for examples in cluster x)
   $CRVy$ (mean CRV_i for examples in cluster y)
   $CRVxy$ (mean CRV_i score in union of clusters x and y)
   $DIFF = mean(CRVx, CRVy) - CRVxy$
4) Select pair of clusters x,y with maximal $DIFF$ value
5) If maximal DIFF is positive then merge clusters x and y
6) Else stop.

The algorithm has a property that at first most similar examples will be merged together. In this way it produces a hierarchy of clusters. It may be noticed that in contrast to most other clustering algorithms, it has a very well defined stopping criteria. The process stops when further merging cannot result in reduction of the example variability measured by the CRV score. It means that the algorithm automatically determines the optimal number of clusters and that some examples may stay unclustered (more precisely, they remain as clusters consisting of only one example).

2.2 Multilayer Algorithm

The basic lesson learnt from redescription mining and multi-view clustering is that the reliability of clustering can be significantly improved by a requirement that the result should be confirmed in two or more attribute layers. The approach for clustering based on example similarity has been presented in the previous section for a single layer case. It can be easily extended to clustering in multilayer domains.

If we have more than one attribute layer then for each of them we compute the example similarity table independently. For each layer we have to construct its own artificial classification problem and execute the supervised learning process in order to determine similarity between examples. Regardless of the number and type of attributes in different layers, the tables will be always matrices of dimension N times N. The reason is that by definition we have the same set of N examples in all layers.

After the computation of similarity tables, we execute the second step of the clustering process. Conceptually it is identical to a single layer approach. The main difference is that merging of two clusters is possible only if there is variability reduction in all layers. For each possible pair of clusters we have to compute potential variability reduction for all attribute layers and to select the smallest value for this pair. If this minimal value is positive it means that merging of the clusters enables variability reduction in all layers. When there are more pairs with positive minimal value, we chose the pair with the largest minimal value and then we merge these clusters in the current iteration.

CRV score based multilayer clustering
1)   Each example is in its own cluster
2)   Iteratively repeat steps 3-8
3)   For each pair of clusters x,y do
4)   For each attribute layer do
                 $CRVx$ (mean CRV_i for examples in cluster x)
                 $CRVy$ (mean CRV_i for examples in cluster y)
                 $CRVxy$ (mean CRV_i score in union of clusters x and y)
                 $DIFF = mean(CRVx, CRVy) - CRVxy$
5)   For the given pair x,y select minimal $DIFF$ for all layers
6)   Select pair of clusters x,y with maximal $DIFF$ value
7)   If maximal DIFF is positive then merge clusters x and y
8)   Else stop.

When we do clustering in two or more layers we have a conjunction of necessary conditions for merging two clusters. A typical consequence is that resulting clusters are smaller than in the case of a single layer clustering. This is illustrated by the experiment presented in the next section.

3   Experimental Data and Results

Our experimental work was conducted on the trading data that are publicly available from UNCTAD [11]. This database contains information for each pair of countries about the value of trade for 106 different commodity types. We have selected 155 countries from the database with relatively small number of unknown values for the year 2012. For them we have computed the total export value for the 106 different commodities. Finally, for each country we normalized indicator values by the value of country's total export in the year 2012. The result is a table with 155 rows and 106 columns. All known values are in the range 0-100 representing the percentage of export that a country has in the respective commodity type. Primary commodities, food and live animals, meat and meat preparations, machinery and transport equipment are some examples of aggregated commodity types. Some of the commodity types overlap. The prepared data table is publicly available from http://lis.irb.hr/DS2014data/ accompanied with the complete list of countries and the list of commodities.

The discovery task is to identify relevant subgroups of countries with a similar export patterns. The results are potentially relevant for understanding global trends, for example, by comparing the current subgroups with those obtained from data in year 2000. Our work has been motivated by the necessity to analyse and predict partial interests of EU countries in respect of a potential free trade agreement with China.

Table 2. Three largest clusters from export data

Cluster with 27 countries: Exporters of primary commodities
Gambia, Seychelles, Zambia, Burkina Faso, Guyana, Ethiopia, Mali, Paraguay, Malawi, Chile, DR Congo, Tajikistan, Afghanistan, Benin, Peru, Belize, Cote d'Ivoire, Mozambique, Guinea, Papua New Guinea, Ghana, Australia, Bolivia, Oman, Russian Federation, Kazakhstan, Romania
Cluster with 24 countries: Exporters of manufactured goods
Germany, Japan, Czech Republic, Slovakia, Italy, Slovenia, Austria, China-Taiwan, R. Korea, Hungary, Poland, Portugal, Turkey, Finland, Sweden, Bangladesh, Cambodia, Luxembourg, France, China, Thailand, USA, Mexico, United Kingdom
Cluster with 17 countries: Fuel exporters
Algeria, Libya, Nigeria, Iraq, Angola, Congo, Brunei, Azerbaijan, Aruba, Gabon, Venezuela, Yemen, Iran, Saudi Arabia, Kuwait, Qatar, Mongolia

Table 2 presents the three largest clusters constructed from the export data layer by the single layer methodology described in Section 2. We have given a name to each cluster based on the common properties of included countries that have been identified by a simple statistical analyses. The largest cluster includes 27 countries that are mainly primary commodity exporters. The other two clusters contain exporters of manufactured goods and fuel exporters. It can be recognized from the lists of countries included in these clusters that the algorithm has been successful in identification of similarities between countries. However, clusters also include some unexpected results such as: Australia, Russia, and Romania being in the cluster of primary commodity exporters together with Guyana and Ethiopia, Bangladesh and Cambodia being in the cluster with Germany and Japan, while Mongolia participates in the cluster of fuel exporters.

Table 3. Three largest clusters from socio-economic data

Cluster with 22 countries: Rural and young population
Ethiopia, Malawi, Uganda, Rwanda, Papua New Guinea, Niger, Burkina Faso, Tanzania, Afghanistan, Kenya, Tajikistan, Mozambique, Yemen, Togo, Zambia, Zimbabwe, DR Congo, Guinea, Madagascar, Mali, Benin, Senegal
Cluster with 18 countries: Modest level of rural population
Estonia, Hungary, Ukraine, Latvia, Austria, Italy, Lithuania, Czech Republic, Germany, Bulgaria, Belarus, Cuba, Spain, Greece, Poland, Croatia, Portugal, Switzerland
Cluster with 16 countries: Urban population
Denmark, France, Sweden, Finland, Netherlands, New Zealand, Iceland, Uruguay, Japan, Belgium, Malta, Australia, Canada, Norway, UK, USA

One possible interpretation is that only export data is insufficient information for effective and very consistent clustering of countries. In order to increase the quality of the results we have prepared the second layer of attributes. It consists of 105 World Bank indicators [12] that describe socio-economic characteristics of countries in the year 2012. We have selected indicators from economic policy, health, agriculture, and gender sets of public World Bank data. Our goal has been to select the most representative indicators from each field. The additional criterion was to use only relative indicators, that do not need normalization, in order to be comparable between countries of different size. We present a small sample of selected indicators for better insight: "Life expectancy at birth", "Percentage of population ages 15-64", "Public health expenditure as percentage of gross domestic product", and "Central government debt as percentage of gross domestic product" etc. The constructed attributes are all numeric and there is a noticeable amount of missing values. The data set is prepared for the same set of 155 countries as in the UNCTAD dataset and it is publicly available from our web site.

Before using both layers, we will present the clustering result obtained with socio-economic data in Table 3. It is interesting because it demonstrates that a dominant socio-economic characteristic of a country is the ratio of rural and urban population. The result is not coherent and it happens that the cluster with moderate number of rural population includes countries like Germany and Switzerland but also Cuba and Belarus. In the same way, the cluster of countries with high percentage of urban population includes USA and Norway together with Uruguay and Malta. From the methodological point of view this is not a bad result but constructed clusters are not very useful for our discovery task because they tend to group economically very different countries.

Next, we have merged the export and socio-economic data into a single layer consisting of 211 attributes. The result obtained by the single layer methodology on this data has been very similar to the result obtained only on export data. Again, the three largest clusters represent primary commodity exporters, manufactured goods exporters, and fuel exporters. The results are now more

Table 4. Clusters detected by the multilayer approach in which export data and socio-economic data are in different layers

Cluster 1 with 8 countries
Coted'Ivoire, Ghana, Guinea, Mozambique, Papua New Guinea, Mali, DR Congo, Zambia
Cluster 2 with 4 countries
Czech Republic, Germany, Austria, Italy
Cluster 3 with 3 countries
Congo, Iraq, Angola
Cluster 4 with 3 countries
Poland, Portugal, Hungary
Cluster 5 with 3 countries
Finland, Sweden, Japan
Cluster 6 with 2 countries
Kuwait, Qatar
Cluster 7 with 2 countries
Ethiopia, Malawi
Cluster 8 with 2 countries
Latvia, Lithuania

consistent, Mongolia is discarded from the fuel exporters cluster and Romania and Russia are not in the cluster of primary commodity exporters. However, Australia and Iceland have been included in this cluster!

Finally, we present the result obtained by the multilayer approach in Table 4. In this approach, export and socio-economic data have been treated as separate layers. At first glance it can be noticed that the constructed clusters are significantly smaller but more coherent. The largest cluster has 8 countries that can be described as a group of countries with rural population that export primary commodities. Their basic common characteristic is that more than 87% of their exports are primary commodities. For Mozambique it is aluminium, beryllium, and tantalum, Ghana exports gold and diamonds, Zambia copper, Mali exports gold and kaolin while Cote d'Ivoire is one important exporter of cocoa. Some other common characteristics of these countries are that they export a low amount of manufactured goods (less than 11%) and a low amount of other food staff excluding tea, coffee, cocoa and spices (less than 24.5%).

For our discovery task, much more relevant result is the identification of a group of four EU countries: Czech Republic, Germany, Austria and Italy. At first, it may be a bit surprising that these countries have been identified as a most coherent group of EU countries. Recognition of their common characteristics is not a simple task because it is a small cluster and each of these four countries share a lot of common characteristics with other developed economies, especially those in EU. A potential solution is a simple statistical comparison of properties with most similar examples *not included* in the cluster. In multilayer methodology most similar examples may be identified as those included in larger clusters constructed for single layers that contain examples from multilayer clusters. Figure 1 illustrates the relations for our domain in which, for example, the cluster consisting of Czech Republic, Germany, Austria and Italy is a subset of the clusters of manufactured goods exporters (layer 1) and the cluster of countries with modest rural population (level 2). In this figure arrows denote superset/subset relation and numbers denote sizes of clusters. Clusters at the basic layers are identified by the given names representing dominant characteristic of included countries while the clusters obtained by the combination of layers are represented by lists of included countries.

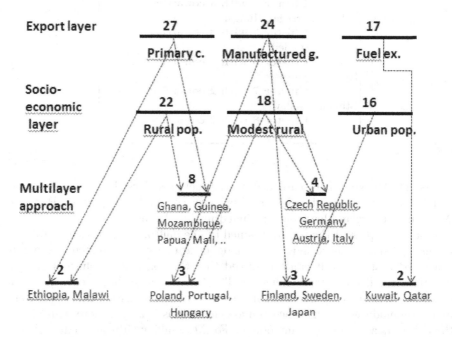

Fig. 1. Approximative superset/subset relations among constructed clusters

By using this approach we have identified the following decisive characteristics for the cluster consisting of Czech Republic, Germany, Austria, and Italy: a) high export of medium-skill and technology-intensive manufactures, b) low export of primary commodities, precious stones and non-monetary gold, c) low but always present export of beverages and tobacco, d) very low percentage of young population, d) low market capitalization of companies relative to gross domestic product. Figure 2 presents distributions of these five characteristics in three different clusters: in the cluster of 24 countries representing manufactured goods exporters, cluster of 18 countries that have modest level of rural population, and finally for the target cluster consisting of four countries. This figure demonstrates that the resulting multilayer cluster has very narrow range of values for some relevant attributes. Furthermore, this fact is also true for some properties which

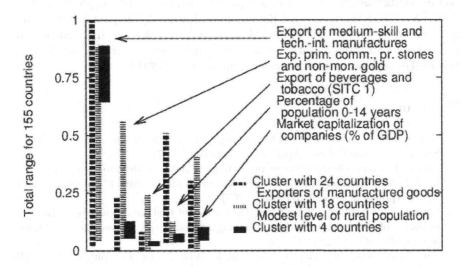

Fig. 2. Distribution of values in three different clusters for five attributes: three from the export layer and two from the socio-economic layer

do not occur in the supersets. In this way we identified that low percentage of young population and low market capitalization of companies as percentage of GDP are additional properties of this cluster of countries. Identification of these properties may present a potentially relevant discovery result.

4 Conclusions

In this work we have presented a novel clustering methodology that may be useful in different discovery tasks. The most decisive advantages are that it may be successfully used on instances described by both numeric and nominal attributes and that it has a well defined stopping criteria. Experimental evaluation of this methodology and its comparison with other known approaches will be a topic of our future work. In this paper we used the country level trading data for the illustration of the results one can expect from this novel methodology. The results are encouraging because we succeeded to get coherent clusters with examples that have narrow ranges of attribute values in some relevant attributes. In the interpretation of the common properties of the included examples, countries in our case, we have used the property that clusters constructed by the multilayer approach are typically subsets of clusters obtained on single layers. This approach enables us to undertake a statistical comparison with most similar examples that are *not* included in the resulting clusters. The most relevant problem of the methodology is that constructed clusters are small and that they will tend to be even smaller if additional data layers are included.

Acknowledgements. This work was partially supported by EU projects MUL-TIPLEX 317532 "Foundational Research on Multilevel Complex Networks and Systems" and and MAESTRA 612944 "Learning from Massive, Incompletely annotated, and Structured Data". It has been supported in part by the Croatian Science Foundation under the project number I-1701-2014.

References

1. Gan, G., Ma, C., Wu, J.: Data Clustering: Theory, Algorithms, and Applications. Society for Industrial and Applied Mathematics (2007)
2. Cha, S.H.: Comprehensive Survey on Distance/Similarity Measures between Probability Density Functions. International Journal of Mathematical Models and Methods in Applied Sciences 1, 300–307 (2007)
3. Kumar, V., Chhabra, J.K., Kumar, D.: Impact of Distance Measures on the Performance of Clustering Algorithms. Intelligent Computing, Networking, and Informatics 243, 183–190 (2014)
4. Sun, S.: A survey of multi-view machine learning. Neural Computing and Applications 23, 2031–2038 (2013)
5. Bickel, S., Scheffer, T.: Multiview clustering. In: Proc. of the Fourth IEEE Int. Conf. on Data Mining, pp. 19–26 (2004)
6. Parida, L., Ramakrishnan, N.: Redescription mining: Structure theory and algorithms. In: Proc. of Association for the Advancement of Artificial Intelligence (AAAI 2005), pp. 837–844 (2005)
7. Galbrun, E., Miettinen, P.: From black and white to full color: extending redescription mining outside the boolean world. In: Statistical Analysis and Data Mining, pp. 284–303 (2012)
8. Caldarelli, G.: Scale-Free Networks: Complex Webs in Nature and Technology. Oxford University Press (2007)
9. Breiman, L.: Random forests. Machine Learning 45(1), 5–32 (2001)
10. Pfahringer, B., Holmes, G., Wang, C.: Millions of random rules. In: Proc. of the Workshop on Advances in Inductive Rule Learning, 15th European Conference on Machine Learning, ECML (2004)
11. UNCTAD database, http://unctadstat.unctad.org/
12. World Bank, http://data.worldbank.org/indicator

Medical Document Mining Combining Image Exploration and Text Characterization

Nicolau Gonçalves, Erkki Oja, and Ricardo Vigário

Department of Information and Computer Science,
Aalto University School of Science, FI-00076 Aalto, Finland
`firstname.lastname@aalto.fi`

Abstract. With an ever growing number of published scientific studies, there is a need for automated search methods, able to collect and extract as much information as possible from those articles. We propose a framework for the extraction and characterization of brain activity areas published in neuroscientific reports, as well as a suitable clustering strategy of said areas. We further show that it is possible to obtain three-dimensional summarizing brain maps, accounting for a particular topic within those studies. After, using the text information from the articles, we characterize such maps. As an illustrative experiment, we demonstrate the proposed mining approach in fMRI reports of default mode networks. The proposed method hints at the possibility of searching for both visual and textual keywords in neuro atlases.

Keywords: Image mining, fMRI, meta-research, default mode network, text mining, neuroscience, brain mapping.

1 Introduction

In the field of neuroscience, research results often take the form of activation/suppression of activity in the brain, as a response to particular stimuli conditions, cognitive tasks, or pathological states, in a variety of image settings, orientations and resolutions. By browsing through those images and looking for specific areas, a neuroscientist is able to find corroborating evidence for his/her own research findings or suggestions of new areas of interest. This manual search stems from the lack of access to the original, raw data sets, resorting then to search the surrogate data itself. With an exponential increase in the number of published studies, that task becomes impractical, in view of the extent of human efforts required, *e.g.* [7,8].

In a field such as neuroscience, any meta-analysis study would clearly benefit from a more automated way to produce summaries of research findings from different articles. Those findings are encoded both in text structures, as well as in image content, providing ample scope for mining information at various levels. Nevertheless, the extraction of such data is not simple, and is subject to intense research in information retrieval and data mining [5].

It is, therefore, important that robust automated means of information extraction are further developed. These methods should allow for comparisons

S. Džeroski et al. (Eds.): DS 2014, LNAI 8777, pp. 99–110, 2014.

between articles, either focusing on particular areas of the brain or reported experimental setups. A recent, fully automated framework was proposed in by Yarkoni et al. [16], based on coordinate information explicitly reported in the scientific manuscripts. By combining text-mining, meta-analysis and machine-learning techniques, they generated probabilistic mappings between cognitive processes and neural states. In addition to requiring the existence of activation coordinates, which are not always reported, none of the vast visual information, such as figures and charts, is used to generate the aforementioned mappings.

In this manuscript, we propose a new framework to retrieve and structure visual information from neuroscientific articles, and use co-occurring text to characterize said information. We focus on the analysis of neuroscientific publications using functional magnetic resonance imaging (fMRI) data. We also assume that fMRI images contain summarizing results of the studies conducted, and that they may provide further information than the one extracted from text alone. To restrict the scope of application to a clear topic, we apply our method to manuscripts focusing on studying the resting state or default mode networks (RSN/DMN). These networks are a very active research topic, which still poses specific conceptual and methodological difficulties [14].

Fig. 1 shows the main areas of activation of the DMN, with axial, sagittal and coronal views from left to right, respectively. The highest areas of activity are the typical five subsystems: the posterior cingulate cortex (PCC), the precuneous, the medial pre-frontal cortex (mPFC) and the lateral parietal cortices. They are active when the individual is not performing any goal-oriented task, and suppressed during activity [11,3].

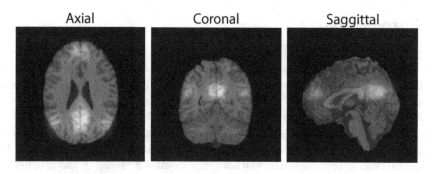

Fig. 1. Average brain activity in the default mode network, superimposed on a Colin-based brain template. From left to right are the axial, coronal and sagittal views of the nodes. The slices selected correspond to the location of highest intensity.

Several methods exist that focus on extracting textual and visual information for content retrieval [9]. Those multi-modal studies allow for an automatic analysis and characterization of documents and their content. Nonetheless, since they mostly restrict themselves to the analysis of complete images, the relation between particular image regions is lost. Furthermore, those studies are often general, and do not attempt to access the data layer behind the reported images. Here, as stated above, we aim at analyzing functional activity regions through their reported images, in a meta-research framework. Such study allows for the analysis of relations between blobs, independently of their reporting origins.

The main goal of this manuscript is to show how information contained in fMRI images of published reports can be extracted and analyzed. In the following sections, we summarize the image extraction procedure used, as well as the subsequent mapping of functional activity patterns onto a common brain template. We then explain how to cluster common areas of activation, gathering information present in different activity maps reported in different articles. The next section explains how those clusters can be characterized, using the textual information in the manuscripts building each cluster. Finally, we conclude the article with some remarks about the proposed approach, its limitations and future directions of research.

This work expands the method proposed in [4], by introducing the textual labeling component, and the clustering image mining strategy, which is used to produce sub-maps of activation of distinct neuroscientific content.

2   Methods

Fig. 2 shows a summarizing flowchart of the proposed methodology. Starting with the database of articles, the first step consists in finding and structuring all activity maps found therein. After, we group those maps according to a similarity measure, based on geometric overlap and proximity between maps. The next step is a hierarchical clustering one, using information about which articles build the group activity maps. Finally, we retrieve the textual information from the articles and use it to characterize each node of the previously created dendrogram.

2.1   Data

We built a database of neuroscientific publications available on-line, with a common research topic. We chose, for this work, to focus on studies of the default mode network. In addition, studies dealing with changes in DMN related to Alzheimer and Schizophrenia were also included. To build this database, methods such as the ones proposed in the context of ImageCLEF [9] benchmark set could be used. Due to the specificity of this work, we opted for a keyword based search, allowing for a more targeted search. This search was carried out using words such as DMN, Alzheimer, fMRI, cognitive impairment, Schizophrenia and resting state. In that way, we collected 183 articles in *pdf* format, from journals such as NeuroImage, Human Brain Mapping, Brain, Magnetic Resonance Imaging, PNAS and PLOS ONE. The time-frame for these articles ranged from early 2000 to June 2013.

The images and text were extracted from each article, using the open source command-line utilities *pdfimages* and *pdftotext*, respectively. This choice was due to both utilities being wide-spread and easy to use in different platforms.

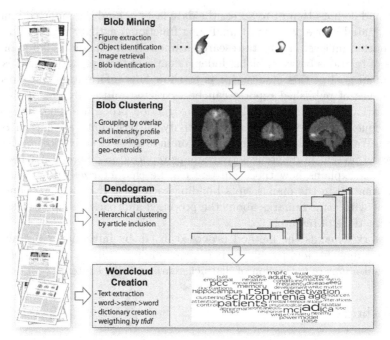

Fig. 2. Flowchart of the methodology proposed. Starting with the blob mining and clustering, the article information is used to group activity maps in a dendrogram, where each node can be characterized using textual information.

2.2   Brain from Blobs

A quick glance at functional MRI brain images permits the identification of several features of relevance, such as the kind of section of the image (axial, sagittal or coronal), various anatomical landmarks, as well as functional activity regions, often represented by colored "blobs" superimposed into gray-level anatomical MRI scans. We do this by relating the observed images to an internal representation of our anatomical and physiological knowledge of the brain.

With that in mind, we devised an image mining method [4], which extracts and maps those images to a common template. After extracting each figure in the article, we detect the various objects present in those figures, such as images, plots, charts or text annotations. To discard the non-interesting objects, and keep only fMRI images, we identify those frames that have properties specific to brains and neural activations, *e.g.* a minimum percentage of color and an aspect ratio typical of a brain image. After a further step to remove undesired annotations, the brain images can then be properly analyzed. This procedure is done automatically, with high accuracy [4]. Note that in fMRI reports, brain images follow a typical standard of activation areas overlayed on an anatomical image. When images differ from the standard, *e.g.* with too many annotations or with a non-standard brain reference, they are discarded.

In order to map the images to a common scale, we need to identify the template and section of the image. This allows for an estimation of the three-dimensional coordinates of the regions with activity changes, and the characterization of those regions in more detail[1]. After a color-map detection procedure, we can then map all blob intensity information to their respective coordinates. To do this, we use a Colin [2] brain template as reference. The result is a four-dimensional intensity map I of size $36 \times 43 \times 36 \times N$, where element $I_n(x, y, z)$ is the intensity of blob n, in the original image, at location (x, y, z) of the summarizing template brain. Tab. 1 presents the number of articles, figures, images and blobs studied in the present work.

For a complete description of the blob information extraction, we refer the reader to [4].

Table 1. Number of articles, figures, images and blobs used in this study

Articles	Figures	Images	Blobs
183	284	1487	6095

2.3 Grouping Similar Blobs

In order to identify the different regions represented in the various articles, one can group the blobs according to their geometric and intensity similarities. The main goal here being to group common activation patterns across various images, figures or even articles. This similarity needs to take into account blob sizes, overlaps and sections.

When considering the grouping of two blobs, we accepted only those with a clear degree of overlap, and for which a considerable similarity in the intensity patterns was evident. We decided to use the *cosine*[15] distance to compare pairs of brains intensity maps, i_{n_1} and i_{n_2}:

$$d_1(n_1, n_2) = \frac{i_{n_1}^* \cdot i_{n_2}^*}{\|i_{n_1}^*\| \|i_{n_2}^*\|} , \tag{1}$$

where $i_{n_1}^*$ and $i_{n_2}^*$ are the vectors forms of I_{n_1} and I_{n_2}, for which both I_n have values greater than 0.

After iteratively joining intensity maps with $d_1(n_1, n_2) > 0.75$, we reached a set of G functional activation groups.

Although these groups already gathered activation in similar locations in the brain, we needed to reduce the number of individual functional basic areas, and join maps from the same regions. Defining the $[1 \times 3]$ vector g_j, as the centroid of group G_j, we clustered the previously found G groups using k-means. The selected k corresponded to the one minimizing:

$$k = \arg\min_k \frac{\sum_{i=1}^k \sum_{g_j \in K_i} \|g_j - \mu_i\|}{\sum_{i=1}^k \sum_{j \neq i} \|\mu_j - \mu_i\|} . \tag{2}$$

---

[1] Note that such processing would allow already for an automated annotation of activation coordinates, essential for the meta-research proposed in [16].

where $\boldsymbol{\mu}_i$ is the centroid of cluster K_i. The number of clusters k has a significant impact in the overall performance. While low values of k force the grouping of distant blobs, large ones create too many singular activity points for a proper study of networks.

At this stage, we have k clusters representing several different regions of brain activity per cluster.

2.4   Clustering Brain Regions through Article Similarity

In many situations, single locations do not fully represent the complete neural response of external stimuli. In fact, most images reported comprise several regions of activation per depicted brains, *e.g.* bilateral activations.

After clearly identifying the various "atomic" regions reported in all the studies, we decided to join those, based on the articles that built each region. The main idea was to consider that different articles reporting the same basic activation sites will have a common research thread.

To find subtle relations between different brain regions, a new processing step needs to be performed. This step uses information from which articles built the previously mentioned clusters K_i. First we create a matrix $\boldsymbol{V}$ of size $k \times A$, where A is the total number of articles in the study (183). We denote $v_{K_i}(j)$ as the element corresponding to the number of blobs from article j, that are found in cluster K_i and $\boldsymbol{v}_{K_i}$ as the corresponding vector for all articles. With this matrix, we performed hierarchical clustering by joining all clusters through similitude of articles building up each cluster of blobs.

Again we group all k clusters using the *cosine* distance:

$$d_2(K_1, K_2) = \frac{\boldsymbol{v}_{K_1} \cdot \boldsymbol{v}_{K_2}}{\|\boldsymbol{v}_{K_1}\| \|\boldsymbol{v}_{K_2}\|} \ . \tag{3}$$

This results in a dendrogram with k end leafs, where the branches are made according to similarity between blob-article vectors.

We can now find those nodes with the least amount of article overlap between them. This is done using with the highest Hamming distance between nodes. For illustration purposes, we decided to select the 4 most disparate nodes, and will refer to them using by the notation M_i. Such nodes should contain interesting regions to analyze and compare. Since they correspond to viewpoint, as well as from the pool of articles and studies that mentioned those regions.

2.5   Text Analysis

After the process described above, the resulting dendrogram nodes still lack an easy way to be analyzed and characterized, besides the visual comparison of the resulting intensity maps. Since we only used images and their originating articles until now, there was still a considerable amount of information that remained unused: the text in those articles.

Therefore, we searched for interesting patterns and regions in the data, by extracting text information from the respective articles in the referred nodes. To avoid confounding topics in the text, we decided to use only the title, the abstract and the conclusion/discussion sections. Furthermore, only the Porter[10] stemmed versions of the words were analyzed. To allow for a human-readable analysis and avoid misleading stems, we substituted them by their most common originating word. We also filtered out common stop words, as they do not contribute significantly to the description of the intensity maps found.

Once the text was processed, we computed the words, bigrams and trigrams for each article using the Ngram Statistics Package[1]. This resulted in a bag-of-words matrix B, with size $A \times W$, where W is the total number of n-grams found in all articles.

We then calculated the *tf-idf* [6,13], which gives more weight to words that appear frequently (term-frequency, *tf*) in a document and less to frequent terms in the corpus (inverse document frequency, *idf*). One could also use BM25 [12], as it seems more robust. Yet, it tends to focus more on scoring documents, whereas our interest lies in scoring terms representing a node. After applying the *tf-idf* to B, we performed a node-by-node analysis on each of the four nodes identified in the dendrogram. Since we searched for words that represented the nodes, we ignored terms present only in a single article per node. Sorting the n-grams for each article inside a cluster and computing the most common ones for the whole cluster, we obtained a sorted dictionary per cluster.

Having identified the n-grams that represent each cluster, we can finally proceed to characterize each of the four interesting nodes. We can do this by finding the most common words in all those clusters, and the ones that differentiate them the most. The former is a logical intersection of the most prevalent terms throughout the nodes, whereas the latter is akin to filtering out the prevalent words of other clusters.

Using the weights obtained by the *tf-idf* on the dictionary of each cluster, we can construct word-clouds, which help visualize the differences and commonalities between the aforementioned clusters.

3 Results

3.1 Blob-Article Dendrogram

Fig. 3 shows the dendrogram obtained when grouping all blob clusters. This dendrogram has $k = 256$ end nodes, and contains $(k - 1)$ leafs.

After calculating the Hamming distance between all nodes, four of them were selected to further analysis, since they had the biggest inter-distance, corresponding to the least amount of article overlap. These nodes are shown in their corresponding leafs in Fig. 3. The number of articles that have contributed for each of the nodes is shown in Tab. 2. Due to the dissimilitude in the articles originating each node, it is not surprising to see that the chosen nodes are located at very different branches of the dendrogram.

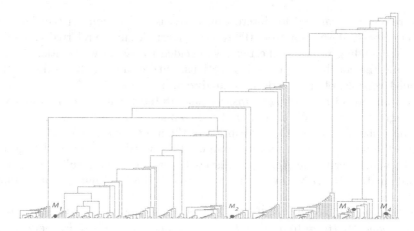

Fig. 3. Dendrogram resulting from the grouping through similitude of articles composing each node. The nodes analyzed in this article, corresponding to the ones with the least overlap of articles between them, are show in their respective location.

Table 2. Number of originating articles for each of the studied nodes

	M_1	M_2	M_3	M_4
Articles	39	81	101	85

3.2　Intensity Maps

Fig. 4 shows a summary of all brain activity changes reported for the clusters M_1 to M_4, from top to bottom, respectively. From left to right, are displayed the axial, coronal and sagittal views of the activity map volumes, centered at their maximum value of intensity.

As expected, they all represent different regions in the brain, since the intensity maps originate from as different as possible sets of articles. For example, see that all volumes have activation in the PCC region, in the central posterior part of the brain, although each with a subtle change in location. Nevertheless, only M_2 and M_4 show considerable frontal activity. Also, when compared to the overall DMN activity map image of Fig. 1, it is clear that none of the chosen nodes focuses on the lateral parietal cortices. Finally, M_4 seems to be the only node highlighting the anterior-posterior cingulate cortex interaction in the DMN.

3.3　Word Clouds

The word-cloud representing node M_1 is show on the top-left corner of Fig. 5. The size of the n-grams in the clouds at the corners of the figure is proportional to their weight in that node, where the biggest word corresponds to the most common word found in that node, but not present in any of the others.

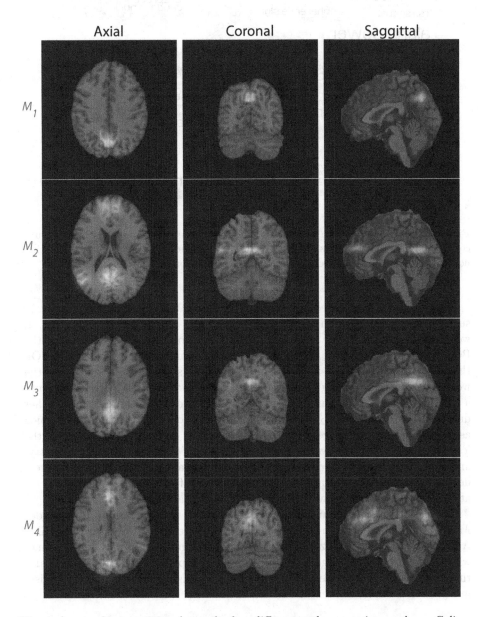

Fig. 4. Average brain activity change for four different nodes, superimposed on a Colin-based brain template. Each row corresponds to a different cluster, from M_1 to M_4. From left to right are the axial, coronal and sagittal views of the nodes.

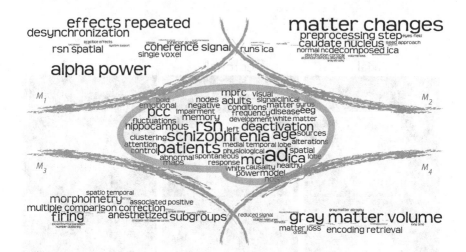

Fig. 5. Word-clouds for the analyzed clusters. On the corners are the word-clouds corresponding to the n-grams representing the corresponding node but not on the other nodes. The cloud in the middle contains the n-grams common to all four clusters.

Node M_1 is characterized by n-grams mainly related with frequency analysis, such as 'desynchronization', 'coherence signal' and 'alpha power'. From Fig. 4, we also know that it is the node that focuses most on one particular part of the PCC, possibly involved in the aforementioned communication mechanisms. On the top-right, bottom-left and bottom-right corners of the same figure are the word-clouds corresponding to M_2, M_3 and M_4 respectively. Without going into a thorough study of these nodes, one may nevertheless note that they are rather distinct in nature, with M_2 containing mostly methodology considerations, such as 'preprocessing' and 'decomposed ICA', as well as distinct regions, namely the 'caudate nucleus' or the 'eyes field'. M_3's n-grams of note include 'morphometry' and 'spatio temporal', and 'Stroop'. M_4 seems to deal with more general matters dealing with Alzheimer and Parkinson's disease effects, such as 'gray matter atrophy', 'matter loss' as well as studies in 'gray matter volume' estimation.

The word-cloud in the middle of Fig. 5 contains words common to all nodes M_i. As can be expected, n-grams referring to the resting state network, 'RSN', Alzheimer, 'AD', and 'schizophrenia' are prevalent in this cloud, as are the somewhat related 'patients', 'age' and mild cognitive impairment, 'MCI'. Notice also that the main region-related term is 'PCC', which is the only region that exists in all studied nodes, as seen in Fig. 4.

4   Discussion and Conclusion

Without having access to the original fMRI raw data, neuroscience meta-research is mostly human-based. Automation methods proposed in literature often rely in heavy curator work, or mining of textual information. Yet, most of the images contained in published articles, the first and foremost source of information used by researchers, is totally discarded.

The methodology here proposed allows for an automated way to gather a wider variety of information from the many neuroscientific articles published. It retrieves fMRI images from articles, and maps the activity reported therein to a template, allowing for an automatic summarization and comparison between images from different studies. The set of articles used in the summarizing maps can then be used to relate all studies, in a more general manner. Finally, the resulting relations and maps are characterized using textual information, retrieved from the original manuscripts. From the example study reported in the current manuscript, it is clear that the proposed method dramatically increases the data available for meta-studies in fMRI, by fully exploiting both image and textual information. Such process opens the door for the construction of a common atlas where to compare imaging results. In addition, a neuroscientist may also search for a particular keyword, either from a functional, methodological or anatomical origin, to find interesting activity maps and their corresponding studies.

Although the illustrative results are very promising, there are some limitations to the proposed methodology. The method scales quadratically with respect to the number of blobs, due to the calculation of pairwise distances. Nevertheless, the addition of new data may require a completely new run. We are currently exploring the use of incremental clustering strategies, to minimize this limitation. In the blob mapping stage, different thresholds and methods are used by the researchers in their publications. This will obviously impair the atlas creation, since currently all images contribute equally for that atlas. On the textual side, the set of stop-words may depend on the studied meta-research topic. The wrong choice of such words may lead to useless labeling of the activity maps. Also, one would need to create a match dictionary, to be able to replace acronyms by their corresponding meanings. This thesaurus would allow for a better dictionary creation stage, where acronyms and their opened versions would correspond to the same entry. While some existing thesaurus deal specifically with medical terms[2], to our knowledge, one specific to general fMRI studies is not available.

The robustness of the methodology proposed in this manuscript is proportional to the number of articles used to create the database. If the database doesn't contain enough information to populate the different regions of the brain, then any search conducted therein will be limited. 183 articles may seem rather limited, in particular when taking into account the vast amounts of published data that are routinely analyzed. Nevertheless, often one is limited to the available data. One example is the current study, where the articles cover most of the available material dealing with DMN, Alzheimer and Schizophrenia.

As future work, we can improve on the text information filtering and develop a system of brain-atlases, or related search engines, capable of finding and displaying information both from an anatomical and functional perspectives. These search engines/atlases would also be able to find links between different studies.

---

[2] http://www.nlm.nih.gov/research/umls/

This would open new ways to create an ontology of the brain, bringing support for further advancement of the neuroscience field.

References

1. Banerjee, S., Pedersen, T.: The design, implementation, and use of the Ngram Statistic Package. In: Proc. of the 4th International Conference on Intelligent Text Processing and Computational Linguistics, Mexico City, pp. 370–381 (2003)
2. Brett, M., Johnsrude, I.S., Owen, A.M.: The problem of functional localization in the human brain. Nature Reviews Neuroscience 3(3), 243–249 (2002)
3. Deco, G., Jirsa, V.K., McIntosh, A.R.: Emerging concepts for the dynamical organization of resting-state activity in the brain. Nature Reviews Neuroscience 12(1), 43–56 (2011)
4. Gonçalves, N., Vranou, G., Vigário, R.: Towards automated image mining from reported medical images. In: Proc. of VipIMAGE 2013 - 4th ECCOMAS Thematic Conference on Computational Vision and Medical Image Processing, pp. 255–261. CRC Press (2013)
5. Hand, D., Mannila, H., Smyth, P.: Principles of Data Mining. MIT Press (2001)
6. Jones, K.S.: A statistical interpretation of term specificity and its application in retrieval. J. Doc. 28(1), 11–21 (1972)
7. Laird, A.R., Lancaster, J.L., Fox, P.T.: Lost in localization? the focus is meta-analysis. Neuroimage 48(1), 18–20 (2009)
8. Levy, D.J., Glimcher, P.W.: The root of all value: a neural common currency for choice. Current Opinion Neurobiology 22(6), 1027–1038 (2012)
9. Müller, H., Clough, P., Deselaers, T., Caputo, B.: Experimental evaluation in visual information retrieval. The Information Retrieval Series, vol. 32 (2010)
10. Porter, M.F.: An algorithm for suffix stripping. Program: Electronic Library and Information Systems 14(3), 130–137 (1980)
11. Raichle, M.E., MacLeod, A.M., Snyder, A.Z., Powers, W.J., Gusnard, D.A., Shulman, G.L.: A default mode of brain function. Proc. National Academy Science U.S.A. 98(2), 676–682 (2001)
12. Robertson, S., Zaragoza, H.: The probabilistic relevance framework: BM25 and beyond. Now Publishers Inc. (2009)
13. Salton, G., Buckley, C.: Term-weighting approaches in automatic text retrieval. Information Processing & Management 24(5), 513–523 (1988)
14. Snyder, A.Z., Raichle, M.E.: A brief history of the resting state: The Washington University perspective. NeuroImage 62(2), 902–910 (2012)
15. Turney, P.D., Pantel, P.: From frequency to meaning: Vector space models of semantics. J. Artificial Intelligent Research 37(1), 141–188 (2010)
16. Yarkoni, T., Poldrack, R.A., Nichols, T.E., Van Essen, D.C., Wager, T.D.: Large-scale automated synthesis of human functional neuroimaging data. Nature Methods 8(8), 665–670 (2011)

Mining Cohesive Itemsets in Graphs

Tayena Hendrickx, Boris Cule, and Bart Goethals

University of Antwerp, Belguim
`firstname.lastname@uantwerp.be`

Abstract. Discovering patterns in graphs is a well-studied field of data mining. While a lot of work has already gone into finding structural patterns in graph datasets, we focus on relaxing the structural requirements in order to find items that often occur near each other in the input graph. By doing this, we significantly reduce the search space and simplify the output. We look for itemsets that are both frequent and cohesive, which enables us to use the anti-monotonicity property of the frequency measure to speed up our algorithm. We experimentally demonstrate that our method can handle larger and more complex datasets than the existing methods that either run out of memory or take too long.

1 Introduction

Graph mining is a popular field in data mining, with wide applications in bioinformatics, social network analysis, etc. Traditional approaches have been largely limited to searching for frequent subgraphs, i.e., reoccurring structures consisting of labelled nodes frequently interconnected in exactly the same way. However, the concept of frequent subgraphs is not flexible enough to capture all patterns. First of all, subgraphs are too strict. If we consider the graph given in Figure 1, we see that items a, b and c make up a pattern that visibly stands out. However, this pattern will not be found by subgraph mining since the three items are never connected in the same way. Subgraph mining approaches are also typically computationally complex. To begin with, they are forced to deal with the graph isomorphism problem. For small graphs, isomorphism checking is not really that hard. However, when we want to mine large graphs, like social networks, the isomorphism checks become computationally very expensive. On top of this, due to the fact that both edges and nodes must be added to the pattern, a large number of candidate subgraphs is generated along the way.

In order to avoid these problems, Cule et al. [5] proposed a *Cohesive Itemset Approach* for mining interesting itemsets in graphs. An interesting itemset is defined as a set of node labels that occur often in the graph and are, on average, tightly connected to each other, but are not necessarily always connected in exactly the same way. Although this method could find previously undetected patterns, there are still a number of drawbacks. The proposed method considers an itemset frequent if a large enough proportion of the graph is covered by items making up the itemset. As a result, large itemsets, sometimes partially consisting of very infrequent items, can be found in the output. This undermines

S. Džeroski et al. (Eds.): DS 2014, LNAI 8777, pp. 111–122, 2014.

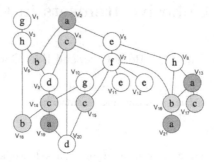

Fig. 1. A graph containing a pattern not discovered by subgraph mining

the attempts to prune the search space and results in prohibitive run-times and memory usage. To overcome this problem, we propose to look for *Frequent Cohesive Itemsets*, where we only consider itemsets consisting of labels that are all, as individual items, frequent, and look for those that on average occur near each other. In this way, we greatly reduce the search space, resulting in a smaller output and a significant reduction in the time and space complexity of our algorithm. We further explore the possibility of pruning candidate itemsets based on the cohesion measure as well. Cohesion is not anti-monotonic, but we develop an upper bound that allows us to prune whole branches of the search tree if certain conditions are satisfied. We experimentally confirm that our algorithm successfully handles datasets on which the existing method fails. In further experiments on an artificial dataset, with a limited alphabet size, we demonstrate exactly where our algorithm outperforms the existing method.

The rest of the paper is organised as follows. In section 2 we discuss the main related work. Section 3 formally describes our problem setting, and Section 4 presents our algorithm. In Section 5 we present the results of our experiments, before ending the paper with our conclusions in Section 6.

2   Related Work

The problem of discovering patterns in graphs is an active data mining topic. A good survey of the early graph based data mining methods is given by Washio and Motoda [19]. Traditionally, pattern discovery in graphs has been mostly limited to searching for frequent subgraphs, reoccurring patterns within which nodes with certain labels are frequently interconnected in exactly the same way.

The first attempts to find subgraph patterns were made by Cook and Holder [4] for a single graph, and by Motoda and Indurkhya [23] for multiple graphs. Both use a greedy scheme that avoids the graph isomorphism problem, but may miss some significant subgraphs. Dehaspe and Toivonen [6] perform a complete search for frequent subgraphs by applying an ILP-based algorithm.

Inokuchi et al. [9] and Kuramochi and Karypis [13] proposed the AGM and FSG algorithms for mining all frequent subgraphs, respectively, using a breadth-first

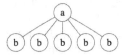

Fig. 2. A graph on which the GRIT algorithm gives counterintuitive results

search. These algorithms suffer from two drawbacks: costly subgraph isomorphism testing and an enormous number of generated candidates (due to the fact that both edges and nodes must be added to the pattern). Yan and Han [20] proposed GSPAN, an algorithm that uses a depth-first search. A more efficient tool, called GASTON, was proposed by Nijssen and Kok [15]. Further attempts at subgraph mining have been made by Inokuchi et al. [10], Yan et al. [21, 22], Huan et al. [8], Kuramochi and Karypis [14] and Bringmann and Nijssen [2].

At first glance, it may seem that itemsets, as patterns, are not as expressive as subgraphs. Nevertheless, Karunaratne and Boström [11] showed that itemset mining algorithms are computationally simpler than their graph mining counterparts and are competitive in terms of results. Recently, Cule et al. [5] proposed the GRIT algorithm for mining interesting itemsets in graphs. An itemset is defined as a set of node labels which often occur in the graph and are, on average, tightly connected to each other. Although the method is more flexible than the traditional approaches, it has some drawbacks. The interestingness of an itemset is defined as the product of its coverage and its cohesion, where the coverage measures what percentage of the graph is covered by items making up the itemset, while the cohesion measures average distances between these items. Due to the small world phenomenon, this approach can result in an item that occurs very infrequently in the dataset being discovered as part of an interesting itemset. Consider the graph given in Figure 2. The GRIT algorithm will discover pattern ab as interesting, because, per definition, the coverage of ab will be larger than the coverage of the individual items a and b. Since each a is connected to a b and vice versa, ab will also score well on cohesion. Although item a is not frequent at all, it has made its way into the output thanks to having many neighbours labelled b. However, itemset ab does not represent a reoccurring pattern and should not be considered more interesting than item b on its own.

In this paper, we build on this work, and focus on mining frequent cohesive itemsets, insisting that each item in a discovered itemset must itself be frequent. A separate cohesion threshold ensures that itemsets we discover are also cohesive. By using a two-step approach of first filtering out the infrequent items, and then using the frequent items to generate candidate itemsets, which are then evaluated on cohesion, we greatly reduce the search space and run-time of our algorithm and produce more meaningful results. By searching for itemsets rather than subgraphs, we also avoid the costly isomorphism testing, and by using a depth-first search algorithm, we avoid the pitfalls of breadth-first search.

Another itemset mining approach has been proposed by Khan et al. [12], where nodes propagate their labels to neighbouring nodes according to given probabilities. Labels are thus aggregated, and can be mined as itemsets in the resulting graph. Silva et al. [16, 17] and Guan et al. [7] introduced methods to identify correlation between node labels and graph structure, whereby the subgraph constraint has been loosened, but not entirely dropped.

3   Frequent Cohesive Itemsets

In this section, we introduce our approach for mining *Frequent Cohesive Itemsets* in a dataset consisting of a single graph. We assume that the graph consists of labelled nodes and unlabelled edges, and we focus on connected graphs with at most one label per node. However, we can also handle input graphs where each node carries a set of labels, as will be shown in Section 5.

To start with, we introduce some notations and definitions. In a graph G, the set of nodes is denoted $V(G)$. Each node $v \in V(G)$ carries a label $l(v) \in S$, where S is the set of all labels. For a label $i \in S$, we denote the set of nodes in the graph carrying this label as $L(i) = \{v \in V(G)|l(v) = i\}$. We define the frequency of a label $i \in S$ as the probability of encountering that label in G, or

$$freq(i) = \frac{|L(i)|}{|V(G)|}.$$

From now on, we will refer to labels as items, and sets of labels as itemsets.

In order to compute the cohesion of an itemset X we first denote the set of nodes labelled by an item in X as $N(X) = \{v \in V(G)|l(v) \in X\}$. In the next step, for each occurrence of an item of X, we must now look for the nearest occurrence of all other items in X. For a node v, we define the sum of all these smallest distances as

$$W(v, X) = \sum_{x \in X} min_{w \in N(\{x\})} d(v, w),$$

where $d(v, w)$ is the length of the shortest path from node v to node w. Subsequently, we compute the average of such sums for all occurrences of items making up itemset X:

$$\overline{W}(X) = \frac{\sum_{v \in N(X)} W(v, X)}{|N(X)|}.$$

Finally, we define the cohesion of an itemset X, where $|X| \geq 2$, as

$$C(X) = \frac{|X| - 1}{\overline{W}(X)}.$$

If $|X| < 2$, we define $C(X)$ to be equal to 1.

Cohesion measures how close to each other the items making up itemset X are on average. If the items are always directly connected by an edge, the sum of these distances for each occurrence of an item in X will be equal to $|X| - 1$, as will the average of such sums, and the cohesion of X will be equal to 1.

Given user defined thresholds for frequency (min_freq) and cohesion (min_coh), our goal is to discover each itemset X if $\forall x \in X : freq(x) \geq min\_freq$ and $C(X) \geq min\_coh$. To allow the user more flexibility, we use two optional size parameters, $minsize$ and $maxsize$, to limit the size of the discovered itemsets.

We will now illustrate the above definitions on the graph given in Figure 1. Assume we are evaluating itemsets abc and ef, with thresholds min_freq and min_coh set to 0.1 and 0.6, respectively. According to our definitions, we first note that $N(abc) = \{v_2, v_4, v_6, v_{13}, v_{14}, v_{15}, v_{16}, v_{17}, v_{18}, v_{19}, v_{21}\}$ and $N(ef) = \{v_5, v_7, v_{11}, v_{12}\}$. It follows that $|N(abc)| = 11$ and $|N(ef)| = 4$. To compute the cohesion of itemset abc, we now search the neighbourhood of each node in $N(abc)$ and obtain the following: $W(v_4, abc) = W(v_6, abc) = W(v_{16}, abc) = W(v_{17}, abc) = W(v_{18}, abc) = W(v_{19}, abc) = 3$, $W(v_{15}, abc) = W(v_{21}, abc) = 4$ and $W(v_2, abc) = W(v_{13}, abc) = W(v_{14}, abc) = 2$. The average of the above sums is $\overline{W}(abc) = \frac{32}{11}$. Doing the same for itemset ef we get $W(v_5, ef) = W(v_7, ef) = W(v_{11}, ef) = W(v_{12}, ef) = 1$. Therefore, $\overline{W}(ef) = 1$. We can now compute the cohesion of the two itemsets abc and ef as

$$C(abc) = \frac{|abc| - 1}{\overline{W}(abc)} = \frac{3 - 1}{\frac{32}{11}} = \frac{22}{32} \approx 0.69 \quad \text{and} \quad C(ef) = \frac{|ef| - 1}{\overline{W}(ef)} = \frac{2 - 1}{1} = 1.$$

We see that both abc and ef are cohesive enough, but, for an itemset to be considered a frequent cohesive pattern, each item in the itemset must be frequent. In our dataset, items a, b, c and e are frequent, but f is not. Therefore, although ef is more cohesive than abc, it will not be discovered as a frequent cohesive itemset. Note that we computed the cohesion of ef above only to illustrate the example. Our algorithm, presented in Section 4, first finds frequent items and then considers only itemsets that consist of these items, so itemset ef would not even be considered, and the above computations would not take place.

4   Algorithm

Our algorithm for mining frequent cohesive itemsets in a given graph consists of two main steps. The first step is to filter out the infrequent items. This can be done while loading the dataset into the memory for the first time, counting the frequency of each item as they occur, and then outputting only the frequent ones. In the second step, given in Algorithm 1, candidates are generated by applying depth-first search, using recursive enumeration. During this enumeration process, a candidate consists of two sets of items, X and Y. X contains those items which make up the candidate, while Y contains the items that still have to be enumerated. The first call to the algorithm is made with X empty and Y containing all *frequent* items appearing in the graph.

At the heart of the algorithm is the *UBC* pruning function (discussed in detail below), which is used to decide whether to prune the complete branch of the search tree, or to proceed deeper. If the branch is not pruned, the next test evaluates if there are still items with which we can expand the candidate. If not, we have reached a leaf node in the search tree, and discovered a frequent

---

Algorithm 1. FCI($\langle X, Y \rangle$) finds frequent cohesive itemsets

---

if $UBC(\langle X, Y \rangle) \geq min\_coh$ **then**
  if $Y = \emptyset$ **then**
    if $|X| \geq minsize$ **then output** X
  else
    Choose a in Y
    if $|X \cup \{a\}| \leq maxsize$ **then** FCI($\langle X \cup \{a\}, Y \setminus \{a\} \rangle$)
    if $|X \cup (Y \setminus \{a\})| \geq minsize$ **then** FCI($\langle X, Y \setminus \{a\} \rangle$)
  end if
end if

---

cohesive itemset which is sent to the output. Otherwise, the first item in Y, for example a, is selected and the FCI algorithm is recursively called twice: once with item a added to X and once without. In both calls, a is removed from Y.

An important property of the cohesion measure is that it is neither monotonic nor anti-monotonic. For example, consider the graph shown in Figure 3, and assume the min_coh threshold is set to 0.6. We can see that $C(ac) < min\_coh < C(abc) < C(a)$, even though $a \subset ac \subset abc$. Therefore, we will sometimes need to go deeper in the search tree, even when we encounter a non-cohesive itemset, since one of its supersets could still prove cohesive. However, traversing the complete search space is clearly unfeasible. In this work, we deploy an additional pruning technique using an upper bound for the cohesion measure, similar to the upper bound for the interestingness measure used by Cule et al. [5]. In a nutshell, we can prune a complete branch of the search tree if we are certain that no itemset generated within this branch can still prove cohesive. To ascertain this, we compute an upper bound for the cohesion measure of all the itemsets in that branch, and if this upper bound is smaller than the cohesion threshold, the whole branch can be pruned.

More formally, recall that a frequent cohesive itemset X, of size 2 or larger, must satisfy the following requirement:

$$C(X) = \frac{(|X|-1)|N(X)|}{\sum_{v \in N(X)} W(v, X)} \geq min\_coh.$$

Assume now that we find ourselves in node $\langle X, Y \rangle$ in the search tree, i.e., at the root of the subtree within which we will generate each candidate itemset Z such that $X \subseteq Z \subseteq X \cup Y$. We know that cohesion is neither monotonic nor anti-monotonic, so we cannot in advance know which such itemset will have the highest cohesion, and we need to develop an upper bound that holds for all of them. What we do know, however, is the cohesion of itemset X, $C(X)$, as shown

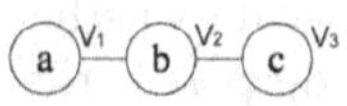

Fig. 3. A small input graph

above. We can now analyse the extent to which this cohesion can maximally grow as items from Y are added to X. Note that if an item is added to X, both the nominator and the denominator in the above expression will grow. In order for the total value to grow maximally, we therefore need to find the case in which the nominator will grow maximally, and the denominator minimally. Let us first examine the case of adding just a single item y to X. Recall that the denominator contains the total sum, for all occurrences of items in X in the graph, of the sums of all minimal distances from such an occurrence to all other items in X. Each such sum of minimal distances will now have to be expanded to include the minimal distance to the new item y. In the worst case, y will be discovered at a distance of 1 from all these occurrences. These sums will therefore grow by exactly 1. Additionally, the total sum will now also include the sums of minimal distances from each occurrence of y to all items in X. In the worst case, from each y we will be able to find each item in X at a distance of 1. Therefore, these sums will be equal to $|X|$. For the case of adding an item y to X, we thus obtain

$$\sum_{v \in N(X \cup \{y\})} W(v, X \cup \{y\}) \geq (\sum_{v \in N(X)} (W(v, X) + 1)) + |N(\{y\})||X|.$$

By induction, for adding the whole of Y to X, we obtain

$$\sum_{v \in N(X \cup Y)} W(v, X \cup Y) \geq (\sum_{v \in N(X)} (W(v, X) + |Y|)) + |N(Y)|(|X \cup Y| - 1).$$

What the above worst-case actually describes is a case of adding maximally cohesive occurrences of $X \cup Y$ to already known occurrences of X. For the overall cohesion to grow as much as possible, we should add as many such occurrences as possible. Clearly, we will obtain this maximal number of occurrences if we add the whole of Y to X. Therefore, we can update the nominator accordingly, and conclude that for any Z, such that $X \subseteq Z \subseteq X \cup Y$, it holds that

$$C(Z) \leq \frac{(|X \cup Y| - 1)|N(X \cup Y)|}{(\sum_{v \in N(X)} (W(v, X) + |Y|)) + |N(Y)|(|X \cup Y| - 1)}.$$

We also need to consider the user-chosen *maxsize* parameter. If $X \cup Y$ is larger than *maxsize*, the worst case will not be obtained by adding the whole of Y to X, but only by adding items from Y to X until *maxsize* is reached. Before proceeding with our analysis, we first abbreviate the new upper bounds for $|Y'|$ and $|N(Y')|$, where $Y' \subset Y$ and $|X \cup Y'| \leq maxsize$, with

$$UBY' = \min(maxsize - |X|, |Y|)$$

and

$$UBNY' = \min(|N(Y)|, \max_{\substack{Y_i \subset Y, \\ |Y_i| = maxsize - |X|}} |N(Y_i)|).$$

Finally, we develop the upper bound for the cohesion of all candidate itemsets generated in the branch of the search tree rooted at $\langle X, Y \rangle$ as

$$UBC(\langle X, Y \rangle) = \frac{(|X| - 1 + UBY')(|N(X)| + UBNY')}{(\sum_{v \in N(X)} (W(v, X) + UBY')) + (|X| - 1 + UBY')UBNY'}.$$

This upper bound is only defined when $|X| \geq 2$. If X is either empty or a singleton, we define $UBC(\langle X, Y \rangle) = 1$.

At first glance, it seems that in order to compute $UBNY'$, we would need to evaluate all possible sets Y_i, such that $Y_i \subset Y$ and $|X \cup Y_i| = maxsize$, which would be computationally expensive. We avoid this problem by enumerating the items in Y sorted on frequency in descending order. For example, if $Y = \{y_1, y_2, \ldots, y_n\}$, given $X = \{a, b, c\}$ and $maxsize$ set to 5, it is obvious that

$$\max_{\substack{Y_i \subset Y, \\ |X \cup Y_i| = maxsize}} |N(X \cup Y_i)| = |N(\{a, b, c, y_1, y_2\})|.$$

Similarly, it may seem that to compute $\sum_{v \in N(X)} W(v, X)$ at each node in our search tree, we would need to traverse the whole graph searching for the minimal distances between all items in X for all relevant nodes, which would be unfeasible. In order to avoid these costly database scans, we adopt the approach that was used in the GRIT algorithm [5] and express the sum of the minimal distances between items making up an itemset and the remaining items in X as a sum of separate sums of such distances for each pair of items individually. Each of these sums between individual items are stored in a matrix which is generated only once at the beginning of the algorithm. Since we are only interested in itemsets consisting of frequent items, it is sufficient to compute the minimal distances only for those frequent items. Consequently, the matrix we generate is of size $|F| \times |F|$, where F is the set of frequent items, which is, depending on the min_freq threshold, considerably smaller than the matrix of size $|S| \times |S|$, where S is the complete alphabet, generated by GRIT.

Thanks to the above two optimisations, given a candidate $\langle X, Y \rangle$, we can compute $UBC(\langle X, Y \rangle)$ in constant time.

5   Experiments

In this section, we experimentally compare the FCI and GRIT algorithms. For our experiments, we used two different datasets — a real-life graph dataset, and a synthetic dataset. The first dataset we used in our experiments is a combination of the yeast protein interaction network available in Saccharomyces Genome Database (SGD) [3] and the yeast regulatory network available in YEASTRACT [1]. In the combined network each node represents a yeast protein and each edge represents an interaction between two proteins. The given interaction network consists of 5 811 protein-nodes and 62 454 edges. The node labels are derived from gene ontology assignments [18], i.e., terms that are assigned to proteins that describe their biological process. The mapping of each of these labels to the list of yeast proteins was obtained from the annotation file provided by the SGD database [3]. In total there are 30 distinct labels. Since each protein-node has multiple labels, we first had to transform the given network. More formally, we expanded the graph by replacing each protein-node with a unique dummy node, surrounded by a set of nodes, each carrying one of the original labels. Our resulting graph consisted of 18 108 nodes and 74 751 edges.

Table 1. Results of the FCI algorithm on the protein interaction network of yeast

min_freq	$|F|$	C-Time (s)	maxsize	min_coh	#itemsets	#candidates	M-Time (ms)
0.001	23	8 974	5	0.40	12	24 774	88
				0.35	154	75 091	224
				0.30	7 989	122 752	586
0.005	15	7 966	5	0.40	9	5 720	37
				0.35	127	9 976	55
				0.30	3 196	13 923	205
0.010	10	6 453	5	0.40	9	1 127	10
				0.35	91	1 290	19
				0.30	612	1 473	70
0.005	15	7 966	∞	0.40	9	23 774	113
				0.35	169	48 916	250
				0.30	25 559	65 518	984

Table 1 reports the number of frequent items $|F|$, the time needed to generate the $|F| \times |F|$ distance matrix (C-Time), the number of discovered itemsets, the number of candidates that were considered, and the time needed for the mining stage (M-Time), for varying values of min_freq and min_coh. In the first three sets of experiments, minsize was set to 2 and maxsize to 5, while we set maxsize to infinity in the fourth set of experiments. C-Time and M-Time are reported separately because the frequent items are fixed for a given frequency threshold, so the distance matrix needs to be computed just once and can then be reused at various cohesion thresholds. The considerable reduction in mining times as the cohesion threshold grows shows that our pruning function has the desired effect. In Section 4, we presented two crucial elements in our pruning function, using the properties of the cohesion measure, and the maxsize parameter. In the fourth set of experiments, we set the maxsize parameter to infinity, to entirely eliminate its effect on pruning. The results show that we still manage to produce output quickly, while pruning large numbers of candidates.

The GRIT algorithm failed to produce output in all of these settings, as the required matrix and the resulting search space proved far too large. Therefore, we can conclude that filtering out the infrequent items is a crucial step if we wish to handle large real-life datasets.

Finally, let us have a closer look at the discovered itemsets. Having shown the output to biologists, they confirmed that the most cohesive patterns were those that could be expected for this type of network. For example, with min_freq set to 0.005 and min_coh to 0.2, the highest scoring itemset consists of two terms that are highly related, namely { cellular metabolic process, organic substance metabolic process }. Indeed, many of the proteins in the studied network are labelled with both terms as they describe overlapping biological processes in yeast. However, besides the expected patterns, biologists discovered some other patterns, such as { cellular metabolic process, organic substance metabolic process, biosynthetic process, catabolic process, regulation of biological quality }, an itemset of size 5 with a cohesion of 0.36. This itemset contains three terms that never occur together on a node, namely biosynthetic process, catabolic process

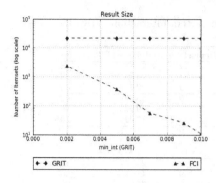
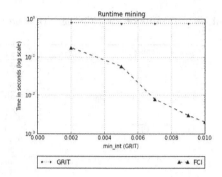

Fig. 4. Comparison of the output size of the GRIT and FCI algorithms

Fig. 5. Comparison of the mining times for the GRIT and FCI algorithms

and regulation of biological quality. Each of these three terms refers to very different, almost opposite, biological processes and thus no proteins exist in yeast that are active in all three categories. However, from a biological point of view, one can expect the nodes with these different terms to be close together in the network due to the regulatory mechanisms that exist in yeast, which propagate through the interactions described in the studied network.

For our second experimental setting, we generated a random graph with 100 000 nodes and 399 495 edges. The labels were randomly allocated and range from 0 to 19. The probability of encountering a label differed for each label, as follows: we defined $p_0 = 1$, $p_1 = 2$ and for $i = 2 \ldots 19$, $p_i = \sum_{j=1 \ldots i} j$. The probability of encountering label i was proportional to p_i. Given that $\sum_{i=0 \ldots 19} p_i = 1332$, the probability of encountering label 0 was $\frac{1}{1332}$ and the probability of encountering label 19 was $\frac{190}{1332}$. We built the graph by, at each step, adding either a new node and connecting it to a random existing node, or adding a new edge between two existing nodes. For this experiment, we set the probability of adding a new node to 25%, and the probability of adding a new edge to 75%, resulting in an average of approximately 8 edges per node

The main goal of this experiment was to be able to compare GRIT and FCI in terms of ouput size and run-time, since GRIT failed to generate output for the yeast protein interaction network. We applied the FCI algorithm with min_coh fixed at 0.1, $minsize$ set to 2, $maxsize$ to 5 and a varying min_freq threshold. The interestingness threshold, min_int, for GRIT was set to the product of min_coh and min_freq, to guarantee that all itemsets found by FCI were also found by GRIT (since GRIT defines interestingness as the product of coverage and cohesion, and coverage is equal to the frequency in case of singletons). Figure 4 shows the number of discovered itemsets for the two approaches. The mining times are reported in Figure 5. The preprocessing stage of GRIT took 2.5 days, while FCI needed between 1 and 50 minutes, depending on the chosen frequency threshold (as reported in Table 2). As we can see in the figures, the output size and the run-time of FCI decrease considerably as we increase the frequency threshold,

Table 2. Experimental results of the FCI algorithm on the artificial dataset

min_coh	maxsize	min_freq	|F|	C-Time (s)	#itemsets	#candidates	M-Time (ms)
0.1	5	0.10	4	62	11	25	2
		0.05	9	643	372	836	58
		0.02	13	2 853	2 366	6 460	175
0.1	∞	0.10	4	62	11	25	2
		0.05	9	643	502	1 012	64
		0.02	13	2 853	8 178	16 368	427

since more items are filtered out to start with, which results in fewer candidate itemsets. On the other hand, as the interestingness threshold, used by GRIT, increases, we see no change in the output size and run-times, since GRIT still generates a huge number of candidates. Finally, Table 2 shows a comparison of using the FCI algorithm with the *maxsize* parameter set to 5 and to infinity, respectively. Once again, we can see that our pruning shows good results, even if we cannot rely on the *maxsize* parameter as a pruning tool.

6   Conclusions

In this paper, we present a novel method for mining frequent cohesive itemsets in graphs. By first filtering out the infrequent items, and only then evaluating the remaining candidate itemsets on cohesion, we achieve much better results than existing algorithms, both in terms of run-times, and the quality of output. Furthermore, by limiting ourselves to itemsets, we avoid the costly isomorphism testing needed in subgraph mining. Using a depth-first search allows us to use an upper bound for the cohesion measure to prune unnecessary candidates, thus further speeding up our algorithm. Experiments demonstrate that the presented method greatly outperforms the existing ones on a variety of datasets.

Acknowledgments. We wish to thank Pieter Meysman, bioinformatician at the University of Antwerp, for helping us interpret the results of the experiments on the biological dataset.

References

[1] Abdulrehman, D., Monteiro, P.T., Teixeira, M.C., Mira, N.P., Lourenço, A.B., dos Santos, S.C., Cabrito, T.R., Francisco, A.P., Madeira, S.C., Aires, R.S., Oliveira, A.L., Sá-Correia, I., Freitas, A.T.: YEASTRACT: providing a programmatic access to curated transcriptional regulatory associations in Saccharomyces cerevisiae through a web services interface. Nucleic Acids Research 39 (2011)

[2] Bringmann, B., Nijssen, S.: What is frequent in a single graph? In: Washio, T., Suzuki, E., Ting, K.M., Inokuchi, A. (eds.) PAKDD 2008. LNCS (LNAI), vol. 5012, pp. 858–863. Springer, Heidelberg (2008)

[3] Cherry, J.: SGD: Saccharomyces Genome Database. Nucleic Acids Research 26(1), 73–79 (1998)

[4] Cook, D.J., Holder, L.B.: Substructure discovery using minimum description length and background knowledge. Journal of Artificial Intelligence Research 1, 231–255 (1994)

[5] Cule, B., Goethals, B., Hendrickx, T.: Mining interesting itemsets in graph datasets. In: Pei, J., Tseng, V.S., Cao, L., Motoda, H., Xu, G. (eds.) PAKDD 2013, Part I. LNCS, vol. 7818, pp. 237–248. Springer, Heidelberg (2013)

[6] Dehaspe, L., Toivonen, H.: Discovery of frequent datalog patterns. Data Mining and Knowledge Discovery 3, 7–36 (1999)

[7] Guan, Z., Wu, J., Zhang, Q., Singh, A., Yan, X.: Assessing and ranking structural correlations in graphs. In: Proc. of the 2011 ACM SIGMOD Int. Conf. on Management of Data, pp. 937–948 (2011)

[8] Huan, J., Wang, W., Prins, J., Yang, J.: Spin: mining maximal frequent subgraphs from graph databases. In: Proc. of the 10th ACM SIGKDD Int. Conf. on Knowledge Discovery and Data Mining, pp. 581–586 (2004)

[9] Inokuchi, A., Washio, T., Motoda, H.: An apriori-based algorithm for mining frequent substructures from graph data. In: Zighed, D.A., Komorowski, J., Żytkow, J.M. (eds.) PKDD 2000. LNCS (LNAI), vol. 1910, pp. 13–23. Springer, Heidelberg (2000)

[10] Inokuchi, A., Washio, T., Motoda, H.: Complete mining of frequent patterns from graphs: Mining graph data. Machine Learning 50, 321–354 (2003)

[11] Karunaratne, T., Boström, H.: Can frequent itemset mining be efficiently and effectively used for learning from graph data? In: Proc. of the 11th Int. Conf. on Machine Learning and Applications (ICMLA), pp. 409–414 (2012)

[12] Khan, A., Yan, X., Wu, K.L.: Towards proximity pattern mining in large graphs. In: Proc. of the 2010 ACM SIGMOD Int. Conf. on Management of Data, pp. 867–878 (2010)

[13] Kuramochi, M., Karypis, G.: Frequent subgraph discovery. In: Proc. of the 2001 IEEE Int. Conf. on Data Mining, pp. 313–320 (2001)

[14] Kuramochi, M., Karypis, G.: Finding frequent patterns in a large sparse graph. Data Mining and Knowledge Discovery 11, 243–271 (2005)

[15] Nijssen, S., Kok, J.: The gaston tool for frequent subgraph mining. Electronic Notes in Theoretical Computer Science 127, 77–87 (2005)

[16] Silva, A., Meira, J.W., Zaki, M.J.: Structural correlation pattern mining for large graphs. In: Proc. of the 8th Workshop on Mining and Learning with Graphs, pp. 119–126 (2010)

[17] Silva, A., Meira, J.W., Zaki, M.J.: Mining attribute-structure correlated patterns in large attributed graphs. Proc. of the VLDB Endowment 5(5), 466–477 (2012)

[18] The Gene Ontology Consortium: Gene Ontology annotations and resources. Nucleic Acids Research 41(Database issue), D530–D535 (2013)

[19] Washio, T., Motoda, H.: State of the art of graph-based data mining. ACM SIGKDD Explorations Newsletter 5, 59–68 (2003)

[20] Yan, X., Han, J.: gspan: Graph-based substructure pattern mining. In: Proc. of the 2002 IEEE Int. Conf. on Data Mining, pp. 721–724 (2002)

[21] Yan, X., Han, J.: Closegraph: Mining closed frequent graph patterns. In: Proc. of the 9th ACM SIGKDD Int. Conf. on Knowledge Discovery in Data Mining, pp. 286–295 (2003)

[22] Yan, X., Zhou, X., Han, J.: Mining closed relational graphs with connectivity constraints. In: Proc. of the 11th ACM SIGKDD Int. Conf. on Knowledge Discovery in Data Mining, pp. 324–333 (2005)

[23] Yoshida, K., Motoda, H., Indurkhya, N.: Graph-based induction as a unified learning framework. Journal of Applied Intelligence 4, 297–316 (1994)

Mining Rank Data

Sascha Henzgen and Eyke Hüllermeier

Department of Computer Science
University of Paderborn, Germany
{sascha.henzgen,eyke}@upb.de

Abstract. This paper addresses the problem of mining rank data, that is, data in the form of rankings (total orders) of an underlying set of items. More specifically, two types of patterns are considered, namely frequent subrankings and dependencies between such rankings in the form of association rules. Algorithms for mining patterns of this kind are proposed and illustrated on three case studies.

1 Introduction

The major goal of data mining methods is to find potentially interesting patterns in (typically very large) data sets. The meaning of "interesting" may depend on the application and the purpose a pattern is used for. Quite often, interestingness is connected to the *frequency* of occurrence: A pattern is considered interesting if its number of occurrences in the data strongly deviates from what one would expect on average. When being observed much more often, ones speaks of a *frequent pattern*, and the problem of discovering such patterns is called *frequent pattern mining* [6]. The other extreme is outliers and exceptional patterns, which deviate from the norm and occur rarely in the data; finding such patterns is called *exception mining* [10].

Needless to say, the type of patterns considered, of measures used to assess their interestingness, and of algorithms used to extract those patterns being highly rated in terms of these measures, strongly depends on the nature of the data. It makes a big difference, for example, whether the data is binary, categorical, or numerical, and whether a single observation is described in terms of a *subset*, like in itemset mining [6], or as a *sequence*, like in sequential pattern mining [9].

In this paper, we make a first step toward the mining of *rank data*, that is, data that comes in the form of *rankings* of an underlying set of items. This idea is largely motivated by the recent emergence of *preference learning* as a novel branch of machine learning [5]. While methods for problems such as "learning to rank" have been studied quite intensely in this field, rank data has not yet been considered from a data mining perspective so far.

To illustrate what we mean by rank data, consider a version of the well-known SUSHI benchmark, in which 5000 customers rank 10 different types of sushi from

S. Džeroski et al. (Eds.): DS 2014, LNAI 8777, pp. 123–134, 2014.

most preferred to least preferred.[1] This data could be represented in the form
of a matrix as follows:

$$
\begin{array}{cccccccccc}
5 & 7 & 3 & 8 & 4 & 10 & 2 & 1 & 6 & 9 \\
6 & 10 & 1 & 4 & 8 & 7 & 2 & 3 & 5 & 9 \\
2 & 7 & 3 & 1 & 6 & 9 & 5 & 8 & 4 & 10 \\
\cdot & \cdot & \cdot & \cdot & \cdot & \cdot & \cdot & \cdot & \cdot & \cdot
\end{array}
$$

In this matrix, the value in row i and column j corresponds to the position of
the jth sushi in the ranking of the ith customer. For example, the first customer
likes the eighth sushi the most, the seventh sushi the second best, and so on.

The above data consists of *complete* rankings, i.e., each observation is a ranking of the complete set of items (the 10 types of sushi). While we do assume
data of that kind in this paper, there are many applications in which rank data
is less complete, especially if the underlying set of items is larger. We shall come
back to corresponding generalizations of our setting in the end of this paper.

The rest of the paper is organized as follows. In the next two sections, we explain more formally what we mean by rank data and rank patterns, respectively.
An algorithm for mining rank patterns in the form of what we call *frequent subrankings* is then introduced in Section 4. Some experimental results are presented
in Section 5, prior to concluding the paper in Section 6.

2   Rank Data

Let $\mathbb{O} = \{o_1, \ldots, o_N\}$ be a set of items or objects. A ranking of these items is a
total order that is represented by a permutation

$$
\pi : [N] \to [N] \ ,
$$

that is, a bijection on $[N] = \{1, \ldots, N\}$, where $\pi(i)$ denotes the position of item
o_i. Thus, the permutation π represents the order relation

$$
o_{\pi^{-1}(1)} \succ o_{\pi^{-1}(2)} \succ \cdots \succ o_{\pi^{-1}(N)} \ ,
$$

where π^{-1} is the inverse of π, i.e., $\pi^{-1}(j)$ is the index of the item on position j.

We assume data to be given in the form of a set $\mathbb{D} = \{\pi_1, \pi_2, \ldots, \pi_M\}$ of
(complete) rankings π_i over a set of items $\mathbb{O}$. Returning to our example above,
$\mathbb{O} = \{o_1, \ldots, o_{10}\}$ could be the 10 types of sushi, and π_i the ranking of these
sushis by the ith customer.

3   Rank Patterns

In the context of rank data as outlined above, there is a large variety of *rank
patterns* that might be of interest. In this section, we introduce two examples of
rank patterns that we shall elaborate further in subsequent sections.

---

[1] http://kamishima.new/sushi/

3.1   Subrankings

An obvious example of such a pattern is a *subranking*. We shall use this term for a ranking π of a subset of objects $O \subset \mathbb{O}$. Here, $\pi(j)$ is the position of item o_j provided this item is contained in the subranking, and $\pi(j) = 0$ otherwise. For example, $\pi = (0, 2, 1, 0, 0, 3)$ denotes the subranking $o_3 \succ o_2 \succ o_6$, in which the items o_1, o_4, o_5 do not occur.

In the following, we will write complete rankings $\boldsymbol{\pi}$ in bold font (as we already did above) and subrankings in normal font. The number of items included in a subranking π is denoted $|\pi|$; if $|\pi| = k$, then we shall also speak of a k-ranking.

We denote by $O(\pi)$ the set of items ranked by a subranking π. The other way around, if $O' \subset O(\pi)$, then $(\pi|O')$ denotes the restriction of the ranking π to the set of objects O', i.e.,

$$(\pi|O')(j) = \begin{cases} \#\{o_i \in O' \,|\, \pi(i) \leq \pi(j)\} & \text{if } o_j \in O' \\ 0 & \text{if } o_j \notin O' \end{cases}$$

If π is a subranking of $O = O(\pi)$, then $\boldsymbol{\pi}$ is a (linear) extension of π if $(\boldsymbol{\pi}|O) = \pi$; in this case, the items in O are put in the same order by $\boldsymbol{\pi}$ and π, i.e., the former is consistent with the latter. We shall symbolize this consistency by writing $\pi \sqsubset \boldsymbol{\pi}$ and denote by $E(\pi)$ the set of linear extensions of π.

Now, we are ready to define the notion of *support* for a subranking π. In analogy to the well-known problem of itemset mining (see also Section 4 below), this is the relative frequency of observations in the data in which π occurs as a subranking:

$$\text{supp}(\pi) = \frac{1}{M} \cdot \#\{\boldsymbol{\pi}_i \in \mathbb{D} \,|\, \pi \sqsubset \boldsymbol{\pi}_i\} \tag{1}$$

A *frequent subranking* is a subranking π such that

$$\text{supp}(\pi) \geq min_{supp} ,$$

where min_{supp} is a user-defined support threshold. A frequent subranking π is *maximal* if there is no frequent subranking π' such that $\pi \sqsubset \pi'$ and $\pi' \not\sqsubset \pi$.

3.2   Association Rules

Association rules are well-known in data mining and have first been considered in the context of *itemset mining*. Here, an association rule is a pattern of the form $A \rightharpoonup B$, where A and B are itemsets. The intended meaning of such a rule is that a transaction containing A is likely to contain B, too. In market-basket analysis, where a transaction is a purchase and items are associated with products, the association $\{\texttt{paper}, \texttt{envelopes}\} \rightharpoonup \{\texttt{stamps}\}$ suggests that a purchase containing paper and envelopes is likely to contain stamps as well.

Rules of that kind can also be considered in the context of rank data. Here, we look at associations of the form

$$\pi_A \rightharpoonup \pi_B , \tag{2}$$

where π_A and π_B are subrankings of $\mathbb{O}$ such that

$$\#\big(O(\pi_A) \cap O(\pi_B)\big) \leq 1 . \tag{3}$$

For example, the rule $b \succ e \succ a \;\rightharpoonup\; d \succ c$ suggests that if b ranks higher than e, which in turn ranks higher than a, then d tends to rank higher than c. Note that this rule does not make any claims about the order relation between items in the antecedent and the consequent part. For example, d could rank lower but also higher than b. In general, the (complete) rankings π that are consistent with a rule (2) is given by $E(\pi_A) \cap E(\pi_B)$.

The condition (3) may call for an explanation. It plays the same role as the condition of empty intersection between items in the rule antecedent A and rule consequent B that is commonly required for association rules $A \rightharpoonup B$ in itemset mining $(A \cap B = \emptyset)$, and which is intended to avoid trivial dependencies. In fact, assuming an item a in the rule antecedent trivially implies its occurrence in all transactions to which this rule is applicable. In our case, this is not completely true, since a subranking is modeling relationships between items instead of properties of single items. For example, a rule like $a \succ b \rightharpoonup a \succ c$ is not at all trivial, although the item a occurs on both sides. A redundancy occurs, however, as soon as two or more items are included both in π_A and π_B. This is why we restrict such occurrences to at most one item.

In itemset mining, the confidence measure

$$\mathrm{conf}(A \rightharpoonup B) = \frac{\mathrm{supp}(A \cup B)}{\mathrm{supp}(A)}$$

that is commonly used to evaluate association rules $A \rightharpoonup B$ can be seen as an estimation of the conditional probability

$$\mathbf{P}(B \,|\, A) = \frac{\mathbf{P}(A \text{ and } B)}{\mathbf{P}(A)} \;,$$

i.e., the probability to observe itemset B given the occurrence of itemset A. Correspondingly, we define the confidence of an association $\pi_A \rightharpoonup \pi_B$ as

$$\mathrm{conf}(\pi_A \rightharpoonup \pi_B) = \frac{\#\{\pi_i \in \mathbb{D} \,|\, \pi_A, \pi_B \subset \pi_i\}}{\#\{\pi_i \in \mathbb{D} \,|\, \pi_A \subset \pi_i\}} = \frac{\mathrm{supp}(\pi_A \oplus \pi_B)}{\mathrm{supp}(\pi_A)} \tag{4}$$

As an important difference between mining itemsets and mining rank data, note that the class of patterns is closed under logical conjunction in the former but not in the latter case: Requiring the simultaneous occurrence of itemset A *and* itemset B is equivalent to requiring the occurrence of their union $A \cup B$, which is again an itemset. The conjunction of two subrankings π_A and π_B, denoted $\pi_A \oplus \pi_B$ in (4), is not again a subranking, however, at least not a single one; instead, it is represented by a *set* of subrankings $\pi_A \oplus \pi_B$, namely the rankings π such that $O(\pi) = O(\pi_A) \cup O(\pi_B)$, $(\pi|O(\pi_A)) = \pi_A$, $(\pi|O(\pi_B)) = \pi_B$. Correspondingly, the *joint support* of π_A and π_B,

$$\mathrm{supp}(\pi_A \oplus \pi_B) = \#\{\pi_i \in \mathbb{D} \,|\, \pi_A, \pi_B \subset \pi_i\} \;, \tag{5}$$

is the support of this subset. As we shall see later on, this has an implication on an algorithmic level.

Finally, and again in analogy with itemset mining, we define a measure of *interest* or *significance* of an association as follows:

$$\text{sign}(\pi_A \rightharpoonup \pi_B) = \text{conf}(\pi_A \rightharpoonup \pi_B) - \text{supp}(\pi_B) \qquad (6)$$

Just like for the measure of support, one is then interested in reaching certain thresholds, i.e., in finding association rules $\pi_A \rightharpoonup \pi_B$ that are highly supported ($\text{supp}(\pi_A \rightharpoonup \pi_B) \geq min_{supp}$), confident ($\text{conf}(\pi_A \rightharpoonup \pi_B) \geq min_{conf}$), and/or significant ($\text{sign}(\pi_A \rightharpoonup \pi_B) \geq min_{sign}$).

3.3   Comparison with Itemset Mining

The connection between mining rank data and itemset mining has already been touched upon several times. Moreover, as will be seen in Section 4, our algorithm for extracting frequent subrankings can be seen as a variant of the basic Apriori algorithm, which has first been proposed for the purpose of itemset mining [1].

Noting that a ranking can be represented (in a unique way) in terms of a *set* of pairwise preferences, our problem could in principle even be reduced to itemset mining. To this end, a new item $o_{i,j}$ is introduced for each pair of items $o_i, o_j \in \mathbb{O}$, and a subranking π is represented by the set of items

$$\{o_{i,j} \mid o_i, o_j \in O(\pi), \ \pi(i) < \pi(j)\} \ .$$

This reduction has a number of disadvantages, however. First, the number of items is increased by a quadratic factor, although the information contained in these items is largely redundant. In fact, due to the transitivity of rankings, the newly created items exhibit (logical) dependencies that need to be taken care of by any mining algorithm. For example, not every itemset corresponds to a valid ranking, only those that are transitively closed.

Apart from that, there are some important differences between the two settings, for example regarding the number of possible patterns. In itemset mining, there are 2^N different subsets of N items, which is much smaller than the $N!$ number of rankings of these items. However, the N we assume for rank data (at least if the rankings are supposed to be complete) is much smaller than the N in itemset mining, which is typically very large. Besides, the itemsets observed in a transaction database are normally quite small and contain only a tiny fraction of all items. In fact, assuming an upper bound K on the size of an itemset, the number of itemsets is of the order $O(N^K)$ and grows much slower in N than exponential.

4   Algorithms

Our algorithm for mining frequent subrankings is based on the well-known Apriori algorithm for mining frequent itemsets [1]. This algorithm constructs itemsets in a level-wise manner, starting with singletons and increasing the size by 1 in each iteration. Candidate itemsets of size k are constructed from the frequent itemsets of size $k-1$ already found. In this step, Apriori exploits an important monotonicity property for pruning candidates: If A is a frequent itemset, then any subset of A must be frequent, too. Thus, by contraposition, any superset of a non-frequent itemset cannot be frequent either.

This monotonicity property also holds for subrankings: If a subranking π is frequent, then all rankings $\pi' \subset \pi$ are frequent, too. Thus, the Apriori approach is in principle applicable to rank data. Nevertheless, since rankings and sets have different properties, the construction and filtering steps need to be adapted. The basic structure of our algorithm for finding frequent subrankings, that will subsequently be explained in more detail, is the following:

1. Initial search for frequent 2-rankings $\mathcal{F}^{(2)}$ (set $k = 3$).
2. LOOP:
   (a) Construct potential frequent k-rankings from the set of frequent $(k-1)$-rankings $\mathcal{F}^{(k-1)}$.
   (b) Filter frequent k-rankings from the potential frequent k-ranking set.
   (c) Stop the LOOP if no k-ranking passes the filtering.
   (d) Set $k = k + 1$.

4.1 Searching Frequent 2-Rankings

While the smallest unit in itemset mining is an item, the smallest unit in the case of subrankings is a preference pair $a \succ b$. Therefore, the initial step is to exhaustively search for all frequent 2-rankings in the data set of rankings.

4.2 Construction of Candidate k-Rankings

Every k-ranking $\pi^{(k)}$ can be decomposed into a set of $(k-l)$-rankings

$$C^{(l)}\left(\pi^{(k)}\right) = \left\{\pi_i^{(k-l)} \mid \pi_i^{(k-l)} \subset \pi^{(k)}\right\}$$

with $0 \le l < k$. A k-ranking $\pi^{(k)}$ is *fully consistent* with $\mathcal{F}^{(k-1)}$ if $C^{(1)}(\pi^{(k)}) \subset \mathcal{F}^{(k-1)}$.

In this step, we search for all k-rankings that are fully consistent with $\mathcal{F}^{(k-1)}$. For this purpose, the algorithm iterates over all pairs $(\pi_i^{(k-1)}, \pi_j^{(k-1)}) \in \mathcal{F}^{(k-1)} \times \mathcal{F}^{(k-1)}$ with $i < j$ and builds *partially consistent* k-rankings. These are rankings $\pi^{(k)}$ such that $\{\pi_i^{(k-1)}, \pi_j^{(k-1)}\} \subset C^{(1)}(\pi^{(k)})$. For example, from the 2-rankings $a \succ b$ and $c \succ d$, we are not able to build any 3-ranking, whereas from $a \succ b$ and $b \succ c$, we can build the 3-ranking $a \succ b \succ c$. Likewise, from the 2-rankings $a \succ c$ and $b \succ c$, we can build the two rankings $a \succ b \succ c$ and $b \succ a \succ c$.

Being partially consistent with a single pair $\{\pi_i^{(k-1)}, \pi_j^{(k-1)}\} \subset \mathcal{F}^{(k-1)}$ is necessary but not sufficient for being fully consistent with $\mathcal{F}^{(k-1)}$. Let us again consider the 2-rankings $a \succ b$ and $b \succ c$, and the only partially consistent 3-ranking $a \succ b \succ c$. In order to assure that this ranking is fully consistent with $\mathcal{F}^{(k-1)}$, we have to check whether the ranking $a \succ c$ is in $\mathcal{F}^{(k-1)}$, too.

Instead of explicitly searching for $a \succ c$ in $\mathcal{F}^{(k-1)}$, we store $a \succ b \succ c$ in a hash map with a key (based on the object sequence abc) and a count as value. This count is set to 1 the first time we put in a key and incremented every time we apply the put(key) operation. The idea is that if a ranking $\pi^{(k)}$ is fully consistent with $\mathcal{F}^{(k-1)}$, we find exactly $|C^{(1)}(\pi^{(k)})|(|C^{(1)}(\pi^{(k)})| - 1)/2 = k(k-1)/2$ pairs of $(k-1)$-rankings from which $\pi^{(k)}$ can be built. After iterating over all pairs, we pass through the entries in our hash map and collect the keys with value $k(k-1)/2$. These rankings form the set of potentially frequent k-rankings. The whole procedure is described in Algorithm 1.

Algorithm 1.

1: **procedure** NCR($\mathcal{F}^{(k-1)}$)
2:     $\mathcal{K} = \emptyset$ ▷ set of candidates
3:     initialize HashMap M
4:     **for** $i = 1$ to $|\mathcal{F}^{(k-1)}| - 1$ **do**
5:         **for** $j = i + 1$ to $|\mathcal{F}^{(k-1)}|$ **do**
6:             $\mathcal{C}_{i,j} \leftarrow$ CONSTRUCT($(\pi_i^{(k-1)}, \pi_j^{(k-1)})$)   ▷ constructs sub consistent
          k-rankings
7:             **for** $\pi^{(k)} \in \mathcal{C}_{i,j}$ **do**
8:                 **if** $M.getValue(\pi^{(k)}) = null$ **then**   ▷ $\pi^{(k)}$ is key
9:                     $M.put(\pi^{(k)}, 1)$
10:                **else**
11:                    $M.incrementValue(\pi^{(k)})$
12:                **end if**
13:            **end for**
14:        **end for**
15:    **end for**
16:    **for** $entry \in M$ **do**
17:        **if** $entry.getValue() = k(k-1)/2$ **then**
18:            $\mathcal{K}.add(entry.getKey())$
19:        **end if**
20:    **end for**
21:    **return** $\mathcal{K}$
22: **end procedure**

The task of CONSTRUCT is to take two $(k-1)$-rankings and construct all partially consistent k-rankings. The way CONSTRUCT works is actually quite easy, although the implementation is a bit intricate. Roughly speaking, the two rankings are compared position by position, and the algorithm has a special handling for the last two positions. CONSTRUCT can be divided into two steps, an alignment step and a construction step. Before we explain the basic principle of operation, let us make a few observations.

| $p_b = 1$ $p_t = 1$ | | $p_b = 2$ $p_t = 2$ | | $p_b = 3$ $p_t = 4$ | | $p_b = 4$ $p_t = 4$ | |
| $u_b = -1$ $u_t = -1$ | | $u_b = -1$ $u_t = -1$ | | $u_b = -1$ $u_t = 2$ | | $u_b = 3$ $u_t = 2$ | |

$$\pi_t : \text{c a f z} \ldots \quad \Longrightarrow \quad \pi_t : \text{c a f z} \ldots \quad \Longrightarrow \quad \pi_t : \text{c a f z} \ldots \quad \Longrightarrow \quad \pi_t : \text{c a f z} \ldots$$
$$\pi_b : \text{c f d z} \ldots \quad\quad\quad \pi_b : \text{c f d z} \ldots \quad\quad\quad \pi_b : \text{c f d z} \ldots \quad\quad\quad \pi_b : \text{c f d z} \ldots$$

	c			c	a	f		c	a	f	–
	c			c	–	f		c	–	f	d

Fig. 1. Illustration of the CONSTRUCT procedure. The second row shows how the object pointers are set after every iteration—the upper arrow is the top object pointer p_t and the lower arrow the bottom object pointer p_b. The alignment, which is implicitly constructed by the algorithm, is shown in the bottom row.

The number of partially consistent k-rankings that can be constructed from two $(k-1)$-rankings is 0, 1 or 2. It is 0 if there is no k-ranking that is consistent with $\pi_i^{(k-1)}$ and $\pi_j^{(k-1)}$. By construction, $\pi_i^{(k-1)} = \pi_j^{(k-1)}$ only if $i = j$, and we only compare rankings with $i \neq j$. Consequently, if there is a partially consistent k-ranking, there is also exactly one object $o_{i,u_i} = O(\pi_i^{(k-1)}) \setminus O(\pi_j^{(k-1)})$ and exactly one object $o_{j,u_j} = O(\pi_j^{(k-1)}) \setminus O(\pi_i^{(k-1)})$. Using terminology from sequence alignment, this means there are exactly two gaps in the alignment of the sequences $\pi_i^{(k-1)}$ and $\pi_j^{(k-1)}$, one in the former and one in the latter (Figure 1). However, the existence of one gap in each sequence is only a necessary but not a sufficient condition. Additionally, we must have $o_{i,u_i} \neq o_{j,u_j}$. For instance, the sequences in Figure 2 contain the same elements but in another order. In the alignment, there is one gap in each sequence, yet there is no consistent $(k-1)$-ranking.

The last observation is that the number of consistent k-rankings which can be constructed from $\pi_i^{(k-1)}$ and $\pi_j^{(k-1)}$ is one if $u_i \neq u_j$ and two if $u_i = u_j$.

Thus, the key tasks consist of finding and counting the gaps—see Algorithm 2 for a description in pseudo code. The meaning of top- and bottom are like in Figure 1.

| $p_b = 1$ $p_t = 1$ | | $p_b = 2$ $p_t = 2$ | | $p_b = 3$ $p_t = 4$ | | $p_b = 4$ $p_t = 4$ | |
| $u_b = -1$ $u_t = -1$ | | $u_b = -1$ $u_t = -1$ | | $u_b = -1$ $u_t = 2$ | | $u_b = 3$ $u_t = 2$ | |

$$\pi_t : \text{c a f z} \ldots \quad \Longrightarrow \quad \pi_t : \text{c a f z} \ldots \quad \Longrightarrow \quad \pi_t : \text{c a f z} \ldots \quad \Longrightarrow \quad \pi_t : \text{c a f z} \ldots$$
$$\pi_b : \text{c f a z} \ldots \quad\quad\quad \pi_b : \text{c f a z} \ldots \quad\quad\quad \pi_b : \text{c f a z} \ldots \quad\quad\quad \pi_b : \text{c f a z} \ldots$$

	c			c	a	f		c	a	f	–
	c			c	–	f		c	–	f	a

Fig. 2. Here, the $(k-1)$-rankings differ by a swap of the objects a and f. Although it is not possible to build a k-ranking, the alignment is the same as in Figure 1. Therefore, the objects o_{b,u_b} and o_{t,u_t} need to be checked for equality.

Algorithm 2.

1: **procedure** CONSTRUCT($\pi_b^{(k-1)}$, $\pi_t^{(k-1)}$)
2:     $u_b \leftarrow -1$; $u_t \leftarrow -1$ ▷ bottom and top gap position
3:     $p_b \leftarrow 1$; $p_t \leftarrow 1$ ▷ bottom and top object pointer
4:     **for** p_b to $|\pi_b^{(k-1)}| - 2$ **do**
5:         **if** $o_{j,p_b} = o_{i,p_t}$ **then**
6:             $p_t \leftarrow p_t + 1$
7:         **else if** $o_{j,p_b} \neq o_{i,(p_t+1)}$ **then**
8:             **if** $u_b > -1$ **then**
9:                 **return** null   ▷ bottom gap already found
10:             **end if**
11:             $u_b \leftarrow p_b$ ▷ bottom gap is found
12:         **else**
13:             **if** $u_b > -1$ **then**
14:                 **return** null   ▷ top gap already found
15:             **end if**
16:             $u_t \leftarrow p_t$ and $p_t \leftarrow p_t + 2$ ▷ the top gap is found
17:         **end if**
18:     **end for**
19:     $\vdots$ ▷ Here the procedure deals with the cases $p_b > |\pi_b^{(k-1)}| - 2$, where you have
    to be careful with the incrementation of p_t
20:     **if** $o_{b,u_b} = o_{t,u_t}$ **then**
21:         **return** null   ▷ necessary gaps are caused by a swap, see Figure 2
22:     **end if**
23:     **return** INNERCONSTRUCT($\pi_b^{(k-1)}$, $\pi_t^{(k-1)}$, u_b, u_t)
24: **end procedure**

While the "outer" CONSTRUCT procedure performs the alignment step, the INNERCONSTRUCT procedure performs the construction step. The idea here is simply to add o_{b,u_b} and o_{t,u_t} to $\pi_t^{(k-1)}$ with the help of the position information u_b and u_t, so that the resulting k-rankings are consistent with $\pi_b^{(k-1)}$ and $\pi_t^{(k-1)}$.

4.3   Filtering Frequent k-Rankings

Like in the original Apriori algorithm, we need to check for every potentially frequent k-ranking whether or not it is indeed frequent. For this purpose, we have to go through all rankings and count the appearance of the k-rankings.

4.4   Association Rule Mining

The mining of association rules of the form (2) is done on the basis of frequent subrankings, just like in itemset mining. However, as mentioned before, the conjunction $\pi_A \oplus \pi_B$ of two subrankings π_A and π_B is not again a subranking. Therefore, the support (5) of a candidate rule $\pi_A \rightharpoonup \pi_B$ cannot simply be looked up.

Instead, for each candidate rules $\pi_A \rightharpoonup \pi_B$, where π_A and π_B are frequent subrankings that meet the consistency constraint (3), we again pass through the data in order to compute the support (5) as well as measures of confidence (4) and interest (6).

5   Experiments

We used three real data sets for our experiments: The SUSHI data, that was already mentioned in the introduction, consists of 5000 rankings of 10 types of sushis. The STUDENTS data [2] comes from a psychological study and consists of 404 rankings of 16 goals (want to get along with my parents, want to feel good about myself, want to have nice things, want to be different from others, want to be better than others, etc.), each one reflecting what a student considers to be more or less important for himself or herself to achieve. Finally, the ESC14 data is derived from the European Song Contest 2014 in Denmark. It consists of rankings of the 26 countries that reached the final. Since each of the 36 participating countries selected 5 jurors, the total number of rankings is 180. There was a need for one adjustment, however: Since jurors are not allowed to rank their own country, we completed such rankings (of length 25 instead of 26) by putting that country on the bottom rank.

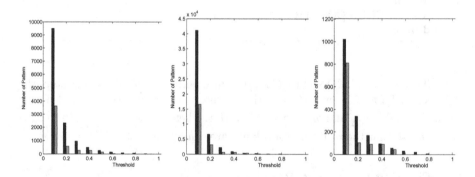

Fig. 3. Number of patterns reaching a threshold min_{supp} (left bar) for the data sets STUDENT, ESC14 and SUSHI (from left to right), compared to the same number for synthetic data sets of the same dimension, in which rankings are generated uniformly at random (right bar)

Figure 3 shows the number of frequent subrankings found in the data sets, compared with the number of frequent subrankings found in synthetic data sets taken from a uniform distribution.

For each of the three data set, we derived a most representative subranking, namely the subranking π that maximizes the relation between its support and the support one would *expect* under a uniform distribution (which is $1/|\pi|!$). Figure 4

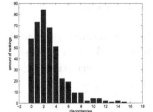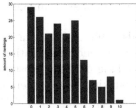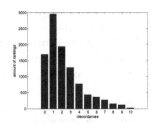

Fig. 4. Distribution of the distance of rankings from the most representative pattern; from left to right: STUDENTS (goal 2 ≻ goal 1 ≻ goal 9 ≻ goal 11 ≻ goal 14 ≻ goal 16), ESC14 (Netherlands ≻ Finland ≻ UK ≻ Italy ≻ Greece), SUSHI (sushi 8 ≻ sushi 3 ≻ sushi 9 ≻ sushi 7 ≻ sushi 10)

shows the distribution of the distances of all rankings from that representative, where the distance is determined as the number of pairwise disagreements. As can be seen, the pattern is indeed representative in the case of STUDENTS and SUSHI, in the sense that the large majority deviates by at most 2-3 pairwise inversions. For ESC14, a representative pattern is more difficult to find, suggesting that the preferences are more diverse in this case.

Finally, we also extracted association rules from the data sets, and found a number of rules with high confidence and interest (for example, the rule goal 10 (material gain) ≻ goal 3 (belongingness) ≻ goal 7 (management) ⇢ goal 10 (material gain) ≻ goal 5 (social responsibility) in the STUDENTS data with confidence 0.9038, interest 0.6544, and support 0.1149). Many of these rules also have a quite interesting semantic interpretation. Due to reasons of space, however, we refrain from a detailed discussion here.

6 Summary and Conclusion

In this paper, we introduced the problem of mining rank data as a novel data mining task—to the best of our knowledge, mining patterns in this type of data has not been studied systematically in the literature so far. Moreover, we have given two concrete examples of rank patterns, namely frequent subrankings and associations between such rankings, and proposed an algorithm for extracting them from rank data. Our algorithm is a rather straightforward generalization of the basic Apriori algorithm for itemset mining. Needless to say, there is much scope for improving this approach, so as to make it scalable to very large data sets. To this end, one may try to adopt ideas of faster algorithms for itemset mining, such as Eclat [11] or FP-growth [7,8], although the data structures used there are not immediately applicable to rank data.

More importantly, however, the problem of mining rank patterns itself can be generalized in various directions:

– For example, as already mentioned, rank information will not always be provided in the form of complete rankings of all items, i.e., the data itself may already be given in the form of subrankings, partial orders or "bags" of order relations.

- In this regard, one may also think of a combination of mining rank and itemset data. For instance, preference information is often given in the form of top-k rankings, i.e., a ranking of the k most preferred alternatives [4]— obviously, information of that kind can be seen as a *ranking* of a *subset* of all items.
- Since the space of rankings is equipped with a natural topology, it would make sense to search for *approximate patterns*, also allowing a ranking to be supported by *similar* rankings, for example [3].
- Yet another direction is the incorporation of quantitative information about rank positions. It could make a difference, for example, whether two objects a and b share adjacent ranks (suggesting that a is only slightly preferred to b), or whether a appears on the top and b on the bottom of the ranking.

Extensions and generalizations of that kind provide interesting challenges for future work.

Acknowledgments. The authors gratefully acknowledge support by the German Research Foundation (DFG). Moreover, they would like to thank Prof. Louis Fono, Université de Douala, Cameroun, for very stimulating discussions on the topic of this paper.

References

1. Agrawal, R., Srikant, R.: Fast algorithms for mining association rules. In: Proc. VLDB, 20th Int. Conf. on Very Large Data Bases, pp. 487–499 (1994)
2. Boekaerts, M., Smit, K., Busing, F.M.T.A.: Salient goals direct and energise students' actions in the classroom. Applied Psychology: An International Review 4(S1), 520–539 (2012)
3. de Sá, C.R., Soares, C., Jorge, A.M., Azevedo, P., Costa, J.: Mining association rules for label ranking. In: Huang, J.Z., Cao, L., Srivastava, J. (eds.) PAKDD 2011, Part II. LNCS, vol. 6635, pp. 432–443. Springer, Heidelberg (2011)
4. Fagin, R., Kumar, R., Sivakumar, D.: Comparing top-k lists. SIAM Journal of Discrete Mathematics 17(1), 134–160 (2003)
5. Fürnkranz, J., Hüllermeier, E. (eds.): Preference Learning. Springer (2011)
6. Han, J., Cheng, H., Xin, D., Yan, X.: Frequent pattern mining: current status and future directions. Data Mining and Knowledge Discovery 15, 55–86 (2007)
7. Han, J., Pei, J., Yin, Y.: Mining frequent patterns without candidate generation. ACM SIGMOD Record 29, 1–12 (2000)
8. Han, J., Pei, J., Yin, Y., Mao, R.: Mining frequent patterns without candidate generation: A frequent-pattern tree approach. Data Mining and Knowledge Discovery 8(1), 53–87 (2004)
9. Srikant, R., Agrawal, R.: Mining sequential patterns: Generalizations and performance improvements. Springer (1996)
10. Suzuki, E.: Data mining methods for discovering interesting exceptions from an unsupervised table. J. Universal Computer Science 12(6), 627–653 (2006)
11. Zaki, M.J.: Scalable algorithms for association mining. IEEE Transactions on Knowledge and Data Engineering 12(3), 372–390 (2000)

Link Prediction on the Semantic MEDLINE Network

An Approach to Literature-Based Discovery

Andrej Kastrin[1], Thomas C. Rindflesch[2], and Dimitar Hristovski[3]

[1] Faculty of Information Studies, Novo mesto, Slovenia
[2] Lister Hill National Center for Biomedical Communications,
National Library of Medicine, Bethesda, MD, USA
[3] Institute of Biostatistics and Medical Informatics, Faculty of Medicine,
University of Ljubljana, Ljubljana, Slovenia
`andrej.kastrin@guest.arnes.si`,
`trindflesch@mail.nih.gov`,
`dimitar.hristovski@mf.uni-lj.si`

Abstract. Retrieving and linking different segments of scientific information into understandable and interpretable knowledge is a challenging task. Literature-based discovery (LBD) is a methodology for automatically generating hypotheses for scientific research by uncovering hidden, previously unknown relationships from existing knowledge (published literature). Semantic MEDLINE is a database consisting of semantic predications extracted from MEDLINE citations. The predications provide a normalized form of the meaning of the text. The associations between the concepts in these predications can be described in terms of a network, consisting of nodes and directed arcs, where the nodes represent biomedical concepts and the arcs represent their semantic relationships. In this paper we propose and evaluate a methodology for link prediction of implicit relationships in the Semantic MEDLINE network. Link prediction was performed using different similarity measures including common neighbors, Jaccard index, and preferential attachment. The proposed approach is complementary to, and may augment, existing LBD approaches. The analyzed network consisted of 231,589 nodes and 10,061,747 directed arcs. The results showed high prediction performance, with the common neighbors method providing the best area under the ROC curve of 0.96.

Keywords: Literature-based discovery, Network analysis, Link prediction, Semantic network.

1 Introduction

The corpus of biomedical papers in online bibliographic repositories, nowadays referred to as the bibliome, is of considerable size and complexity. Increase in knowledge greatly depends on the synthesis of information from the existing scientific literature. It is a challenging task to link diverse scientific information

S. Džeroski et al. (Eds.): DS 2014, LNAI 8777, pp. 135–143, 2014.

into coherently interpretable knowledge. Computer-based methods complement manual literature management and knowledge discovery from biomedical data. Common text mining tasks in biomedicine include information extraction from the literature, document summarization, question answering and literature-based discovery (LBD) [1].

LBD is a methodology for automatically generating hypotheses for scientific research by uncovering hidden, previously unknown relationships from existing knowledge. The LBD methodology was pioneered by Swanson [2], who proposed that dietary fish oils might be used to treat Raynaud's disease because they lower blood viscosity, reduce platelet aggregation and inhibit vascular reactivity. Swanson's approach is based on the assumption that there exist two nonintersecting scientific domains. Knowledge in one may be related to knowledge in the other, without the relationship being known. The methodology of LBD relies on the idea of concepts relevant to three literature domains: X, Y, and Z. For example, suppose a researcher has found a relationship between disease X and a gene Y. Further suppose that a different researcher has studied the effects of substance Z on gene Y. The use of LBD may suggest an XZ relationship, indicating that substance Z may potentially treat disease X (Fig. 1). For a recent review of LBD tools and approaches, see [3].

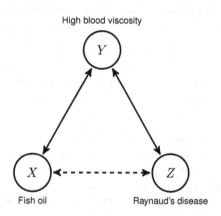

Fig. 1. Swansons XYZ discovery model

Effective retrieval crucially underpins knowledge management applications, such as LBD. Widely used document retrieval systems such as Google or PubMed typically have no access to the meaning of the text being processed [4]. To fill the gap between raw text and its meaning Rindflesch [5] introduced the SemRep system. SemRep is a rule-based, symbolic natural language processing system that recovers semantic propositions from the biomedical research literature. The system relies on domain knowledge in the Unified Medical Language System (UMLS) [6] to provide partial semantic interpretation in the form of predications consisting of UMLS Metathesaurus concepts as arguments and UMLS Semantic

Network relations as predicates. SemRep uses a partial syntactic analysis based on the SPECIALIST Lexicon [7] and MedPost tagger [8]. Each noun phrase in this analysis is mapped to a concept in the Metathesaurus using MetaMap [9]. Both syntactic and semantic constraints are employed to identify propositional assertions. For example, three predications are extracted from the text 'dexamethasone is a potent inducer of multidrug resistance-associated protein expression in rat hepatocytes':

1. Dexamethasone STIMULATES Multidrug Resistance Associated Proteins
2. Multidrug Resistance-Associated Proteins PART_OF Rats
3. Hepatocytes PART_OF Rats

SemRep extracts 30 predicate types expressing assertions in clinical medicine (e.g., TREATS, ADMINISTERED_TO), substance interactions (e.g., INTERACTS_WITH, STIMULATES), genetic etiology of disease (e.g. CAUSES, PREDISPOSES), and pharmacogenomics (e.g., AUGMENTS, DISRUPTS). The program has been run on all of MEDLINE and the extracted predications deposited in a MySQL database (SemMedDB [10]) updated quarterly and available to researchers.

Knowledge of a particular biomedical domain can be viewed as a set of concepts and the relationships among them [11]. For example relations among genes, diseases, or chemical substances constitute an important part of biomedical knowledge. These associations can be represented as a graph consisting of nodes and edges, where the former represent concepts and the latter relationships.

Link (association) prediction is a novel research field at the intersection of network analysis and machine learning. Understanding the mechanisms of link formation in complex networks is a long-standing challenge for network analysis. Link prediction refers to the discovery of associations between nodes that are not directly connected in the current snapshot of a given network [12]. Seen in this way, the link prediction problem is similar to LBD. Several techniques have been proposed to predict new links by estimating the likelihood of link formation between two nodes on the basis of the observed network topology.

In this paper we examine link prediction from the novel perspective of literature-based discovery. We propose a method for predicting and evaluating implicit or previously unknown connections between biomedical concepts. Our approach is complementary to traditional LBD. To evaluate the link prediction techniques for LBD, we investigated performance on a network obtained from Semantic MEDLINE.

2 Methods

2.1 Basic Terminology

A network can be represented as a graph $G(V, A)$ that consists of a set of nodes V representing concepts and a set of directed arcs A representing relationships between the nodes [13].

The link prediction problem can be formally represented as follows. Suppose we have network $G[t_1, t_2]$ which contains all interactions among nodes that take place in the time interval $[t_1, t_2]$. Further suppose that $[t_3, t_4]$ is a time interval occurring after $[t_1, t_2]$. The task of link prediction is to provide a list of edges that are present in $G[t_3, t_4]$ but absent in $G[t_1, t_2]$. We refer to $G[t_1, t_2]$ as the training network and $G[t_3, t_4]$ as the testing network.

2.2   Data Preparation

We prepared the experimental network based on SemMedDB distribution [10]. The current snapshot of SemMedDB (version 24.2) contains 32,505,704 arguments and 69,333,420 semantic predications from 23,659,049 MEDLINE citations published between January 1 1990 and December 31 2012. We further selected only those predications which refer to UMLS Metathesaurus concepts.

We applied Pearson's chi-square (χ^2) test for independence [14] for each relation pair to obtain a statistic which indicates whether a particular pair of concepts occurs together more often than by chance. To the best of our knowledge, this technique is novel in the network analysis community. In the following paragraphs, we provide a detailed description of the χ^2 test for independence and its application to network reduction.

For each co-occurrence pair (u, v) we are interested in co-occurrence frequency and also in the co-occurrences of u and v with other terms. Complete frequency information is summarized in a contingency table and yields four cell counts (Table 1). O_{11} is the joint frequency of the co-occurrence, the number of times the terms u and v in a co-occurrence are seen together. The cell O_{12} is the frequency of pairs in which term u occurs, but term v does not occur. Likewise, the O_{21} is the frequency of pairs in which term v occurs, but term u does not occur. The cell O_{22} is the frequency of pairs in which neither term u nor term v occurs. The marginal totals are denoted with Rs and Cs with subscripts corresponding to the rows and columns. The grand total N is the total of all four frequencies (i.e., $O_{11} + O_{12} + O_{21} + O_{22}$).

Table 1. Contingency table of observed frequencies for pairs of concepts

	$V = v$	$V \neq v$	
$U = u$	O_{11}	O_{12}	R_1
$U \neq u$	O_{21}	O_{22}	R_2
	C_1	C_2	

Next we calculated the corresponding expected frequencies E_{ij} for each table cell, as demonstrated in Table 2.

Given the observed and expected frequencies for each concept pair, the χ^2 statistic was calculated as

$$\chi^2 = \sum_{i=1}^{2} \sum_{j=1}^{2} \frac{(O_{ij} - E_{ij})^2}{E_{ij}}. \tag{1}$$

Table 2. Calculation of expected frequencies for pairs of concepts

	$V = v$	$V \neq v$
$U = u$	$E_{11} = (R_1 \times C_1)/N$	$E_{11} = (R_1 \times C_2)/N$
$U \neq u$	$E_{11} = (R_2 \times C_1)/N$	$E_{11} = (R_2 \times C_2)/N$

If an expected value was less than five, we applied Yates's correction for continuity by subtracting 0.5 from the difference between each observed frequency and its expected frequency. The limiting distribution of χ^2 statistic for 2×2 contingency table is a χ^2 distribution with one degree-of-freedom. If the χ^2 is greater than the critical value of 3.84 ($p \leq 0.05$), we can be 95% confident that a particular concept relation occurs more often than by chance.

2.3 Experimental Setup

The link prediction approach we used follows the procedure first introduced by Liben-Nowell and Kleinberg [15]. We performed link prediction using proximity measures, which are used to find similarity between a pair of nodes. For each node pair (u, v), a link prediction method gives score $s(u, v)$, an estimate of the likelihood of link formation between nodes u and v. Link prediction stated in this form is a binary classification task in which links that form constitute the positive class and links that do not form constitute the negative class. From among various proximity measures proposed in the literature we selected common neighbors, Jaccard index, and preferential attachment.

Common Neighbors. For node u, let $\Gamma_{\text{out}}(u)$ denote the set of outgoing neighbors of u and similarly $\Gamma_{\text{in}}(u)$ denote the set of incoming neighbors. The common neighbors measure is the simple count of common neighbors of nodes u and v, formally defined as

$$s_{uv}^{CN} = |\Gamma_{\text{out}}(u) \cap \Gamma_{\text{in}}(v)|. \qquad (2)$$

Two nodes are more likely to have a relation if they have many common neighbors. In a directed network, a common neighbor exists if there is a neighbor relationship between the source and target node.

Jaccard Index. This measure produces a similarity score corresponding to the quotient of the intersection of the two neighbor sets and the union of the two neighbor sets. Formally this is defined as

$$s_{uv}^{JI} = \frac{|\Gamma_{\text{out}}(u) \cap \Gamma_{\text{in}}(v)|}{|\Gamma_{\text{out}}(u) \cup \Gamma_{\text{in}}(v)|}. \qquad (3)$$

Preferential Attachment. The basic premise underling this measure is that the probability that a new link has node u as a terminal point is proportional to the current number of neighbors of u. The probability that a new link will connect u and v is thus proportional to

$$s_{uv}^{PA} = |\Gamma_{\text{out}}(u)| \cdot |\Gamma_{\text{in}}(v)|. \tag{4}$$

Performance Evaluation. We examined how accurately we could predict which node pairs will connect between times t_3 and t_4 despite not having any connections before time t_3. The training network was built on nodes and appropriate links for the time period 2000–2005. Similarly, test network was constructed for the period 2006–2011. Next, we computed the link prediction score $s(u, v)$ for each node pair that is not associated with any interaction before year 2006 by using one of the similarity measures introduced in the previous paragraphs. We assigned the class label 'positive' to this node pair if the connection occurs in the test network and 'negative' otherwise.

As a measure of prediction performance we used area under the ROC curve (AUC). The AUC can be interpreted as the probability that a randomly selected link is given a higher link prediction score than a randomly selected non-existent link.

3   Results

The semantic network consists of 231,589 nodes with 10,061,747 arcs. The global density of the network was 2e-04. We filtered out all arcs with a χ^2 statistic lower than 3.84. After filtering non-useful relations, the number of arcs decreased to 7,801,995. The density of the reduced network decreased to 1e-04.

The network exhibited relatively short average path length between all pairs of nodes; on average there are only about 3.41 hops from the selected node to any other node. The clustering coefficient of the exploiting network was 0.04. The network exhibits the small world property because of small average path length and relatively high clustering.

Figure 2 shows the frequency distribution of co-occurrence pairs in the network (i.e., number of sentences in which the concepts co-occur). Distribution is highly right asymmetric with the majority of the counts falling in the first quartile. Only about 9% of the co-occurrence pairs have a frequency greater than one. The mean frequency of pairs between two concepts is 1.12 ($\pm$0.44), with a maximal frequency of 81. We further computed the degree distribution of concept nodes. The shape of the degree distribution is similar to Fig. 2. Mean and maximal degree are 67.38 ($\pm$339.36) and 50,33, respectively. About 67% of the nodes have degree greater than 1,000.

The prediction performance for all three similarity measures is summarized in Fig. 3. Figure plots the false positive rate on the x axis vs. the true positive rate on the y axis for different cutoff values. The point $(0, 1)$ is a perfect predictor. Point $(0, 0)$ represents a predictor that predicts all links to be negative (i.e., no relationship between concept A and concept B), while the point $(1, 1)$ corresponds to a predictor that predicts each link to be positive (i.e., a link between concept A and concept B is present). The AUC was 0.94, 0.96, and 0.94

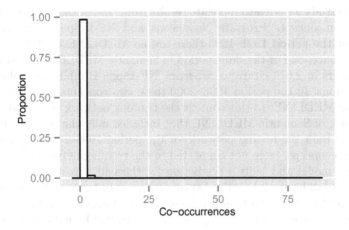

Fig. 2. Frequency distribution of co-occurrence counts in Semantic MEDLINE

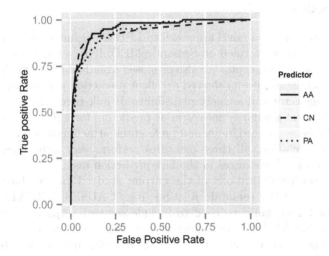

Fig. 3. ROC curve. Note: AA – Adamic/Adar, CN – common neighbor, PA – preferential attachment.

for common neighbor, Jaccard index, and preferential attachment, respectively. According to Swets's guidelines for qualitative interpretation of the AUC scale [17], our results demonstrate excellent prediction performance. Standard performance measures (i.e., accuracy, precision, recall) are not suitable in our setting, because class distribution is highly asymmetric (e.g., the ratio between positive and negative predictions was only 2e-04 for all three predictors). We also repeated the prediction procedure without χ^2 filtering, and AUC values decreased slightly: the AUC was 0.88, 0.92, and 0.90 for common neighbor, Jaccard index, and preferential attachment, respectively.

As a case study we tried to replicate results from [18]; authors found an association between concepts 'Prostatic Neoplasms' and 'NF-kappa-B inhibitor alpha' protein. For the period 1991–1995 there are no MEDLINE citations which include both concepts. For the same period in Semantic MEDLINE 'Prostatic Neoplasms' occurs in 2,673 citations, whereas 'NF-kappa-B inhibitor alpha' occurs in 102 citations. In the period 1996–2000 these two concepts co-occur 13 times in Semantic MEDLINE. In this context the training network was constructed as a subnetwork of Semantic MEDLINE that included only the period 1991–1995. Similarly we built the testing network for the period 1996–2000. We made sure that the two concepts were not connected in the training network. Then we ran the learning model and built a prediction for the pair 'Prostatic Neoplasms' – 'NF-kappa-B inhibitor alpha'. Prediction estimates in terms of probability were 0.98, 0.60, and 0.72 for common neighbors, Jaccard coefficient, and preferential attachment. All three predictors suggesting a strong link between target concepts in the period 1996–2000.

4   Discussion

In this work we studied a novel approach to LBD using link discovery methods. Link prediction was performed on Semantic MEDLINE, a large-scale relational dataset of biomedical concepts. We also proposed a method for network reduction using the χ^2 statistic. Results showed excellent prediction performance in terms of the AUC measure and suggest plausibility of link prediction for LBD. We used three different similarity measures as predictors for link discovery, namely common neighbors, Jaccard index and preferential attachment score. In contrast to common expectation, all three measures perform very well. They correctly predict practically all instances in the link prediction task.

There are several limitations of the current study. First, we have not considered the type of the semantic relation (e.g., CAUSES, TREATS) between concepts. For future work we need to include semantic type as a covariate in the prediction model. It is well known that nodes with similar attributes tend to create connections to each other [19]. Second, our analysis was based on a static view of the network, not taking temporal characteristics of the network into account. The semantic network exploited is an evolving system in which new concepts and relations are constantly added.

There are also many possible opportunities for future work. First, we will implement a greater number of similarity measures and include them in the link prediction task (e.g., PageRank, SimRank, Katz measure). In future settings we also plan to include arc weights as an additional attribute. We expect that this will significantly improve prediction performance. Our long-standing research interest is also to develop a Web application that will exploit modern network analysis methods for LBD.

Acknowledgments. This work was supported by the Slovenian Research Agency and by the Intramural Research Program of the U.S. National Institutes of Health, National Library of Medicine.

References

1. Rebholz-Schuhmann, D., Oellrich, A., Hoehndorf, R.: Text-mining solutions for biomedical research: Enabling integrative biology. Nat. Rev. Genet. 13, 829–839 (2012)
2. Swanson, D.R.: Fish oil, Raynaud's syndrome, and undiscovered public knowledge. Perspect. Biol. Med. 30, 7–18 (1986)
3. Hristovski, D., Rindflesch, T., Peterlin, B.: Using literature-based discovery to identify novel therapeutic approaches. Cardiovasc. Hematol. Agents Med. Chem. 11, 14–24 (2013)
4. Rindflesch, T., Kilicoglu, H.: Semantic MEDLINE: An advanced information management application for biomedicine. Inf. Serv. Use. 31, 15–21 (2011)
5. Rindflesch, T.C., Fiszman, M.: The interaction of domain knowledge and linguistic structure in natural language processing: interpreting hypernymic propositions in biomedical text. J. Biomed. Inform. 36, 462–477 (2003)
6. Bodenreider, O.: The Unified Medical Language System (UMLS): integrating biomedical terminology. Nucleic Acids Res. 32, D267–D270 (2004)
7. McCray, A.T., Srinivasan, S., Browne, A.C.: Lexical methods for managing variation in biomedical terminologies. In: Ozbolt, J.G. (ed.) Proceedings of the Eighteenth Annual Symposium on Computer Application in Medical Care, pp. 235–239. Hanley & Belfus, Washington, DC (1994)
8. Smith, L., Rindflesch, T., Wilbur, W.J.: MedPost: a part-of-speech tagger for bioMedical text. Bioinformatics 20, 2320–2321 (2004)
9. Aronson, A.R., Lang, F.-M.: An overview of MetaMap: historical perspective and recent advances. J. Am. Med. Inform. Assoc. 17, 229–236 (2010)
10. Kilicoglu, H., Shin, D., Fiszman, M., Rosemblat, G., Rindflesch, T.C.: SemMedDB: a PubMed-scale repository of biomedical semantic predications. Bioinformatics 28, 3158–3160 (2012)
11. Bales, M.E., Johnson, S.B.: Graph theoretic modeling of large-scale semantic networks. J. Biomed. Inform. 39, 451–454 (2006)
12. Lü, L., Zhou, T.: Link prediction in complex networks: A survey. Phys. A Stat. Mech. its Appl. 390, 1150–1170 (2011)
13. Newman, M.E.J.: The structure and function of complex networks. SIAM Rev. Soc. Ind. Appl. Math. 45, 167–256 (2003)
14. Manning, C.D., Schuetze, H.: Foundations of statistical natural language processing. MIT Press, Cambridge (1999)
15. Liben-Nowell, D., Kleinberg, J.: The link-prediction problem for social networks. J. Am. Soc. Inf. Sci. Technol. 58, 1019–1031 (2007)
16. Sarkar, P., Chakrabarti, D., Moore, A.W.: Theoretical justification of popular link prediction heuristics, pp. 2722–2727 (2011)
17. Swets, J.A.: Measuring the accuracy of diagnostic systems. Science 240, 1285–1293 (1988)
18. Katukuri, J.R., Xie, Y., Raghavan, V.V., Gupta, A.: Hypotheses generation as supervised link discovery with automated class labeling on large-scale biomedical concept networks. BMC Genomics 13(suppl. 3), S5 (2012)
19. Liu, Z., He, J.-L., Kapoor, K., Srivastava, J.: Correlations between community structure and link formation in complex networks. PLoS One 8, e72908 (2013)

Medical Image Retrieval Using Multimodal Data

Ivan Kitanovski, Ivica Dimitrovski, Gjorgji Madjarov, and Suzana Loskovska

Faculty of Computer Science and Engineering, University "Ss. Cyril and Methodius",
Skopje, Macedonia
{ivan.kitanovski,ivica.dimitrovski,gjorgji.madjarov,
suzana.loskovska}@finki.ukim.mk
http://www.finki.ukim.mk

Abstract. In this paper we propose a system for medical image retrieval
using multimodal data. The system can be separated in an off-line and
on-line phase. The off-line phase deals with modality classification of
the images by their visual content. For this part we use state-of-the-art
opponentSIFT visual features to describe the image content, as for the
classification we use SVMs. The modality classification labels all images
in the database with their corresponding modality. The off-line phase,
also, implements the text-based retrieval structure of the system. In this
part we index the text associated with the images using the open-source
search engine Terrier IR. In the on-line phase the retrieval is performed.
In this phase the system receives a text query. The system processes the
query and performs the text-based retrieval with Terrier IR and the initial results are generated. Afterwards, the images in the initial results are
re-ranked based on their modality and the final results are provided. Our
system was evaluated against the standardized ImageCLEF 2013 medical dataset. Our system reported results with a mean average precision
of 0.32, which is state-of-the-art performance on the dataset.

Keywords: medical image retrieval, retrieval in medical texts, image
modality classification, visual image descriptors.

1 Introduction

Medical image collections are a valuable source of information and play an important role in clinical decision making, medical research and education. Their
size poses a serious technical problem, especially as technology advances and
imaging equipment is more accessible to medical institutions. The growth of the
size of the image collections is exponential, which in turn creates huge repositories of valuable information which is difficult to maintain and process in an
appropriate manner. This underlines the need for tools for efficient access to this
kind of information.

The Picture Archiving and Communication System (PACS) [1] is a standard way of accessing medical image databases by means of textual information.
PACS is a central entity which integrates imaging modalities and interfaces with
medical institutions to appropriately store and distribute the images to medical

S. Džeroski et al. (Eds.): DS 2014, LNAI 8777, pp. 144–155, 2014.
© Springer International Publishing Switzerland 2014

professionals. Organizing these collections was usually done manually. But, as the collections grow it becomes a very expensive task and also prone to human-made errors. Hence, this task needs to be automated with the end goal to organize and retrieve large medical image collections [2].

The retrieval of medical images has usually been text-based i.e. the retrieval relied on text annotations/descriptions of the images and by extent the queries were also of textual nature. There is another approach where the retrieval is depended on the actual visual content of the images which is referred as Content-Based Image Retrieval (CBIR) [3]. The systems which follow these approaches are called CBIR systems. CBIR systems have been proven to be effective in somewhat narrowed medical domains such as lung CTs [4], head MRIs [5] and mammography images [6]. Additionally, utilizing both visual and textual data in the retrieval should increase overall retrieval performance [2]. In that situation, the queries are composed of a text part (keywords) and/or visual part (couple of sample images). The text part may contain relevant medical information about the patient such demographics, symptoms etc. So, the goal of this task is for a given query (text and/or visual) to retrieve the most relevant medical images from a given image collection. This task is referred as *ad-hoc image-based retrieval* [7].

General medical image collections used for research or education purposes, contain images from different modalities such as MRI, CT, x-ray etc. Image modality is a crucial characteristic of an image and it can be used for the benefit of the retrieval. Research has shown that physicians often prefer to filter the retrieval by the image modality [8]. However, the annotations of the images usually do not contain that information. This is set afterwards, frequently manually by a physician or radiologist, and that exposes this process to potential human errors [9]. Hence, automated modality classification would be a useful feature, which can be incorporated in a medical image retrieval system.

In this paper, we propose a system for medical image retrieval that utilizes, both, text and visual features. As visual features we use the current state-of-the-art Scale Invariant Feature Transform (SIFT) in opponent color space [10]. The text features are resented with the bag-of-words model. The text-based retrieval is further improved by incorporating query expansion. We have made additional improvement over traditional medical image retrieval systems by adding information for the medical modality of the images as input during the retrieval phase.

The rest of the paper is organized as follows. Section 2 presents the relevant state-of-the-art work in the field. The overall architecture of our proposed system is described in Section 3. The textual part of the retrieval is described in more details in Section 4. Section 5 contains the details of the modality classification part. The experimental design and evaluation methodology is presented in Section 6. Section 7 contains the results and discussion. Finally, the concluding remarks and directions for further work are given in Section 8.

2   Related Work

Typical approaches to medical image retrieval are text-based i.e. the images are retrieved based on their annotations/descriptions. So, one part of the problem is acquiring appropriate text-based representation of the images. Often, medical images are used to provide better description of the content of medical articles and are an essential piece of information in the context. Hence, the text data from the medical articles can be used to make the image representation.

Medical image retrieval systems, which search for images within a collection of medical articles usually represent and retrieve the images according to their captions in the appropriate article [11]. For example, BioText uses Lucene [1] to index over 300 journals and retrieves images based on their captions.

Stathopoulos et al. [12] make a field-based representation of the images and index it using Lucene. In the retrieval phase they add different weights to the fields of representation based on the part of the article they were extracted from.

Goldminer [13] is a search engine that retrieves images by looking at figure captions of journal articles from the Radiological Society of North America (RSNA). They map keywords from the captions to UMLS [2] concepts. The Yale Image Finder (YIF) [14] retrieves images by looking at title, captions and abstract of medical journal articles. It uses optical character recognition (OCR) to recognize the text in the images.

The above methods use primarily text data for the retrieval. Methods that rely on visual data are also developed. Ozturkmenoglu et al [15] propose a retrieval framework based on the Terrier IR search engine. The framework consists on two parts. The first part is a classification mechanism, which classifies the modality of the potential target image and filters out the images which do not belong in that modality class. The retrieval is then performed within that filtered subset. The drawback of this approach is the relatively lower retrieval performance compared to the current best result on the given database (by roughly 10%).

Rahman et al. [16] also propose a similar filtering approach. Their system uses multimodal data for the retrieval phase. The text-based part focuses on retrieving the image representation with Essie search engine developed by the National Library of Medicine (NLM). The visual part of the retrieval uses modality classification to filter out images that do not belong to the query image class and then retrieves in that subset. The results from the text-based and visual search are merged using weighted linear combination and final results are provided. The major issues in their approach is scalability and efficiency, because they are using many low level visual descriptors to obtain the medical modalities for each image.

In our previous work we have concluded that BM25 is the best performing weighting model for this type of problem [17]. Also, it has shown that merging text-based and content-based retrieval does not directly provide a significant boost in performance [18]. Hence, we were motivated to add value by using

---

[1] http://lucene.apache.org
[2] http://www.nlm.nih.gov/research/umls/

text-based retrieval and modality classification (based on visual content of the images) to perform the retrieval.

3   System for Medical Image Retrieval

The architecture of the system for multimodal medical image retrieval is presented in Figure 1. The system consists of an off-line and on-line phase. The off-line phase implements the algorithm for medical modality classification and the data retrieval structure. The on-line phase implements the image querying and results presentation.

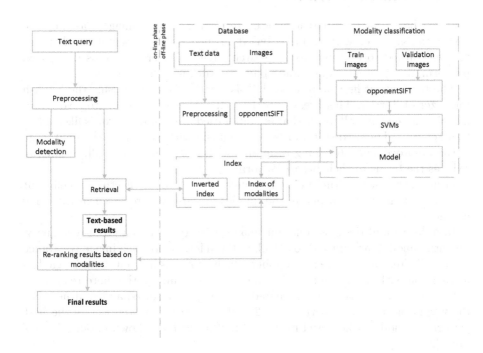

Fig. 1. Architecture of the proposed system for multimodal medial image retrieval

The off-line phase starts with the generation, preprocessing and indexing of the text-based representation (data) for each image. This representation based on the journal/article text the image belongs to. The proposed method for medical modality classification is as follows. First, we generate visual features (opponentSIFT) for a given dataset of training images. The generated features are used to train the classifier (in our case SVMs). The obtained classifier is validated by a dataset of validation images with known modalities (ground-truth). This model is capable of determining the modality of a given previously unseen image based on its visual features. Once we have the model, we extract the same visual features for all images in the database and pass them through the model. The label obtained from the classifier represents the image-modality index.

In the on-line phase the input in the system are textual queries. These queries are preprocessed and retrieval is performed. Because, the queries often contain keywords related to the modality of the images that need to be retrieved, in stage we extract the modalities of the query based on simple keyword matching. After the retrieval is performed the retrieved images are re-ranked using the image-modality index and the final results are presented.

A more detailed explanation for each of these parts (mainly modality classification and text retrieval) is given in the following sections.

4   Text-Based Retrieval

The text-based retrieval part focuses on retrieving medical images based on the text data associated with them i.e. based on their text representations. For this part we turned to the open-source search engine Terrier IR [19] as a retrieval platform, due to its flexibility and ability to work with large datasets.

The text data is first preprocessed. Tokenization is applied using the English tokenizer of Terrier. This breaks up the text into words. Then, stop-words removal is applied and all words that are often used are removed (words like: a, for, from, of etc.). The used stop-words set is built-in with Terrier. In the final stage of the preprocessing, stemming is applied with Porter stemmer [20]. Stemming reduces the terms to their base/root form.

From the resulting terms an inverted index is created for effective and efficient retrieval. In the retrieval phase, the queries are also preprocessed in the same manner.

In order to find the most relevant images with respect to a given text query, weighting models are applied to calculate the relevancy of each image (relevance score). The relevance score is a number which shows us how much an image has in common with a given query. The higher is the number the more relevant is the image. Once the score is calculated the images are sorted and returned. For the weighting model, we turn to BM25 [21], which has proven as one the best performing models when used in this context as we have shown in our previous work [22].

The text-based retrieval is further boosted by applying query expansion. Query expansion is a process of reformulating a query with the aim to improve retrieval performance. We implemented with pseudo-relevance feedback in Terrier. Pseudo-relevance feedback analyzes the top n number of initially retrieved documents and finds the m most informative terms. These terms are then added to the query and another retrieval is performed. The results from this retrieval are the final text-based results.

5   Modality Classification

Image modality is an important aspect of the image for medical retrieval purposes. In user studies, clinicians have indicated that modality is one of the most important filters that they would like to have to limit their search [2]. The usage of modality information often significantly improves the retrieval results.

5.1 Scale-Invariant Feature Transform Descriptors

Collections of medical images typically contain various images obtained using different imaging techniques. To properly represent the images, different feature extraction techniques that are able to capture the different aspects of an image (e.g., texture, shapes, color distribution...) need to be used [23]. Texture is especially important, because it is difficult to classify medical images using shape or gray level information. Effective representation of texture is necessary to distinguish between images with equal modality and layout. Furthermore, local image characteristics are fundamental for image interpretation: while global features retain information on the whole image, the local features capture the details. Thus, they are more discriminative concerning the problem of inter and intra-class variability [24].

Our approach to medical modality classification uses scale-invariant feature transform (SIFT) image descriptors in combination with the bag of features approach commonly used in many state-of-the-art image classification problems [2]. The basic idea of this approach is to sample a set of local image patches using some method (densely, randomly or using a key-point detector) and to calculate a visual descriptor on each patch (SIFT descriptor, normalized pixel values). The resulting set of descriptors is then matched against a pre-specified visual codebook, which converts it to a histogram. The main issues that need to be considered when applying this approach are: sampling of patches, selection of visual patch descriptors and building a visual codebook.

We use dense sampling of the patches, which samples an image grid in a uniform fashion using a fixed pixel interval between patches. We use an interval distance of 6 pixels and sample at multiple scales ($\sigma = 1.2$ and $\sigma = 2.0$ for the Gaussian filter [10]). Due to the low contrast of some of the medical images (e.g., radiographs), it would be difficult to use any detector for points of interest. We calculate opponentSIFT (OSIFT) descriptors for each image patch [25], [10]. OpponentSIFT describes all channels in the opponent color space using SIFT descriptors. The information in the O_3 channel is equal to the intensity information, while the other channels describe the color information in the image. These other channels do contain some intensity information, but due to the normalization of the SIFT descriptor they are invariant to changes in light intensity [10].

The crucial aspects of a codebook representation are the codebook construction and assignment. An extensive comparison of codebook representation variables is given by van Gemert et al. [26]. We employ k-means clustering (a custom implementation in the *Python* programming language was used) on 250K randomly chosen descriptors from the set of images available for training. k-means partitions the visual feature space by minimizing the variance between a predefined number of k clusters. Here, we set k to 1000 and thus define a codebook with 1000 codewords [24].

5.2   Classifier Setup

The training database is taken from the ImageCLEF 2013 medical modality classification task which is a standardized benchmark for systems that automatically classify medical image modality from PubMed journal articles [2]. The provided dataset consists of images from 31 modalities. Figure 2 shows sample images from the dataset. The number of images in the provided train set is 2901, and the number of validation images is 2582.

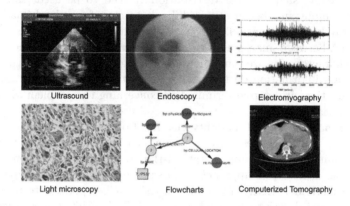

Fig. 2. Sample images from the ImageCLEF 2013 database

For classification, we used the LIBSVM implementation of support vector machines (SVMs) [27] with probabilistic output [28]. To solve the multi-class classification problems, we employ the *one-vs-all* approach. Namely, we build a binary classifier for each modality/class: the examples associated with that class are labeled positive and the remaining examples are labeled negative. This results in an imbalanced ratio of positive versus negative training examples. We resolve this issue by adjusting the weights of the positive and negative classes [10]. In particular, we set the weight of the positive class to $\frac{\#pos+\#neg}{\#pos}$ and the weight of the negative class to $\frac{\#pos+\#neg}{\#neg}$, with $\#pos$ the number of positive instances and $\#neg$ the number of negative instances in the train set.

We used SVMs with a precomputed χ^2 kernel. We optimize the cost parameter C of the SVMs using an automated parameter search procedure. For the parameter optimization, we used the validation set. In the final stage, using the learned model we have predicted the modalities for 300K images which are part of the retrieval system. These 300K images are a part of the ImageCLEF 2013 dataset for the same medical task.

5.3   Evaluation of the Classifier

To assess the performance of the classifiers, we use the overall recognition rate/accuracy. This is a very common and widely used evaluation measure. It is calculated as the fraction of the validation images whose class/modality was predicted correctly.

We can note that the opponentSIFT descriptor offers very good predictive performance with accuracy of 76.21%. The opponentSIFT descriptor is able to capture specific details from the images and is robust to noise, illumination, scale, translation and rotation changes. The detailed performance per modality (class) is given in Figure 3. The figure shows the confusion matrix for the validation set and the main diagonal contains the per modality classification accuracy.

From the results we can note the high predictive performance for most of the modalities. If we analyze the confusion matrix we can note that the classifier is biased towards the class/modality denoted as compound figure (COMP).

6   Experimental Design

6.1   Dataset Description

ImageCLEF [2] has positioned itself as a valid platform for benchmarking of medical image retrieval techniques, hence we turned to its dataset for our experiments. Namely, we used the dataset from the ad-hoc image-based retrieval subtask. The subtask is defined in the following manner. The input is a set of keywords and/or sample medical images and the output is a sorted list of relevant medical images. The goal is to find the most relevant images for a given query. For evaluation purposes the top 1000 image are taken into consideration as it was in the task.

The provided dataset contains sample images and medical journal articles that contain the images. A set of queries is also provided, so they can be used to evaluate the algorithms. The total number of images is 306,539 and the total number of queries is 35.

Each test query consists of a textual part and come with 2-3 sample images. The textual queries are usually short and contain a couple of keywords. Examples queries: **1.**osteoporosis x-ray, **2.**nephrocalcinosis ultrasound images, **3.**lymphoma MRI images

The medical articles from which the images are extracted are organized in a XML format and contain the following fields: title,abstract,article text - referred as full-text,captions of images present in the article, Medical Subject Headings (MeSH ®terms) of the article. The best runs from previous years [2] reported that the best image representation consisted of the title, abstract, MeSH terms, image caption and the snippets (sentences) of the text where the image is mentioned (referred as image mention). Hence, we choose the same representation as well.

6.2   Evaluation Metrics

For evaluation of our system we applied the evaluation metrics used by Image-CLEF: mean average precision (MAP), precision at first 10 (P10) and first 30 (P30) images retrieved [2].

7   Results and Discussion

The goal of this study is to answer the following question: *Does integrating modality information in the retrieval stage improve the regular retrieval performance
using only text data?*

First, we conduct pure text-based retrieval, where only the text part of the
query is used. Then for each image in the initial results we check the image-
modality index to determine whether the image has the same modality as the
ones extracted from the query. If that is the case the relevance score of the

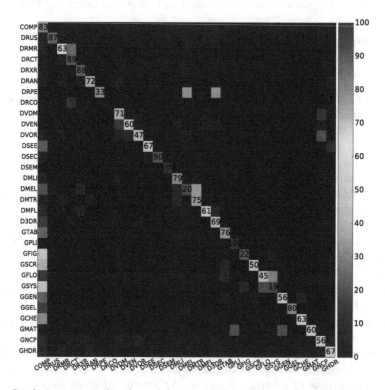

Fig. 3. Confusion matrix for the validation set. The image modalities are as follows:
Tables and forms (GTAB), Fluorescence microscopy (DMFL), Chromatography/Gel
(GGEL), Statistical figures/graphs/charts (GFIG), Other organs (DVOR), Chemi-
cal structure (GCHE), Light microscopy (DMLI), Angiography (DRAN), Screenshots
(GSCR), Endoscopy (DVEN), Hand-drawn sketches (GHDR), Gene sequence (GGEN),
System overviews (GSYS), Compound or multipane images (COMP), Ultrasound
(DRUS), Combined modalities in one image (DRCO), Electromyography (DSEM),
Program listing (GPLI), Electroencephalography (DSEE), Mathematics/formulae
(GMAT), Electrocardiography (DSEC), X-Ray/2D Radiography (DRXR), Transmis-
sion microscopy (DMTR), Flowcharts (GFLO), Dermatology/skin (DVDM), Electron
microscopy (DMEL), Computerized Tomography (DRCT), PET (DRPE), Non-clinical
photos (GNCP), Magnetic Resonance (DRMR), 3D reconstructions (D3DR)

image is multiplied by a certain factor. We then re-rank the images based on this modification and get the final results (this is done for the results from each query separately). This should provide us with an answer on whether modality improves the retrieval.

The results from the experiments are presented in Table 1.

Table 1. Results from ad-hoc image-based retrieval over ImageCLEF 2013 using text and multimodal data with/without query expansion (denoted as qe in the table)

	MAP	P10	P30
text	0.26	0.32	0.22
text + modality	0.29	0.35	0.24
text + qe	0.30	0.38	0.24
text + qe + modality	0.32	0.39	0.25

We did four experiments to determine whether our proposed architecture improves the retrieval process. The first run *text* is a pure textual retrieval over the ImageCLEF 2013 collection. The run *text + modality* is the run from our proposed architecture, which takes into account the image modality and the text-based query. The proposed architecture provides improved MAP for roughly 3%. The run *text + qe* is the baseline textual retrieval with added query expansion. Query expansion by itself adds improvements over the pure textual retrieval. The run *text + qe + modality* is the textual retrieval with added query expansion and modality re-ranking at the end. The highest performance is provided by this run. It provides a roughly 2% increase in performance.

Incorporating modality increases performance in both cases. That confirms our hypothesis that introducing modality information in this kind of retrieval process can improve the performance. Furthermore, the obtained results are so far the best on this database by our knowledge. Namely, the Natural Library of Medicine (NLM) group at ImageCLEF 2013 reported a MAP of 0.32 on the same dataset (with the same queries) as well.

8   Conclusion

In this paper we proposed an approach for retrieval of medical images, which uses multimodal (textual and visual) features. For the visual features we used the state-of-the-art opponentSIFT features, whereas, for the textual features we referred to the standard bag-of-words representation. We applied query expansion to further improve the text-based retrieval. At the end, we included the medical modality of the images as input to the retrieval.

The results confirmed our hypothesis that introducing modality information to the retrieval process can improve the overall performance. The modality classification in the retrieval process boosts the retrieval by roughly 2-3%. We obtained state-of-the-art results over the ImageCLEF 2013 dataset.

Our future work goes into the lines of implementing the proposed architecture in a form of a publicly accessible website and including different databases.

Acknowledgement. We would like to acknowledge the support of the European Commission through the project MAESTRA - Learning from Massive, Incompletely annotated, and Structured Data (Grant number ICT-2013-612944).

References

1. Choplin, R., Boehme, J., Maynard, C.: Picture archiving and communication systems: an overview. Radiographics 12(1), 127–129 (1992)
2. de Herrera, A.G.S., Kalpathy-Cramer, J., Fushman, D.D., Antani, S., Müller, H.: Overview of the imageclef 2013 medical tasks. In: Working notes of CLEF 2013 (2013)
3. Lehmann, T.M., Wein, B.B., Dahmen, J., Bredno, J., Vogelsang, F., Kohnen, M.: Content-based image retrieval in medical applications: a novel multistep approach. In: Proceedings of SPIE: Storage and Retrieval for Media Databases, vol. 3972, pp. 312–320 (2000)
4. Shyu, C.-R., Brodley, C.E., Kak, A.C., Kosaka, A., Aisen, A.M., Broderick, L.S.: Assert: A physician-in-the-loop content-based retrieval system for HRCT image databases. Computer Vision and Image Understanding 75(12), 111–132 (1999)
5. Simonyan, K., Modat, M., Ourselin, S., Cash, D., Criminisi, A., Zisserman, A.: Immediate ROI search for 3-D medical images. In: Greenspan, H., Müller, H., Syeda-Mahmood, T. (eds.) MCBR-CDS 2012. LNCS, vol. 7723, pp. 56–67. Springer, Heidelberg (2013)
6. El-Naqa, I., Yang, Y., Galatsanos, N.P., Nishikawa, R.M., Wernick, M.N.: A similarity learning approach to content-based image retrieval: application to digital mammography. IEEE Transactions on Medical Imaging 23(10), 1233–1244 (2004)
7. Müller, H., de Herrera, A.G.S., Kalpathy-Cramer, J., Demner-Fushman, D., Antani, S., Eggel, I.: Overview of the imageclef 2012 medical image retrieval and classification tasks. In: CLEF (Online Working Notes/Labs/Workshop) (2012)
8. Medical retrieval task, http://www.imageclef.org/node/104/ (accessed: July 03, 2014)
9. Guld, M.O., Kohnen, M., Keysers, D., Schubert, H., Wein, B.B., Bredno, J., Lehmann, T.M.: Quality of DICOM header information for image categorization. In: Medical Imaging 2002: PACS and Integrated Medical Information Systems: Design and Evaluation, SPIE, vol. 4685, pp. 280–287 (2002)
10. van de Sande, K., Gevers, T., Snoek, C.: Evaluating color fescriptors for object and scene recognition. IEEE Transactions on Pattern Analysis and Machine Intelligence 32(9), 1582–1596 (2010)
11. Hearst, M.A., Divoli, A., Guturu, H., Ksikes, A., Nakov, P., Wooldridge, M.A., Ye, J.: Biotext search engine: beyond abstract search. Bioinformatics 23(16), 2196–2197 (2007)
12. Kyriakopoulou, A., Stathopoulos, S., Lourentzou, I., Kalamboukis, T.: Ipl at clef 2013 medical retrieval task. In: CLEF (Online Working Notes/Labs/Workshop) (2013)
13. Kahn Jr., C.E., Thao, C.: Goldminer: a radiology image search engine. American Journal of Roentgenology 188(6), 1475–1478 (2007)
14. Xu, S., McCusker, J., Krauthammer, M.: Yale image finder (yif): a new search engine for retrieving biomedical images. Bioinformatics 24(17), 1968–1970 (2008)
15. Ceylan, N.M., Ozturkmenoglu, O., Alpkocak, A.: Demir at imageclefmed 2013: The effects of modality classification to information retrieval. In: CLEF (Online Working Notes/Labs/Workshop) (2013)

16. Rahman, M.M., You, D., Simpson, M.S., Antani, S.K., Demner-Fushman, D., Thoma, G.R.: Multimodal biomedical image retrieval using hierarchical classification and modality fusion. International Journal of Multimedia Information Retrieval 2(3), 159–173 (2013)

17. Kitanovski, I., Trojacanec, K., Dimitrovski, I., Loshkovska, S.: Merging words and concepts for medical articles retrieval. In: Proceedings of the 10th Conference on Open Research Areas in Information Retrieval, pp. 25–28. Le Centre De Hautes Etudes Internationales D'Informatique Documentaire (2013)

18. Kitanovski, I., Trojacanec, K., Dimitrovski, I., Loskovska, S.: Multimodal medical image retrieval. In: Markovski, S., Gushev, M. (eds.) ICT Innovations 2012. AISC, vol. 207, pp. 81–89. Springer, Heidelberg (2013)

19. Ounis, I., Amati, G., Plachouras, V., He, B., Macdonald, C., Johnson, D.: Terrier information retrieval platform. In: Losada, D.E., Fernández-Luna, J.M. (eds.) ECIR 2005. LNCS, vol. 3408, pp. 517–519. Springer, Heidelberg (2005)

20. Macdonald, C., Plachouras, V., He, B., Lioma, C., Ounis, I.: University of Glasgow at webclef 2005: Experiments in per-field normalisation and language specific stemming. In: Peters, C., Gey, F.C., Gonzalo, J., Müller, H., Jones, G.J.F., Kluck, M., Magnini, B., de Rijke, M., Giampiccolo, D., et al. (eds.) CLEF 2005. LNCS, vol. 4022, pp. 898–907. Springer, Heidelberg (2006)

21. Amati, G., Van Rijsbergen, C.J.: Probabilistic models of information retrieval based on measuring the divergence from randomness. ACM Transactions on Information Systems (TOIS) 20(4), 357–389 (2002)

22. Kitanovski, I., Dimitrovski, I., Loskovska, S.: Fcse at medical tasks of imageclef 2013. In: CLEF (Online Working Notes/Labs/Workshop) (2013)

23. Dimitrovski, I., Kocev, D., Loskovska, S., Dzeroski, S.: Hierarchical annotation of medical images. Pattern Recognition 44(10-11), 2436–2449 (2011)

24. Tommasi, T., Orabona, F., Caputo, B.: Discriminative cue integration for medical image annotation. Pattern Recognition Letters 29(15), 1996–2002 (2008)

25. Lowe, D.G.: Distinctive image features from scale-invariant keypoints. International Journal of Computer Vision 60(2), 91–110 (2004)

26. van Gemert, J.C., Veenman, C.J., Smeulders, A.W.M., Geusebroek, J.M.: Visual word ambiguity. IEEE Transactions on Pattern Analysis and Machine Intelligence 99(1)

27. Chang, C.-C., Lin, C.-J.: LIBSVM: a library for support vector machines (2001), software available at http://www.csie.ntu.edu.tw/~cjlin/libsvm

28. Lin, H.-T., Lin, C.-J., Weng, R.C.: A note on Platt's probabilistic outputs for support vector machines. Machine Learning 68, 267–276 (2007)

Fast Computation of the Tree Edit Distance between Unordered Trees Using IP Solvers

Seiichi Kondo*, Keisuke Otaki**, Madori Ikeda, and Akihiro Yamamoto

Graduate School of Informatics, Kyoto University
Yoshida-Honmachi, Sakyo-ku, Kyoto, 606-8501, Japan
{s.kondo,ootaki,m.ikeda}@iip.ist.i.kyoto-u.ac.jp, akihiro@i.kyoto-u.ac.jp

Abstract. We propose a new method for computing the tree edit distance between two unordered trees by problem encoding. Our method transforms an instance of the computation into an instance of some IP problems and solves it by an efficient IP solver. The tree edit distance is defined as the minimum cost of a sequence of edit operations (either substitution, deletion, or insertion) to transform a tree into another one. Although its time complexity is NP-hard, some encoding techniques have been proposed for computational efficiency. An example is an encoding method using the clique problem. As a new encoding method, we propose to use IP solvers and provide new IP formulations representing the problem of finding the minimum cost mapping between two unordered trees, where the minimum cost exactly coincides with the tree edit distance. There are IP solvers other than that for the clique problem and our method can efficiently compute ariations of the tree edit distance by adding additional constraints. Our experimental results with Glycan datasets and the Web log datasets CSLOGS show that our method is much faster than an existing method if input trees have a large degree. We also show that two variations of the tree edit distance could be computed efficiently by IP solvers.

Keywords: tree edit distance, unordered tree, IP formulation.

1 Introduction

Computing similarities between tree structured data (e.g., RNA secondary structures [10], Glycan structures [12], markup documents) is an important task in machine learning applications, and many efficient algorithms for the *tree edit distance* (*distance*, for short) of *rooted labeled ordered trees* have been proposed [13]. However, computing the distance between unordered trees is known to be NP-hard [19] and MAX SNP-hard [20], which indicates that the problem does not have any polynomial time approximation scheme (PTAS) unless P = NP.

---

* Seiichi Kondo's current affiliation is Ricoh Inc., Japan. This work was conducted while he was a student at Graduate School of Informatics, Kyoto University.
** Keisuke Otaki is supported as a JSPS research fellow (DC2, 26·4555).

S. Džeroski et al. (Eds.): DS 2014, LNAI 8777, pp. 156–167, 2014.
© Springer International Publishing Switzerland 2014

In this paper, we propose a new method to compute the *tree edit distance* for *rooted labeled unordered trees*. The distance is formulated as the minimum cost to transform a tree into another by applying *edit operations* (either substitution, deletion, or insertion). It is known that the distance coincides with the minimum cost of all possible *Tai mappings* between two trees [16]. Our key idea is to transform every instance of the problem into an IP problem finding the minimum cost mapping, and solve it using IP solvers. Though the time complexity of IP problems is also NP-hard, there exist several IP solvers (e.g., SCIP [1] and CPLEX [9]) that run in practical time, which is roughly 1000 times faster than those in 90s without progresses of hardwares [4].

Existing methods for unordered trees are roughly classified into three groups. The first group adopts heuristic search. Horesh et al. developed an A* algorithm for *unlabeled* unordered trees [8] and Higuchi et al. extended it for *labeled* trees [7]. The second one concerns parameterized algorithms. Shasha et al. proposed $O(4^{l_1+l_2} poly(n_1, n_2))$ time algorithm [15], where l_i and n_i are the numbers of leaves and nodes in input trees respectively. Another example is that proposed by Akutsu et al., which is an $O(1.26^{n_1+n_2})$ time algorithm [2]. The third one performs problem reductions. Fukagawa et al. proposed a method to use an maximum vertex weighted clique solver after transforming every instance of the distance problem into some instance of the maximum vertex weighted clique problem [5]. They showed that the the clique-based method is as fast as an A* based method. Mori et al. improved it by introducing several heuristics [14]. They showed that their method is much faster than the previous study. Although these clique-based methods are faster than the other previous methods, yet their method can only solve moderate size trees. Moreover, there is a problem that there are few available maximum vertex weighted clique solvers.

Our approach is classified into the third category, which uses IP solvers instead of clique solvers. An IP problem is a linear programming (LP) problem with additional restrictions on variables [3]. Compared to the clique-based methods, we have two advantages: There exist IP solvers other than those for the clique problem, and IP formulations could represent variations of the distance easily by adding additional constraints. To explain our method, we give preliminaries in Section 2 and propose our IP formulations in Section 3. In Section 4, we evaluate the proposed method using Glycan datasets from KEGG [12] and the Web log data CSLOGS [18]. We also evaluate our method for computing variations of the tree edit distance. We conclude our work in Section 5.

2   Preliminaries

Let V be a (finite) set of *nodes* and $\leq$ be an order between nodes. A *rooted unordered tree* T is a poset $(V, \leq)$ satisfying

1. There exists a unique element $r \in V$ such that $x \leq r$ for all $x \in V$, and
2. for all $x, y, z \in V$, if $x \leq y$ and $x \leq z$, then y and z are comparable.

The node r is called the *root* of T and denoted by $root(T)$. For two nodes x and y, $x < y$ means $x \leq y$ and $x \neq y$. If $x \leq y$, x is an *descendant* of y and y is an

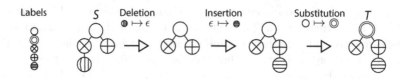

Fig. 1. An example of a sequence of edit operations from S to T

ancestor of x. For a node $x \in V \setminus \{r\}$, the minimum ancestor is called the *parent* and denoted by $par(x)$ and x is called a *child* of $par(x)$. The set of all children of x is denoted by $ch(x)$. The degree $deg(x)$ of a node x is defined as $|ch(x)|$ and that $deg(T)$ of a tree T is also defined as $\max_{x \in T} deg(x)$, respectively. A node x is called *leaf* if $deg(x) = 0$, and $leaves(T)$ denotes the set of all leaves in T.

With a labeling function $l_T : V \to \Sigma$, a tuple (T, l_T) with $T = (V, \le)$ is called a *labeled tree*, where Σ is the *alphabet*. In the following, we often represent labeled trees by rooted trees without labeling functions.

2.1   Tree Edit Distance

The tree edit distance between two trees is defined by using the cost of sequences of *edit operations* which are required to transform a tree into another.

Definition 1 (Edit Operations). Let $\Sigma_\epsilon = \Sigma \cup \{\epsilon\}$ where ϵ is a special blank symbol not in Σ. An edit operation is either

substitution replacing the label of a node in T with a new label,
deletion deleting a non-root node s of T, making all children of s be the children
    of $par(s)$, or
insertion inserting a new node t as a child of some node v in T, making some
    children of v be the children of t.

For each of the edit operations, we define a pair in $\Sigma_\epsilon \times \Sigma_\epsilon \setminus \{(\epsilon, \epsilon)\}$ as follows. For substituting a label n of a node t in T by another label m, the pair is (n, m). For deleting a node having a label n, the pair is (n, ϵ). For inserting a new node having a label m, the pair is (ϵ, m). A *cost function* $d : \Sigma_\epsilon \times \Sigma_\epsilon \setminus \{(\epsilon, \epsilon)\} \to \mathbb{R}$ defines the cost of the operations on trees according to pairs representing the edit operations. In the following, we write $d(s, t)$ for $(s, t) \in S \times T$ to mean $d(l_S(s), l_T(t))$, where l_S and l_T are labeling functions of two trees S and T. In this paper we adopt the *unit cost*, which is defined by using the Kronecker's delta $\delta_{i,j}$ as:

$$d(n_1, n_2) = 1 - \delta_{l_S(n_1), l_T(n_2)}$$

It is important that our IP formulation adopts not only the unit cost but also various cost functions. Because of the space limitation, we discuss only the unit cost in this paper. The cost of a sequence $E = \langle e_1, \ldots, e_n \rangle$ of edit operations is defined as $cost(E) = \sum_i cost(e_i)$. Figure 1 shows an example of such a sequence transforming S into T. With the cost of the edit operations, the tree edit distance is defined as follows.

Definition 2 (Edit Distance [16]). Let S and T be trees and $\mathcal{E}(S,T)$ be the set of all possible sequences of edit operations transforming S into T. The edit distance $\mathrm{D}(S,T)$ is defined as $\mathrm{D}(S,T) = \min_{E \in \mathcal{E}(S,T)} \mathrm{cost}(E)$.

The tree edit distance is closely related to *Tai mappings*.

Definition 3 (Tai Mapping [16]). Let S and T be trees. A *Tai mapping* M is a subset of $S \times T$ satisfying constraints below for any $(s_1, t_1), (s_2, t_2)$ in M:

One-to-one mapping : $s_1 = s_2 \iff t_1 = t_2$, and
Preserving ancestor : $s_1 < s_2 \iff t_1 < t_2$.

The set of all possible Tai mappings between S and T is denoted by $\mathcal{M}^{\mathrm{Tai}}(S,T)$. For a mapping $M \subseteq S \times T$, we let $M^{(1)} = \{s \in S \mid (s,t) \in M\}$ and $M^{(2)} = \{t \in T \mid (s,t) \in M\}$. With a cost function d, we define the cost of a Tai mapping M, denoted by $\mathrm{cost}(M)$, as follows.

$$\mathrm{cost}(M) = \sum_{(s,t) \in M} d(s,t) + \sum_{s \in S \setminus M^{(1)}} d(s,\epsilon) + \sum_{t \in T \setminus M^{(2)}} d(\epsilon, t).$$

Based on the cost of Tai mappings, Tai showed the following theorem.

Theorem 1 ([16]). For trees S and T, $\mathrm{D}(S,T) = \min_{M \in \mathcal{M}^{\mathrm{Tai}}(S,T)} \mathrm{cost}(M)$.

This well-known theorem means that the tree edit distance can be computed by finding the minimum cost Tai mapping instead of finding directly the minimum cost sequence of the edit operations.

2.2   Mixed Integer Linear Programming

We compute the tree edit distance by finding a tree mapping with the minimum cost based on Theorem 1 by transforming the original problem into an instance of IP problems instead of computing the distance directly. An IP problem is a linear programming problem with constraints, where some or all variables must be integer values. In this paper, we deal with only *Mixed Integer Linear Programming* problems and call them *IP problems*.

Definition 4 (Mixed Integer Linear Programming). A *Mixed Integer Linear Programming* (MILP) is defined as follows:

$$\begin{aligned}
&\text{maximize } \textstyle\sum_{j=1}^{n} c_j x_j, \\
&\text{subject to } \textstyle\sum_{j=1,2,\ldots,n} a_{ij} x_j \leq b_i && (1 \leq i \leq m), \\
&\qquad\quad x_j \geq 0 && (1 \leq j \leq n), \\
&\qquad\quad \text{some or all } x_j \text{ must be integer,}
\end{aligned}$$

where c_j and a_{ij} are constant integers.

Because a large number of problems can be formulated as IP problems, the expressiveness of IP formulations has been recognized since early times [6]. However, they had not been used in practice because their computational complexity is NP-hard [4]. Recently several IP solvers have been improved drastically, and they can handle IP problems with several hundred thousand constraints.

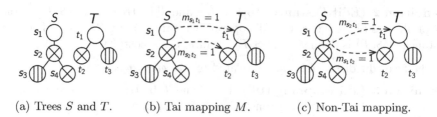

(a) Trees S and T.   (b) Tai mapping M.   (c) Non-Tai mapping.

Fig. 2. A example of representing mappings by using binary variables m_{s_i,t_j}

3   Two IP Formulations for the Tree Edit Distance

As an example case, let we consider the trees $S = \{s_1, s_2, s_3, s_4\}$ and $T = \{t_1, t_2, t_3\}$ illustrated in Figure 2a. For representing mappings, we first prepare binary variables $m_{s_i,t_j} \in \{0,1\}$ for each pair of nodes $(s_i, t_j) \in S \times T$. We use these binary variables as indicators: $m_{s,t} = 1$ if and only if $(s,t) \in M$ for $s \in S, t \in T$. In Figure 2b, a Tai mapping $M = \{(s_1,t_1),(s_2,t_2)\} \subseteq S \times T$ is represented by $m_{s_1,t_1} = m_{s_2,t_2} = 1$. A mapping in Figure 2c is not a Tai mapping.

We here design an objective function for computing $D(S,T)$ with binary variables by transforming a $\text{cost}(M)$ into an objective function as follows:

$$\text{cost}(M) = \sum_{(s,t)\in M} d(s,t) + \sum_{s\in S\setminus M^{(1)}} d(s,\epsilon) + \sum_{t\in T\setminus M^{(2)}} d(\epsilon,t),$$

$$= \sum_{(s,t)\in S\times T} d(s,t)m_{s,t}$$

$$+ \sum_{s\in S} d(s,\epsilon)\left\{1-\sum_{t\in T}m_{s,t}\right\} + \sum_{t\in T} d(\epsilon,t)\left\{1-\sum_{s\in S}m_{s,t}\right\}, \quad (1)$$

$$= \sum_{(s,t)\in S\times T}\{d(s,t)-d(s,\epsilon)-d(\epsilon,t)\}\,m_{s,t} + \sum_{s\in S}d(s,\epsilon)+\sum_{t\in T}d(\epsilon,t). \quad (2)$$

where terms in Equation 1 indicate the costs of pairs in M, those of nodes only in S which are removed, and those only in T which are inserted, respectively. In IP formulations, we minimize $\text{cost}(M)$ under constraints on binary variables $m_{s,t}$ representing conditions of Tai mappings in Definition 3. The first condition (one-to-one mapping) can be transformed into two types of linear constraints.

$$\text{For all } s \in S, \sum_{t\in T} m_{s,t} \le 1 \text{ and for all } t \in T, \sum_{s\in S} m_{s,t} \le 1. \quad (3)$$

The first constraint means that for a node $s \in S$ (resp. $t \in T$) we allow at most one pair in M containing s (resp. t).

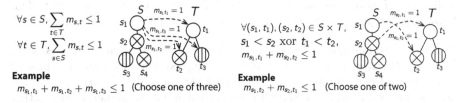

$$\forall s \in S, \sum_{t \in T} m_{s,t} \le 1$$

$$\forall t \in T, \sum_{s \in S} m_{s,t} \le 1$$

Example
$m_{s_1,t_1} + m_{s_1,t_2} + m_{s_1,t_3} \le 1$ (Choose one of three)

(a) One-to-one mapping constraint.

$$\forall (s_1, t_1), (s_2, t_2) \in S \times T,$$

$$s_1 < s_2 \text{ xor } t_1 < t_2,$$

$$m_{s_1,t_1} + m_{s_2,t_2} \le 1$$

Example
$m_{s_1,t_2} + m_{s_2,t_1} \le 1$ (Choose one of two)

(b) Preserving ancestor constraint.

Fig. 3. Examples of constraints representing both one-to-one mapping constraint and preserving ancestor constraint

The second condition (preserving ancestor) is transformed into linear constraints in the same manner:

$$\text{For all } (s_1, t_2), (s_2, t_2) \in S \times T \text{ s.t. } s_1 < s_2 \not\Leftrightarrow t_1 < t_2, m_{s_1,t_1} + m_{s_2,t_2} \le 1. \quad (4)$$

This means that a mapping M cannot contain two pairs whose mapping from S to T violates preserving ancestor relationships. Figure 3 shows examples for these two constraints.

Putting all together, our formulation is given as follows.

Proposition 1 (IP Formulation for the Tree Edit Distance)

$$\begin{aligned}
\text{minimize} \quad & \sum_{(s,t) \in S \times T} \{d(s,t) - d(s,\epsilon) - d(\epsilon,t)\} m_{s,t} \\
& + \sum_{s \in S} d(s,\epsilon) + \sum_{t \in T} d(\epsilon,t) \\
\text{subject to} \quad & m_{s,t} \in \{0,1\} \qquad \text{(for all } (s,t) \in S \times T) \\
& \sum_{t \in T} m_{s,t} \le 1 \qquad \text{(for all } s \in S) \\
& \sum_{s \in S} m_{s,t} \le 1 \qquad \text{(for all } t \in T) \\
& m_{s_1,t_1} + m_{s_2,t_2} \le 1 \\
& \text{(for all } (s_1, t_2), (s_2, t_2) \in S \times T \text{ s.t. } s_1 < s_2 \not\Leftrightarrow t_1 < t_2)
\end{aligned}$$

As a result, given a pair of trees S and T, our IP formulation requires $O(|S||T|)$ binary variables and $O(|S|^2|T|^2)$ constraints.

In general, there are many ways to obtain IP formulations, which depends on how to represent constraints on variables. For example, we can transform the problem also into the following form:

Proposition 2 (IP Formulation Using Big-M Method)

$$\begin{aligned}
\text{minimize} \quad & \sum_{(s,t) \in S \times T} \{d(s,t) - d(s,\epsilon) - d(\epsilon,t)\} m_{s,t} \\
& + \sum_{s \in S} d(s,\epsilon) + \sum_{t \in T} d(\epsilon,t) \\
\text{subject to} \quad & m_{s,t} \in \{0,1\} \qquad\qquad\qquad \text{(for all } (s,t) \in S \times T) \\
& \sum_{t \in T} m_{s,t} \le 1 \qquad\qquad\qquad \text{(for all } s \in S) \\
& \sum_{s \in S} m_{s,t} \le 1 \qquad\qquad\qquad \text{(for all } t \in T) \\
& \sum_{\substack{(s',t') \in S \times T \\ \text{s.t. } s<s' \not\Leftrightarrow t<t'}} m_{s',t'} \le M(1 - m_{s,t}) \quad \text{(for all } (s,t) \in S \times T)
\end{aligned}$$

Note that M is a large integer and the difference is the last constraint. This formulation requires $O(|S||T|)$ binary variables and $O(|S||T|)$ constraints.

The IP formulations using a large constant integer M such as Proposition 2 are called the *Big-M method* [6], which are adopted when we want to represent disjunction of constraints. We can interpret the inequality $\sum m_{s',t'} \leq M(1 - m_{s,t})$ as follows: If $m_{s,t} = 1$, the r.h.s. becomes 0 and it consequently means that M cannot contain a pair (s', t') which violates the preserving ancestor condition which we want to represent.

3.1   Constraints for Variations of the Tree Edit Distance

Since we have IP formulations for Tai mappings, we could easily extend them to deal with other mappings. We provide two such mappings; *segmental mappings* and *bottom-up segmental mappings*, which are seen in previous studies on variations of the tree edit distance [17, 11]. These types of mapping are useful when we focus on mappings between connected trees as long as possible.

Definition 5 (Segmental Mapping). Let S and T be trees. A Tai mapping M between S and T is a *segmental mapping* if it satisfies the following condition: for any (s_1, t_1) and (s_2, t_2) in M such that $s_1 \neq root(S)$ and $t_1 \neq root(T)$, $(s_1 < s_2 \wedge t_1 < t_2) \Rightarrow (par(s_1), par(t_1)) \in M$.

This new constraint can be represented as follows: for all $(s_1, t_1), (s_2, t_2) \in S \times T$ such that $s_1 \neq root(S)$, $t_1 \neq root(T)$, $s_1 \leq s_2$, and $t_1 \leq t_2$,

$$m_{s_1,t_1} + m_{s_2,t_2} \leq m_{par(s_1),par(t_1)} + 1. \tag{5}$$

Definition 6 (Bottom-up Segmental Mapping). Let S and T be trees. A segmental mapping M between S and T is a *bottom-up segmental mapping* if it satisfies the following conditions: for any (s_1, t_1) in M,

$$((s_1 \in leaves(S)) \wedge (t_1 \in leaves(T))) \vee$$
$$\exists (s_2, t_2) \in M. (s_2 \leq s_1 \wedge t_2 \leq t_1 \wedge s_2 \in leaves(S) \wedge t_2 \in leaves(T)).$$

It is also possible to design constraints for bottom-up segmental mappings as follows: for all $(s_1, t_1) \in S \times T$ satisfying $s_1 \notin leaves(S)$ and $t_1 \notin leaves(T)$,

$$m_{s_1,t_1} \leq \sum_{s \in leaves(S), t \in leaves(T), s < s_1, t < t_1} m_{s,t}. \tag{6}$$

4   Experiments

The purposes of our experiments are comparing performances of our method with an existing algorithm and evaluating our IP formulations including variations of the tree edit distance. We selected as our competitor the DpCliqueEdit-E developed by Mori et al. [14], which is much faster than the other existing methods on the third approach and methods based on A* algorithm. Note that we added an additional constraint: $m_{root(S),root(T)} = 1$ to our IP formulation for fairness. because the minimum cost mapping computed by the DpCliqueEdit-E always contains a pair $(root(S), root(T))$,

We implemented our IP-based method in C++ and used the IP solver CPLEX 12.5 [9][1]. These methods run on a service provided by the Sakura Internet Cloud[2]. We set the timeout to 60 minutes and adopted the unit cost.

Datasets. We used two datasets: Glycan datasets from KEGG [12] and CSLOGS consisting of Web browsing log files [18]. We separated CSLOGS into two sub-datasets: SUBLOG3 and SUBLOG49. Each sub-dataset has 15,000 trees in which every tree T satisfies $|T| \leq 80$. Every tree T in SUBLOG3 (resp. SUBLOG49) is restricted to be $deg(T) \leq 3$ (resp. $deg(T) \leq 49$). Following the ways of experiments in the previous work [14], we randomly selected 100 pairs of trees structured data from each database with a specified range of the total number of nodes (i.e., $|S| + |T|$). We measured the average CPU time, the standard deviation of the time, and the number of problems causing timeout per pair. Unbalanced cases on size in which the size of one structure was smaller than one-third of the other structure were excluded.

4.1 Experimental Results and Discussions

Tables 1, 2, and 3 show the average of CPU time (in "avg."), the standard deviation of CPU time (in "s.d."), and the number of timeout (in "t.o.") for each dataset Glycan, SUBLOG3, and SUBLOG49, respectively.

Experimental results show that the clique-based method DpCliqueEdit-E is much faster than our method in Glycan (shown in Table 1) and SUBLOG3 (shown in 2). While, Table 3 shows that our method works faster in SUBLOG49, where our method is much more stable with respect to computational time shown in columns of s.d., and there are no timeout pairs of trees in use of our method. Note that for many pairs the DpCliqueEdit-E cannot complete the computations by the timeout. From these observations, we can conclude that our method is more efficient and reliable in computations of the tree edit distance if the average of degrees of trees in a dataset is large.

Since there exists many heuristics to improve clique-based methods, it is more efficient for databases in which trees have small degrees. For example, they deal with nodes that have only one child in a special way [14]. This suggests that we also could improve the performance of our method by introducing heuristics and additional techniques to our IP formulations. Unfortunately, in general it is not easy to analyze behaviors of IP solvers and IP formulations. We conjecture

---

[1] We set `IloCplex::Threads` to 1, `IloCplex::EpGap` to 0.0, and `IloCplex::EpAGap` to 0.0, respectively.

[2] http://cloud.sakura.ad.jp/. We performed our experiments using 10 PCs with Intel(R) Xeon(R) 3.07GHz CPU and 4GB RAM on the Ubuntu Server 13.10 (64bit).

Table 1. Experimental results with Glycan

# of nodes	# of instances	IP-based method			DpCliqueEdit-E		
		avg.	s.d.	t.o.	avg.	s.d.	t.o.
[50, 54]	100	4.488	3.030	0	0.656	0.086	0
[55, 59]	100	12.513	11.933	0	0.695	0.126	0
[60, 64]	88	68.616	180.805	0	0.778	0.208	0
[65, 69]	36	96.492	74.772	0	0.809	0.180	0
[70, 74]	100	40.321	72.647	0	1.508	3.002	0
[75, 79]	29	48.990	71.934	0	1.339	0.430	0
[80, 84]	9	113.092	90.738	0	2.171	0.797	0
[85, 89]	5	204.978	88.713	0	3.716	2.027	0
[90, 94]	4	1586.188	1083.090	0	15.793	10.954	0

Table 2. Experimental results with SUBLOG3

# of nodes	# of instances	IP-based method			DpCliqueEdit-E		
		avg.	s.d.	t.o.	avg.	s.d.	t.o.
[50, 54]	100	2.991	1.863	0	0.610	0.080	0
[55, 59]	100	5.774	4.136	0	0.905	2.238	0
[60, 64]	100	10.947	12.373	0	0.950	0.420	0
[65, 69]	100	21.023	21.026	0	1.078	0.548	0
[70, 74]	100	24.614	27.077	0	1.513	1.299	0
[75, 79]	100	73.684	165.768	0	1.714	0.840	0
[80, 84]	100	165.068	266.840	0	21.289	130.041	0
[85, 89]	100	212.862	406.964	0	6.024	14.441	0
[90, 94]	100	435.462	631.875	2	7.901	25.401	1
[95, 99]	100	527.586	673.865	11	18.018	74.920	1

that our approach is more useful than the existing approaches because we have many possible choices for IP solvers and formulations, particularly for the case databases consist of trees whose degree are relatively large.

In general there are many possible IP formulations for computing the tree edit distance. Here we compare two IP Formulations in Section 3.

The results are shown in Table 4, which shows that Proposition 1 is faster than Proposition 2 even though the former has more constraints than the latter. This result coincides with a well-known statement: we should avoid to adopt the Big-M methods for the numerical stability [4].

Table 3. Experimental results with SUBLOG49

# of nodes	# of instances	IP-based method			DpCliqueEdit-E		
		avg.	s.d.	t.o.	avg.	s.d.	t.o.
[50, 54]	100	1.309	1.148	0	41.369	327.451	0
[55, 59]	100	2.077	1.854	0	77.050	273.540	4
[60, 64]	100	4.750	7.725	0	136.286	473.147	9
[65, 69]	100	6.333	6.405	0	151.853	474.503	18
[70, 74]	100	9.486	12.985	0	194.986	537.871	36
[75, 79]	100	10.732	11.198	0	366.557	818.226	40
[80, 84]	100	20.261	25.272	0	427.151	848.688	48
[85, 89]	100	39.072	72.938	0	426.430	853.436	60
[90, 94]	100	78.213	203.843	0	424.456	927.602	63
[95, 99]	100	163.556	385.869	0	204.049	376.761	71

Table 4. CPU time for comparing two IP Formulations in Proposition 1 and Proposition 2 (using the Big-M method) with Glycan from KEGG

# of nodes	# of instances	Proposition 1			Proposition 2 (the Big-M method)		
		avg.	s.d.	t.o.	avg.	s.d.	t.o.
[50, 54]	100	5.820	3.326	0	51.371	276.219	0
[55, 59]	100	13.197	16.608	0	88.716	129.957	0
[60, 64]	88	49.258	60.757	0	536.320	787.060	3
[65, 69]	36	91.514	41.501	0	797.209	735.258	1
[70, 74]	100	40.661	66.718	0	244.962	317.952	3
[75, 79]	25	41.759	19.081	0	442.778	450.224	0

4.2   Comparison of Two IP Formulations

4.3   Variations of the Tree Edit Distance

We computed two variations: segmental and bottom-up mappings based on our IP formulation (Proposition 1). As far as we know, there are no competitive counterparts for computing the distances of those mappings. We thus only compare our results of those mappings with that of Tai mapping and show that our algorithm works efficiently for variations of the tree edit distance.

The result is shown in Table 5. On the viewpoint of computational complexity, computing the distance based on Tai mappings is NP-hard, and those based on segmental mappings and bottom-up segmental mappings are MAX SNP-hard [11]. In opposite to these computational intractability, Table 5 shows that the computational time of the tree edit distance based on bottom-up segmental mappings is shorter than one based on Tai mapping by using our algorithm.

Table 5. Comparison computation time for three mappings: Tai mappings, segmental mappings, and bottom-up segmental mappings with Glycan from KEGG

# of nodes	# of instances	Tai			Segmental			Bottom-up		
		avg.	s.d.	t.o.	avg.	s.d.	t.o.	avg.	s.d.	t.o.
[50, 54]	100	5.820	3.326	0	19.401	23.934	0	1.917	0.905	0
[55, 59]	100	13.197	16.608	0	51.857	72.340	0	3.308	1.971	0
[60, 64]	88	49.258	60.757	0	275.356	349.770	0	11.687	12.121	0
[65, 69]	36	91.514	41.501	0	357.015	202.331	1	24.781	10.312	0
[70, 74]	100	40.661	66.718	0	550.508	526.035	1	11.227	17.452	0
[75, 79]	25	41.759	19.081	0	716.843	556.473	0	20.914	36.057	0

5　Conclusion and Future Work

We proposed a new encoding method to compute the tree edit distance between (rooted labeled) unordered trees based on IP solvers and IP formulations representing the problem of finding the minimum cost mapping between two trees. The proposed method has several advantages against existing methods. First, once an IP formulation of the tree edit distance problems is found, the variations of the tree edit distance could be modeled conveniently by introducing additional constraints to the base formulation as seen in Section 3 and its experimental evaluation in Section 4.3. Second, many IP solvers are available that can solve even NP-hard problems in practice.

We performed experiments using real tree-structured datasets, the Glycan and the Web log data CSLOGS. They showed that the our method outperforms the DpCliqueEdit-E when the maximum degrees of input trees are large. However, in simple problems such that nodes in input trees have small degrees, the proposed method is less effective than the existing method.

In some applications, especially in computational biology, it is common that degrees of most nodes are small. Thus there is need for more speed up against such data by redesigning IP formulations or introducing heuristics used in the existing methods. In addition, more experiments for other datasets, using other IP solvers are our future work to clarify behaviors of IP-based methods.

Acknowledgments. The authors would like to thank both the anonymous reviewers for their valuable comments and Dr. Tetsuji Kuboyama, Gakushuin University, Japan for his introduction and valuable discussion. This study is partially supported by the JSPS KAKENHI Grant Number 26280085.

References

[1] Achterberg, T.: Scip: Solving constraint integer programs. Mathematical Programming Computation 1(1), 1–41 (2009),
http://mpc.zib.de/index.php/MPC/article/view/4

[2] Akutsu, T., Tamura, T., Fukagawa, D., Takasu, A.: Efficient exponential time algorithms for edit distance between unordered trees. In: Kärkkäinen, J., Stoye, J. (eds.) CPM 2012. LNCS, vol. 7354, pp. 360–372. Springer, Heidelberg (2012)

[3] Bixby, E.R., Fenelon, M., Gu, Z., Rothberg, E., Wunderling, R.: MIP: theory and practice-closing the gap. In: Powell, M.J.D., Scholtes, S. (eds.) System Modelling and Optimization: Methods, Theory, and Applications. IFIP, vol. 46, pp. 19–49. Springer, Boston (2000)

[4] Bixby, R.E., Fenelon, M., Gu, Z., Rothberg, E., Wunderling, R.: Mixed integer programming: a progress report. In: The Sharpest Cut: The Impact of Manfred Padberg and His Work. MPS-SIAM Series on Optimization, vol. 4, pp. 309–326 (2004)

[5] Daiji, F., Takeyuki, T., Atushiro, T., Etsuji, T., Tatsuya, A.: A clique-based method for the edit distance between unordered trees and its application to analysis of glycan structures. BMC Bioinformatics 12 (2011)

[6] Griva, I., Nash, S.G., Sofer, A.: Linear and Nonlinear Optimization, 2nd edn. Society for Industrial Mathematics (2008)

[7] Higuchi, S., Kan, T., Yamamoto, Y., Hirata, K.: An A* algorithm for computing edit distance between rooted labeled unordered trees. In: Okumura, M., Bekki, D., Satoh, K. (eds.) JSAI-isAI 2012. LNCS, vol. 7258, pp. 186–196. Springer, Heidelberg (2012)

[8] Horesh, Y., Mehr, R., Unger, R.: Designing an A* algorithm for calculating edit distance between rooted-unordered trees. Journal of Computational Biology 13(6), 1165–1176 (2006)

[9] IBM: IBM ILOG CPLEX Optimizer (2010), http://www-01.ibm.com/software/integration/optimization/cplex-optimizer/

[10] Jiang, T., Lin, G., Ma, B., Zhang, K.: A general edit distance between rna structures. Journal of Computational Biology 9(2), 371–388 (2002)

[11] Kan, T., Higuchi, S., Hirata, K.: Segmental mapping and distance for rooted labeled ordered trees. In: Chao, K.-M., Hsu, T.-S., Lee, D.-T. (eds.) ISAAC 2012. LNCS, vol. 7676, pp. 485–494. Springer, Heidelberg (2012)

[12] Kanehisa, M., Goto, S.: Kegg: kyoto encyclopedia of genes and genomes. Nucleic Acids Research 28(1), 27–30 (2000)

[13] Kuboyama, T.: Matching and learning in trees. Ph.D Thesis (The University of Tokyo) (2007)

[14] Mori, T., Tamura, T., Fukagawa, D., Takasu, A., Tomita, E., Akutsu, T.: An improved clique-based method for computing edit distance between rooted unordered trees. SIG-BIO 2011(3), 1–6 (2011)

[15] Shasha, D., Wang, J.L., Zhang, K., Shih, F.Y.: Exact and approximate algorithms for unordered tree matching. IEEE Transactions on Systems, Man and Cybernetics 24(4), 668–678 (1994)

[16] Tai, K.C.: The tree-to-tree correction problem. Journal of the ACM (JACM) 26(3), 422–433 (1979)

[17] Valiente, G.: An efficient bottom-up distance between trees. In: Proceedings of the 8th International Symposium of String Processing and Information Retrieval, pp. 212–219. Press (2001)

[18] Zaki, M.J.: Efficiently mining frequent trees in a forest: Algorithms and applications. IEEE Transactions on Knowledge and Data Engineering 17(8), 1021–1035 (2005)

[19] Zhang, K., Statman, R., Shasha, D.: On the editing distance between unordered labeled trees. Information Processing Letters 42(3), 133–139 (1992)

[20] Zhang, K., Shasha, D., Wang, J.T.L.: Approximate tree matching in the presence of variable length don't cares. Journal of Algorithms 16(1), 33–66 (1994)

Probabilistic Active Learning: Towards Combining Versatility, Optimality and Efficiency

Georg Krempl, Daniel Kottke, and Myra Spiliopoulou

Knowledge Management and Discovery Lab, University Magdeburg, Germany
{georg.krempl,daniel.kottke,myra}@iti.cs.uni-magdeburg.de,
kmd.cs.ovgu.de

Abstract. Mining data with minimal annotation costs requires efficient active approaches, that ideally select the optimal candidate for labelling under a user-specified classification performance measure. Common generic approaches, that are usable with any classifier and any performance measure, are either slow like error reduction, or heuristics like uncertainty sampling. In contrast, our Probabilistic Active Learning (PAL) approach offers versatility, direct optimisation of a performance measure and computational efficiency. Given a labelling candidate from a pool, PAL models both the candidate's label and the true posterior in its neighbourhood as random variables. By computing the expectation of the gain in classification performance over both random variables, PAL then selects the candidate that in expectation will improve the classification performance the most. Extending our recent poster, we discuss the properties of PAL and perform a thorough experimental evaluation on several synthetic and real-world data sets of different sizes. Results show comparable or better classification performance than error reduction and uncertainty sampling, yet PAL has the same asymptotic time complexity as uncertainty sampling and is faster than error reduction.

1 Introduction

Recently, the application of machine learning to large data pools and fast data streams has gained attention. This application often requires classification of data where features are cheap but labels are costly [8]. Examples are applications where features are obtained from an automated process but labels require human annotation efforts. Active learning (AL) [15, p. 4] addresses such applications, where the machine learning system can actively select instances for labelling, rather than passively processing a given set of labelled instances. Its tasks are to decide a) for which instance to request a label, and b) whether to continue labelling at all, given some labels have already been acquired.

The ideal active learning strategy should select those instances first that, once incorporated into the training data, will result in the highest gain in terms of a classification performance measure. Furthermore, it provides a quantification of this performance gain, needed for a sound answer to the stop-criterion related second question. It therefore considers the already acquired amount of training data. Finally, it is fast, requiring solely linear asymptotic computational time per instance with respect to the pool size, in order to enable its application in large data pools and fast data streams. Active learning strategies that are usable in conjunction with any classifier technology provide some

S. Džeroski et al. (Eds.): DS 2014, LNAI 8777, pp. 168–179, 2014.

of the above qualities. However, as discussed further in Section 2, they do not offer a *combination of all these qualities* in one single approach.

We address this challenge by a novel, probabilistic active learning (PAL) technique for classification that combines the above qualities and constitutes an alternative to other generic strategies like error reduction or uncertainty sampling. It is not limited to a particular classifier technology, and usable with any point [12] performance performance measure. Given a pool of candidates, it computes for each candidate the expected gain in classification performance from obtaining its label. This expectation models the candidate's label *and* the true posterior at its location as a random variables, and uses likelihood weights according to the already obtained labels in the candidate's neighbourhood. Subsequently, it selects the optimal candidate under this expected overall performance gain for labelling. This active selection from a pool requires asymptotic computational time that is solely linear in the size of the pool, as fast uncertainty sampling approaches do. While deriving stop-criteria is not within the scope of this paper, but our quantification of a label's expected impact provides a fundamental first step.

This paper is a full-version of our recent poster [10], extending it by a more detailed discussion of related work, an additional discussion of PAL's properties, and additional experiments. It is structured as follows: In the next section, we provide the necessary background and discuss related approaches. In section 3, we present our probabilistic active learning approach. In section 4, we report on our evaluation results, where we compare *PAL* to the strategy considered to be optimal for minimising classification error (error reduction), and to a popular fast heuristic strategy (uncertainty sampling). [1]

2 Background and Related Work

This paper addresses *pool-based* active learning (AL) for binary classifiers, as described in [15, p. 9] and [4]. In this scenario, an active classifier has access to a pool of unlabelled instances $\mathcal{U} = \{(x, .)\}$. From this pool of labelling candidates it repeatedly selects an instance $(x^*, .)$ for labelling. Upon receiving its label y^*, the instance (x^*, y^*) is moved to a pool of labelled instances $\mathcal{L} = \{(x, y)\}$, the classifier is retrained, and the process is repeated. There exist various approaches for this scenario, recent surveys are provided in [15], [6], [4] and [14]. We will focus on popular families of approaches that are usable with any classification technique, and discuss the ones most related to our approach: error reduction, uncertainty sampling and query-by-transduction.

Expected error reduction (ER) is a decision-theoretic approach. It considers the improvement in classification performance by selecting the candidate, that has the minimal expected classification error if incorporated into the training pool. The seminal work of [5], which coined the term "statistically optimal active learning", derived closed-form solutions for optimal data selection for two specific learning methods. In contrast, the approach suggested in [13] is generic, both with respect to arbitrary performance measures and classifiers: using a Monte Carlo sampling approach, it estimates the performance on a labelled validation sample $\mathcal{V}$, rather than integrating over the full feature distribution $Pr(x)$. It uses the posterior estimate $\hat{p} = \hat{Pr}(y|x)$ provided by the current classifier as proxy for the true posterior $Pr(y|x)$ that is required for the expectation over

---

[1] For additional resources please consult http://kmd.cs.ovgu.de/res/pal/.

the label realisations y. However, as discussed in [2], this proxy is not reliable if solely few labels are available, requiring regularisation approaches such as using Beta priors. Furthermore, the labelled (or self-labelled) validation sample $\mathcal{V}$ must be representative of the data. Not only is this difficult, in particular at the beginning with few available labels and a still unreliable classifier, but it also makes error reduction prohibitively slow [14] for using it in applications that require fast processing of big amounts of data, as even for incremental classifiers its asymptotic time complexity is $O(|\mathcal{V}| \cdot |\mathcal{U}|)$.

In comparison, a faster method [15, p.64] is uncertainty sampling (US), introduced in [11]. It uses simple uncertainty measures, like sample margin, confidence, or entropy as proxies for a candidate's value, and selects the candidate with maximal uncertainty. However, these proxies do not consider the number of similar instances on which posterior estimates are made. This is problematic, as Figure 2 (next page) illustrates on four exemplary active learning situations. These situations could, for example, occur simultaneously in different regions of a feature space such that the next label must be actively requested in either of them[2] The first (in Roman numeral) and second situation differ in the number of obtained labels (6 vs. 1), but lead to the same posterior estimate $\hat{Pr}(+|x) = 1$, as all obtained labels are positive. Uncertainty sampling is indifferent between them, as both entropy and confidence are zero. This indicates equal and absolute certainty, which is not justified as in II the single positive label can simply be due to chance, even if the true posterior of the positive class is actually smaller than 0.5 and the classifier is wrong. In contrast, in I a high true positive posterior is indeed very likely, and additional labels have less impact on the classifier. Similarly, in IV the classifier's prediction is quite reliable, but uncertainty according to measures like entropy or confidence is maximal. This leads to sampling in regions of high Bayesian error rate, even if the classifier can not be further improved there.

Some of the many existing *classifier-specific* AL approaches offer high processing speeds for particular applications. However, they require classifier selection to be made with respect to the available active learning strategy, as sample reusability between different types of classifiers for selector and consumer strategies is an open question [16]. Finally, even recently proposed classifier-specific approaches are mostly either information-theoretic (i.e. agnostic to the decision task at hand) or use the most likely or most pessimistic posterior under the current model, thus ignoring the reliability associated with this estimate, as for example [7]. A very recent information-theoretic approach that considers the reliability of a predictive model is Query-By-Transduction (QbT) [9]. QbT is based on conformal prediction and selects the instances with respect to the p-values obtained using transduction. This quantification of the reliability using p-values is related to ours, although we use the likelihood weights of the posterior estimates and follow a decision-theoretic Bayes-optimal active learning approach that directly optimises a classification performance measure.

---

[2] For simplicity, this illustration assumes conditional independence of the posterior from the feature given the region, i.e. $Pr(y|x, z) = Pr(y|z)$, where y is the class, x the feature vector, and z the region. Thus no further differentiation can be made within a region. We also assume equal numbers of instances in all regions, making accuracy everywhere equally important.

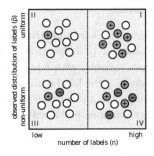

Fig. 1. Different AL situations, where entropy- or confidence-based uncertainty measures differentiate only on a class' *relative* (vert.) but not on all classes' *total* (horiz.) number of labels

3 Probabilistic Active Learning

Following the common smoothness assumption [3], we consider that an instance x influences the classification the most in its neighbourhood. Thus, the impact of an additional label primarily depends on the already obtained labels in its neighbourhood. We summarise these by their *absolute* number n, and the share of positives $\hat{p}$ therein, yielding the *label statistics* $ls = (n, \hat{p})$. Here, n is obtained by counting the similar labelled instances for pre-clustered or categorical data (as for the partitions in Figure 2), or approximated by frequency estimates such as kernel frequency estimates for smooth, continuous data. Thus, in x's neighbourhood, n expresses the absolute quantity of labelled information, whereas the density d_x of unlabelled instances quantifies the importance of this neighbourhood, i.e. the share of future classifications that will take place therein compared to other regions of the feature space.

Given a labelling candidate $(x, .)$ from a pool of unlabelled instances $\mathcal{U}$ for a user-specified point classification performance measure [12] like accuracy, we want to compute the expected overall gain in classification performance if requesting its label. This requires knowledge of its label statistics ls, but also of its label y and the true posterior p of the positive class within its neighbourhood. As the latter values of y and p are not directly accessible, we use a probabilistic approach and model Y and P as random variables. This allows us to compute the *expected value* of the gain in performance over all different true posteriors and label realisations, which we denote as *probabilistic gain*[3] (pgain). Finally, we weight it by the neighbourhood's density d_x (over labelled and unlabelled data) to consider the importance of the neighbourhood on the whole data set, quantifying the overall expected performance change. Comparing the overall expected performance change of all candidates, we select the optimal candidate for labelling.

We now first provide the modelling and derive the necessary equations, present the framework of *Probabilistic Active Learning (PAL)* with its pseudo-code, and close with discussing its properties.

---

[3] We do this to differentiate it from the expected gain as in expected error reduction methods like [2], where expectation is solely over label outcomes, but not over the true posterior.

3.1 Probabilistic Gain Calculation

Given a candidate $(x, .)$, the label statistics ls summarise the obtained labels in its neighbourhood. We model the true posterior P of the positive class $(y = 1)$ in this neighbourhood as a Beta-distributed random variable, whose realisation p is itself the parameter of the Bernoulli distribution controlling the label realisation $y \in \{0, 1\}$ of any instance within the neighbourhood. Consequently, the number of positives $n \cdot \hat{p}$ among the n already obtained labels in the neighbourhood is the realisation of a Binomial-distributed random variable:

$$P \sim \text{Beta}_{n \cdot \hat{p}+1, n \cdot (1-\hat{p})+1} \tag{1}$$

$$Y \sim \text{Bernoulli}_p = \text{Ber}_p \tag{2}$$

$$(n \cdot \hat{p}) \sim \text{Binomial}_{n,p} \tag{3}$$

The true posterior's Beta distribution above results from its normalised likelihood given the already observed labels, that is

$$\omega_{ls}(p) = \frac{L(p|ls)g(p)}{\int_0^1 L(\psi|ls)g(\psi)d\psi} = (1 + n) \cdot L(p|ls) \tag{4}$$

$$= \frac{\Gamma(n+2) \cdot p^{n \cdot \hat{p}} \cdot (1 - p)^{n \cdot (1-\hat{p})}}{\Gamma(n \cdot \hat{p} + 1) \cdot \Gamma(n \cdot (1 - \hat{p}) + 1)} = \text{Beta}_{\alpha,\beta}(p) \tag{5}$$

where the parameters $\alpha = n \cdot \hat{p} + 1$ and $\beta = n \cdot (1 - \hat{p}) + 1$ of the Beta-distribution's pdf $\text{Beta}_{\alpha,\beta}(p)$ are obtained by following a Bayesian approach under a uniform prior for P such that $g()$ is a constant function, and by using the probability mass function according to Eq. 3 for the likelihood $L(p|ls)$, and $(1 + n) \cdot \Gamma(n + 1) = \Gamma(n + 2)$.

We take the expectation on the performance gain over these two random variables, yielding the candidate's probabilistic gain (pgain), that defines the expected change of the performance measure for its neighbourhood:

$$\text{pgain}(ls) = \text{E}_p \left[\text{E}_y \left[\text{gain}_p(ls, y) \right] \right] \tag{6}$$

$$= \int_0^1 \text{Beta}_{\alpha,\beta}(p) \cdot \sum_{y \in \{0,1\}} \text{Ber}_p(y) \cdot \text{gain}_p(ls, y) \, dp \tag{7}$$

Here, $\text{gain}_p(ls, y)$ is the candidate's $(x, .)$ performance gain given its label realisation y and the neighbourhood's true posterior p:

$$\text{gain}_p(ls, y) = \text{perf}_p\left(\frac{n\hat{p} + y}{n + 1}\right) - \text{perf}_p(\hat{p}) \tag{8}$$

The definition of Eq. 7 and 8 allow the use of any point performance measure (see e.g. [12]) for perf. An example is accuracy (acc), defined as

$$\text{perf}_p(\hat{p}) = 1 - \text{err}_p(\hat{p}) = 1 - \begin{cases} p & \hat{p} < 0.5 \\ 1 - p & otherwise \end{cases} \tag{9}$$

where $\mathrm{err}_p(\hat{p})$ is the error rate under Bayes' optimal classification, given a true posterior p and observed posterior $\hat{p}$ of the positive class.

Plugging this in Eq. 7 yields the probabilistic accuracy gain

$$\mathrm{pgain}_{\mathrm{acc}}(\textit{ls}) =$$
$$= \int_0^1 \mathrm{Beta}_{\alpha,\beta}(p) \sum_{y \in \{0,1\}} \mathrm{Ber}_p(y) \left(\mathrm{err}_p(\hat{p}) - \mathrm{err}_p \left(\frac{n\hat{p}+y}{n+1} \right) \right) dp$$

which we compute by trapezoidal numerical integration over p.

Finally, we weight each candidate's probabilistic gain with the density d_x over *labelled* and *unlabelled* instances in its neighbourhood, and select the candidate with the highest density-weighted probabilistic gain for labelling:

$$x^* = \arg\max_{x \in \mathcal{U}} \left(d_x \cdot \mathrm{pgain}_{\mathrm{acc}}(\textit{ls}_x) \right) \qquad (10)$$

3.2 PAL Algorithm

The pseudo-code for the resulting probabilistic, pool-based active learning algorithm is given in Figure 2. Iterating over the candidate pool $\mathcal{U}$ (Lines 2-6), for each labelling candidate x one computes its label statistics $\textit{ls}_x = (n_x, \hat{p}_x)$, its density weight d_x, and using numerical integration its probabilistic gain, which is weighted by its density weight to obtain g_x. Finally, the candidate with the highest g_x is selected (Line 7).

```
1: function POOLBASEDPAL(U,L)
2:     for x ∈ U do
3:         (n_x, p̂_x) ← labelstatistics(x, L)
4:         d_x ← densityweight(x, L ∪ U)
5:         g_x ← pgain((n_x, p̂_x)) · d_x
6:     end for
7:     return arg max_{x∈U}(g_x)
8: end function
```

Fig. 2. The PAL Algorithm

3.3 PAL's Properties

Statistical Optimality in Disjoint Neighbourhoods. For a disjoint neighbourhood concept, like in pre-clustered or categorical data, where instances are partitioned such that instances having an influence on each others' classification belong to the same subset, the density-weighted probabilistic gain of a candidate corresponds precisely to the expected change in overall performance from acquiring the candidate's label. Thus selecting the candidate with highest probabilistic gain is statistically optimal.

For smooth, continuous neighbourhoods, the density-weighted probabilistic gain is the expected change at the candidate's location, serving as an approximation of the overall performance gain. We use this latter concept in our evaluation, as it applies to more data sets and is better comparable to the baseline active learning algorithms.

Computational Efficiency. In this subsection, we discuss the asymptotic (with respect to data set size) computational time complexity of PAL and related algorithms for active learning of binary, incremental classifiers. For selecting a candidate from a pool $\mathcal{U}$ of labelling candidates, the PAL algorithm above needs to iterate over all candidates in the pool (Lines 2 – 6). Each iteration consists of 1) querying labelstatstics, 2) querying density weights, and 3) computing the probabilistic gain. The first step requires absolute frequency estimates of labels in the candidate's neighbourhood, similar to the relative frequency estimates needed by entropy or confidence uncertainty measures. These are obtained in constant time by probabilistic classifiers. The second step requires density estimates over all instances, that is over labelled $\mathcal{L}$ and unlabelled $\mathcal{U}$ ones. Precomputing these density estimates once for all later calls of PAL leads to constant query time, as in the pool-based setting the union $\mathcal{L} \cup \mathcal{U}$ is constant. The third step consists of a numeric integration over the true posterior p and a summing over possible label realisations y. Both factors do not depend on the data set size. We used fifty numeric integration steps in all our experiments to get highly precise estimates for expected classification accuracy gain, resulting in a constant factor of $O(50 \cdot 2)$ per probabilistic gain computation. Overall, the iteration over the pool is done in $O(|\mathcal{U}|)$ time.

Selecting the candidate with highest density-weighted probabilistic gain in Line 7 is done in constant time, by using a sweep line approach and storing the maximal value and its corresponding candidate in the previous for-loop.

Overall, PAL requires $O(|\mathcal{U}|)$ time for selecting a candidate from the pool. Uncertainty sampling, using probabilistic classifiers and entropy or confidence uncertainty measures, requires asymptotically the same time, but due the simplicity of its computation with a smaller constant factor involved. In contrast, error reduction as discussed in [15], requires $O(|\mathcal{U}| \cdot |\mathcal{V}|)$ time, where $|\mathcal{V}| \approx |\mathcal{U}|$, as $\mathcal{V}$ needs to be a representative sample of the data.

Characteristics of the Probabilistic Gain. For a better understanding of the probabilistic gain function, Figure 3.3 shows the computed probabilistic gain (in terms of accuracy) for different label statistics, i.e. combinations of different numbers of already obtained labels n and observed posteriors $\hat{Pr}(+|x)$. The following main characteristics of the curve underline its reasonable behaviour:

Monotonicity with variable n: With increasing n and a fixed $\hat{Pr}(+|x)$ the probabilistic gain decreases, because it is more likely that the posterior already is correct.

Symmetry with respect to $\hat{Pr}(+|x) = 0.5$: Evaluating accuracy, pos. and neg. labels count the same, i.e. the probabilistic gain is equal for $\hat{Pr}(+|x)$ and $\hat{Pr}(-|x)$.

Zero for irrelevant candidates: If one label would not change the decision in its neighbourhood, the accuracy remains the same. Thus, gain and probabilistic gain are 0.

This figure is inspired by an illustration of Settles, where different uncertainty measures are plotted as functions of the posterior of a class (see figure 2.4 in [15, p. 15]). Comparing the least confident curve (plot (a) in [15]), it behaves nearly similarly as our probabilistic gain for $n = 1$, but does not change with n.

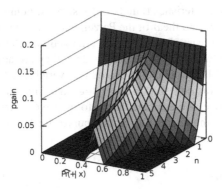

Fig. 3. Illustration of the probabilistic gain (pgain) as a function of $\hat{Pr}(+|x)$, which is the observed posterior of the positive class, and of n, which is the number of already obtained labels

4 Experimental Evaluation

From its theoretical characteristics, we expect PAL to be comparable to error-reduction in terms of classification performance, yet faster, and we expect PAL to be better than uncertainty sampling. This section will now verify these characteristics empirically. After outlining the experimental setup, we will discuss the results in the second subsection.

4.1 Evaluation Settings

We compare our new base method PAL with expected error-reduction (in the extended variant proposed by Chapelle in [2], denoted Chap), with uncertainty sampling (using confidence [15] as uncertainty measure, den. Uncer), and with random sampling (den. Rand). While error-reduction is considered as one of the best available AL-methods [15, p. 64], uncertainty sampling is fast and very popular for large or streaming data.

We used Gaussian kernels for frequency estimation, and a Parzen window classifier as in [2] for ensuring comparability with [2]. So, the estimated label frequencies $labelFreq_c, c \in \{+, -\}$ at an instance x for the the positive $\mathcal{L}_+$ and the negative class $\mathcal{L}_-$ are calculated by an unnormalised Gaussian function. These frequencies build the label statistics $n = labelFreq_+ + labelFreq_-$ and $\hat{p} = labelFreq_+/n$.

$$labelFreq_c(x) = \sum_{x' \in \mathcal{L}_c} \exp\left(-\frac{\|x' - x\|^2}{2\sigma^2}\right)$$

Our framework starts *without* initial labels, and finishes after 40 label requests. The classifiers, implemented in Octave/MATLAB and run separately on a cluster, use the same pre-tuned, data set-specific bandwidth, and are re-evaluated in each of the 40 steps on the same, dedicated (labelled) test sample. This ensures that only the difference in the active learning strategy is influencing the performance. For better performance assessment, we generated 100 random training and test subsets for each data set, and averaged the results. Evaluation is done on 2 synthetic (based on [2]) and 6 real-world data sets (from [1]). The main characteristics (number of instances, number of attributes), such as training and test set size and the σ of the Parzen window, are summarised in Table 4. The synthetic data sets consist of $4x4$ clusters, arranged in a checker-board formation. While the clusters are low-density-separated in Che, they are adjoined in Che2. The real-world data sets are Mammographic mass (Mam), Vertebral (Ver), Haberman's survival (Hab), Blood transfusion (Blo), Seeds (See) and Abalone (Aba). All attributes are scaled to a $[0; 1]$-range. We evaluate the performance over the first 40 active label acquisitions and provide the results as learning curves for the optimised performance measure accuracy for all data sets and algorithms.

4.2   Evaluation Results

In accordance to [2] and [15], we provide learning curves in the subfigures of Figure 6. These curves depict the progress in the active classifier's accuracy as 40 training instances are selected one after another for training. This allows to evaluate the performance based on several criteria, and is more informative than tables of the performance at arbitrarily selected learning stages.

(1) When does a curve become flat, i.e. when does the learner converge? On subfigure g) for data set Seeds, the curves become flat already after reading 10 labels, while the curves for data set Checkboard 2 (b) do not converge. Convergence indicates that additional labels do not provide additional use to the classifier, ideally a classifier converges fast and to a high level of performance. This is seen on subfigures a and c, where PAL in contrast to Random Sampling quickly converges to a high performance level.

(2) At what accuracy does a learner stop improving? Clearly, a learner that achieves a 99% accuracy after reading 10 labels is better than one that needs 40 labels to reach the same accuracy value, and also better than one that converges at 75%. Hence, PAL outperforms all other algorithms except on Blood (f), Seeds (g) Abalone (h). The moment of convergence gives also indication on the appropriateness of the data set for

Dataset	Inst	Attr	$Pr(+)$	|Train|	|Test|	σ
See	210	7	33 %	160	50	0.1
Che	308	2	44 %	200	108	0.08
Che2	392	2	49 %	250	142	0.08
Hab	306	3	73 %	256	50	0.1
Ver	310	6	32 %	260	50	0.1
Aba	4177	8	50 %	400	1177	0.06
Blo	748	4	24 %	600	148	0.1
Mam	830	11	51 %	630	200	0.1

Fig. 4. Dataset characteristics and parameters (number of instances, number of attributes, proportion of positive instances, training set size, test set size, bandwidth for Parzen window classifier)

Dataset	PAL	Chap	Uncer	Rand
See	0.50	0.93	0.03	0.01
Che	0.61	1.16	0.03	0.01
Che2	0.92	1.54	0.03	0.02
Hab	0.89	1.72	0.03	0.02
Ver	0.91	1.84	0.04	0.02
Aba	1.51	3.82	0.07	0.04
Blo	2.34	6.14	0.1	0.05
Mam	2.56	8.48	0.25	0.12

Fig. 5. Average execution time (in seconds), ordering of rows is in ascending training dataset size

active learning. If we contrast subfigures b and g, we must assert that data set Seeds is not truly interesting in terms of active learning: after reading the labels of 5 or at most 10 instances, all learners converge to an accuracy very close to 1. Thus, comparative performance of the active learners on Seeds is not truly informative; this data set is not very appropriate for experiments on active learning (except as a counterexample). The curves on the Blood Transfusion data set (cf. subfigure f) also indicate that active learning is not truly beneficial on this data set.

(3) Does a learner recover from previous errors?   If a curve becomes flat early, then the learner might be trapped in low accuracy values. This is the case for the algorithm Chapelle on data set Mammographic Mass (c). In contrast, PAL recovers on this data set, as well as on data sets Checkboard, Vertebral and Habermans Survival (a, d, e). Random Sampling never recovers from earlier choices: its performance curves are either flat or go upwards, indicating that an early poor choice cannot be amended. Uncertainty Sampling recovers in some data sets, while Chapelle and PAL always manages

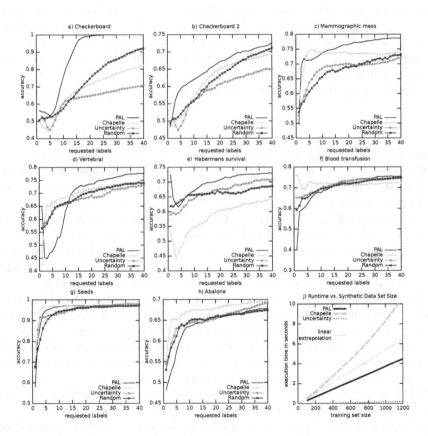

Fig. 6. a-h: accuracy curves for the algorithms on each dataset; early convergence to very high values is best; improvement after a performance drop is better than a flat curve on low accuracy values; **j**: runtime of PAL on a synthetic data set of varying size (100–1200 candidate instances)

to recover if they err in their early choices of label. Summarising the results on accuracy progress, PAL exhibits high performance in all data sets, manages to recover from poor choices and makes best use of available labels, as long as needed (i.e. longer for Checkboard 2 than for Seeds). PAL reaches the best accuracy values on 5 of the data sets, achieves comparable accuracy to the other learners on two data sets (Seeds and Blood Transfusion). PAL is only outperformed once on the Abalone data set.

(4) Execution time The execution time of PAL is shown in Table 5 and plot j of figure 6. Table 5 indicates the execution times of all active learning algorithms on each dataset. We see that PAL achieves better accuracy curves with lower (up to $1/2.5$ times) execution time than the error-reduction algorithm of Chapelle. Nevertheless, the execution time is still significantly higher than that of uncertainty sampling, but like the former its time increases solely linearly with the training set size, i.e. the number of labelling candidates. This is also shown in plot j) of Figure 6, where the execution times on various training set sizes of the same synthetic dataset are plotted. Overall, the uniformly low execution time of uncertainty sampling is accompanied by a stronger variance among the accuracy curves (cf. Figure 6): while PAL has very high performance on all data sets, escapes from earlier errors and exploits well all labels (whenever reasonable, see counterexample on Subfigure 6g), the accuracy curves of Uncertainty Sampling and Random Sampling vary in dependence on the data set. Thus, PAL exhibits stable performance at lower execution time than the expensive error-reduction mechanism, while the simpler algorithms are affected stronger by the idiosyncrasies of the data sets.

5 Conclusion

In this paper, we introduced the probabilistic active learning approach (PAL). It uses probabilistic estimates (label statistics) calculated within the neighbourhood of a labelling candidate. In contrast to Monte-Carlo-based error reduction approach proposed in [13], it models both the true posterior and the candidate's label as random variables. Given a user-specified performance measure, PAL computes the probabilistic gain, that is the expected performance gain over *both* random variables by numeric integration. It subsequently selects the candidate with highest density-weighted probabilistic gain. Like uncertainty sampling [11], PAL requires asymptotically linear time with respect to the pool size, in contrast to quadratic time required by error reduction in [13].

Thus PAL combines two previously incompatible qualities: being fast, and computing and optimising directly a point-performance measure. Given such a user-specified performance measure and the label statistics as input, no additional parameters are required. Our experimental evaluation shows that PAL yields comparable or better classification performance than error-reduction, uncertainty-sampling or random active learning strategies, while requiring less computational time than error-reduction.

Future work will comprise deriving *specific* closed-form solutions for some point-performance measures such as misclassification loss, as this promises further improvements in speed. Further research is also needed to address *non-myopic* scenarios, where optimising the resulting performance gain from acquiring several labels is required. Finally, as PAL is fast and requires only label statistics but no samples to be kept, its application in *data streams* seems a promising direction for future research.

Acknowledgements. We thank Vincent Lemaire from Orange Labs, France, for the insightful discussions.

References

1. Asuncion, A., Newman, D.J.: UCI ML repository (2013)
2. Chapelle, O.: Active learning for parzen window classifier. In: Proc. 10th Int. Workshop on AI and Statistics, pp. 49–56 (2005)
3. Chapelle, O., Schölkopf, B., Zien, A. (eds.): Semi-Supervised Learning. MIT Press (2006)
4. Cohn, D.: Active learning. In: Sammut, C., Webb, G.I. (eds.) Encyclopedia of ML, pp. 10–14. Springer (2010)
5. Cohn, D.A., Ghahramani, Z., Jordan, M.I.: Active learning with statistical models. J. of AI Research 4, 129–145 (1996)
6. Fu, Y., Zhu, X., Li, B.: A survey on instance selection for active learning. Knowledge and Inf. Syss. 35(2), 249–283 (2012)
7. Garnett, R., Krishnamurthy, Y., Xiong, X., Schneider, J.G., Mann, R.: Bayesian optimal active search and surveying. In: Proc. of the 29th ICML (2012)
8. Gopalkrishnan, V., Steier, D., Lewis, H., Guszcza, J.: Big data, big business: Bridging the gap. In: Workshop on Big Data, Streams and Heterogeneous Source Mining, pp. 7–11 (2012)
9. Ho, S.S., Wechsler, H.: Query by transduction. IEEE Trans. on Pattern A. & Mach. Int. 30(9), 1557–1571 (2008)
10. Krempl, G., Kottke, D., Spiliopoulou, M.: Probabilistic active learning: A short proposition. In: Proc. 21st Europ. Conf. on AI (ECAI 2014). IOS Press (2014)
11. Lewis, D.D., Gale, W.A.: A sequential algorithm for training text classifiers. In: Proc. of the 17th ACM SIGIR, pp. 3–12 (1994)
12. Parker, C.: An analysis of performance measures for binary classifiers. In: Proc. of the 11th ICDM, pp. 517–526. IEEE (2011)
13. Roy, N., McCallum, A.: Toward optimal active learning through sampling estimation of error reduction. In: Proc. of the 18th ICML, pp. 441–448 (2001)
14. Settles, B.: Active Learning literature survey. CS Tech. Rep. 1648, U. Wisconsin (2009)
15. Settles, B.: Active Learning. in Synth. Lect. AI and ML. Morgan Claypool, vol. 18 (2012)
16. Tomanek, K., Morik, K.: Inspecting sample reusability for active learning. In: Guyon, I., et al. (eds.) AISTATS workshop on Act. Learning and Exp. Design., vol. 16, pp. 169–181 (2011)

Incremental Learning with Social Media Data to Predict Near Real-Time Events

Duc Kinh Le Tran[1,2], Cécile Bothorel[1],
Pascal Cheung Mon Chan[2], and Yvon Kermarrec[1]

[1] UMR CNRS 3192 Lab-STICC
Département LUSSI – Télécom Bretagne, France
{duc.letran,cecile.bothorel,yvon.kermarrec}@telecom-bretagne.eu
[2] Orange Labs, France
{duckinh.letran,pascal.cheungmonchan}@orange.com

Abstract. In this paper, we focus on the problem of predicting some particular user activities in social media. Our challenge is to consider real events such as message posting to friends or forwarding received ones, connecting to new friends, and provide near real-time prediction of new events. Our approach is based on *latent factor models* which can exploit simultaneously the timestamped interaction information among users and their posted content information. We propose a simple strategy to learn incrementally the latent factors at each time step. Our method takes only recent data to update latent factor models and thus can reduce computational cost. Experiments on a real dataset collected from Twitter show that our method can achieve performances that are comparable with other state-of-the-art non-incremental techniques.

Keywords: social media mining, incremental learning, latent factor models, matrix factorization.

1 Introduction

Recent years have witnessed the explosion of social media on the Internet. Vast amounts of user-generated content are created on social media sites every day. Social media data are often characterized as vast, noisy, distributed, unstructured and dynamic [7]. These characteristics make it difficult or impossible to apply conventional data mining techniques on social media data.

One of the challenges in mining social media is how to leverage the *interaction information* (or *relation*) in the data. Interaction information in social media can be any type of interactions between two users (e.g send a message, write a comment) or relations between them (e.g friendship declared in a social network). These interactions and relations are heterogeneous (can be different in nature) and very rich in volume. In general, interaction information is worthy to consider. Conventional machine learning techniques relying on attribute-value data representation (i.e *content information*) cannot fully exploit this kind of information.

S. Džeroski et al. (Eds.): DS 2014, LNAI 8777, pp. 180–191, 2014.

Another challenge of mining social media data lies in the fact that these data are vast and continuously evolving. Social media provide a continuous stream of data. Some applications in social media mining require building prediction model to periodically extract useful information. Using offline learning techniques (*batch learning*), we have to consider all data available from the past until the present. This approach is not suitable for mining social media because (1) as new data come, the size of the dataset grows, it gets more and more expensive to learn and to apply the model (2) this approach treats old data in the past and recent data the same way; intuitively, this may not be a good choice because old data often contain less relevant information (in the context of predicting future events).

These two challenges have not often been considered together. Recently, there have been some works on mining social media stream, for example [1,9], but they mostly concern topic extraction or trending topic detection on social media. We are interested in predicting actions or attributes on each user. For this problem, there have been a lot of works on exploring relational information in data. These techniques are often referred to as *statistic relational learning* [5]. Unfortunately they can only deal with static datasets. On the other hand, the second challenge can be overcome by using *incremental learning* techniques [8], which are capable of incrementally updating the model with new data. However, most incremental learning algorithms only deal with attribute-value data.

This paper aims to tackle both these two challenges. We are interested in predicting some particular users' actions in social media: post a message mentioning a telecommunication brand on Twitter. The problem is described in details in the next sections. We show that our proposed method based on *latent factor models* achieves comparable or better performances than other learning techniques in leveraging simultaneously interaction information and attribute-value information in social media. The basic idea of our method has been introduced in our previous work [14], but here we test it for a different task and in a different context. We also show that incremental learning is more appropriate for mining social media: it is at least as good as batch learning in terms of prediction performance and can gain a lot in computational time.

2   Data Representation and Problem Statement

For reasons of convenience, we adopt the concept of the *social attribute network (SAN)* [6] to represent the data from social media. A *SAN* is a social network $G_s = (V_s, E_s)$ where V_s is the set of nodes and E_s is the set of (undirected) edges. The social graph is augmented with a bipartite graph $G_a = (V_s \cup V_a, E_a)$ connecting the *social nodes* in V_s with *attribute nodes* in V_a. The edges in E_s are *social links* and the edges in E_a (connecting social nodes and attribute nodes) are *attribute links*. The value of an attribute a for a social node u is represented by the weight of the link (u, a). Social media data can be represented by a SAN as follows: social nodes represent the users, social links represent their relations or interactions and attribute links represent known values of attributes (profiles, user-generated content) on these users.

The advantage of the SAN model is that it can easily represent data stream from social media. When data come in stream, at each new time step, we can have new users and new attributes. We can also have new social relations and interactions between users and new values of attributes for each user. All of these elements can be represented by an "incremental" SAN: at each time step new nodes (social nodes or attribute nodes) and new links (social links or attribute links) are added. The new links can be social links or attribute link and concern both existing nodes and new nodes. To be clear, we do not consider node and link disappearance.

Our objective is to predict a target variable on the users in the next time step $t + 1$ using data up to (including) t. In this paper, we consider a binary target variable (label) which concerns some particular real-time action of the users. In each time step, it takes the value 1 (positive) if the user take the action and 0 (negative) otherwise. This is a near real-time prediction problem (i.e requires building prediction model at each time step). We aim to design prediction models that can be learned incrementally from an "incremental" SAN . Our problem is, *with new nodes and new links added at each time step, how to adapt the model built at the previous time step to get a new model.*

3   Related Work

The problem stated above is concerned with building classifiers from both attribute-value data and the social graph. To use attribute-value data, any conventional machine learning technique can be employed. Among these techniques, *support vector machine* (SVM) [4] is one of the most robust.

To explore the social graph, many techniques of statistical relational learning have been proposed. We cite here some interesting graph-based approaches. The neighbor-based approach [12] infers the target attribute of a node from that of its neighbors as follows:

$$P(y_i = 1) = \frac{\sum_{j \in N_i} y_j S_{ij}}{\sum_{j \in N_i} S_{ij}} \qquad (1)$$

where y_i denotes the attribute value of the node t (0, 1 where 1 corresponds to a positive label), S_{ij} denotes the weight of the social link (i, j) and N_i is the set of neighbors of t. This is a very simple approach it was proven to be better than other relational techniques in some particular datasets [12].

Another approach of using the social graph for classification is *Social Dimension* [13]. The basic idea of this method is to transform the social network into features of nodes using a graph clustering algorithm (where each cluster, also called a *social dimension*, corresponds to a feature) and then train a discriminative classifier (SVM) using these features. Any graph clustering algorithm can be used to extract social dimensions but *spectral clustering* [11] was shown to be the best. This approach helps exploiting the graph globally, not just the neighborhood of a node. It was shown in [13] that the *Social Dimension* outperforms other well-known methods of graph-based classification on many social media datasets.

There have also been efforts to use both the social graph and the attributes to improve prediction performances. For example, the *Social Dimension* approach [13] was extended to handle attribute on nodes. It is a simple combination of features extracted from the social graphs with attributes on nodes to learn a SVM classifier.

Another class of interesting techniques for mining both the social graph and the attributes are based on *latent factor models(LFM)* [2]. LFMs represent data points (in this case social nodes and attribute nodes) as vectors of unobserved variables (latent factors or latent features). All observation on nodes (in this case, links between nodes) depend on their latent factors. When we only have attribute-value data, latents factors can be learned using *matrix factorization (MF)* techniques, which consists of decomposing the data matrix into two matrices: one contains latent features and one contains those of attributes. To use the relational information in the social graph, in [10] the authors proposed an extension of MF, called *relation regularized matrix factorization (RRMF)*. RRMF simultaneously exploits the social graph and the attribute graph. Suppose that we have a dataset represented by a SAN $\mathcal{G}$, RRMF learns latent factors by minimizing:

$$
Q\left(U, P, \mathcal{G}\right) = \alpha \sum_{(i,j) \in E_s} S_{ij} \left\|u_i - u_j\right\|^2 + \sum_{(i,k) \in E_a} \left(A_{ik} - u_i p_k^T\right)^2
$$
$$
+ \lambda \left(\sum_{i=1}^{n_s} \left\|u_i\right\|^2 + \sum_{k=1}^{n_a} \left\|p_k\right\|^2 \right) \tag{2}
$$

where E_s is the set of social links, E_a is the set of attribute links; S is the adjacent matrix of the social graph and A is the adjacent matrix of the bipartite attribute graph; U is the matrix constituted of the latent vectors of all the social nodes and similarly, P is the matrix constituted of the latent vectors of all the attribute nodes of $\mathcal{G}$. The parameter α allows to adjust the relative importance of the social network in the model. The third term is a regularization term to penalize complex models with large magnitudes of latent vectors. λ is a regularization parameter. We can see that this is in fact the factorization of the attribute matrix A when adding regularization term $\alpha \sum_{(i,j) \in E_s} S_{ij} \left\|u_i - u_j\right\|^2$.

This term is called the *relational regularization term* which allows to minimize the distances between connected social nodes in the latent space. The RRMF approach assumes that connected social actors tend to have similar profiles. In some cases, it is better to use the *normalized Laplacian* of the social graph and the regularization term becomes $\alpha \sum_{(i,j) \in E_s} S_{ij} \left\|u_i/\sqrt{d_i} - u_j/\sqrt{d_j}\right\|^2$ where d_i is the degree of the node i in the social graph (see [10] for more details). The latent factors learned with RRMF are then used to train a classifier for label prediction problem.

All techniques mentioned above concern batch learning, i.e learning from a static dataset. In their problem setting, they assume that the set of social nodes is partially labeled and the problem is to infer labels of unlabeled nodes. Our problem (near real-time prediction problem) is not in same context but we can use the same idea : using latent factors to train a classifier at each time step. In the next sections, we describe our method in which we learn LFM (more precisely RRMF) incrementally and then use these factors to predict labels at each time step. We describe our strategy (based on least squares regularization) and show that it works well in a real world problem with data collected via Twitter.

4   Incremental Learning with Latent Factor Model

In the incremental learning context defined in Section 2, we need to learn a model (i.e the latent features of nodes) at each time step. The *batch learning* approach suggests that we learn the latent features at each time step using the whole snapshot of the SAN $\mathcal{G}(t)$ (which contains all nodes and links collected up to t)

$$U^{\star}(t), P^{\star}(t) = \arg\min_{U,P} Q(U, P, \mathcal{G}(t)) \qquad (3)$$

where Q is the objective function defined above (Equation 2).

The incremental method learns a model (latent factors of nodes) only from new data (i.e the incremental part of the SAN, denoted by SAN $\Delta\mathcal{G}(t)$) when reusing the old model, i.e latent features of nodes calculated in the previous time step. To do this, we minimize the following objective function:

$$Q_{inc}(U, P, t) = Q(U, P, \Delta\mathcal{G}(t))$$

$$+\mu\left(\sum_{i\in V_s(t-1)}\|u_i - u_i^{\star}(t-1)\|^2 + \sum_{k\in V_a(t-1)}\|p_k - p_k^{\star}(t-1)\|^2\right)$$

$$(4)$$

where $V_s(t-1)$ and $V_a(t-1)$ are respectively the set of social nodes and the set of attribute nodes in the previous time step; $u_i^{\star}(t-1)$ and $p_k^{\star}(t-1)$ are respectively the latent vectors of the social node i and the attribute node k learned in the previous time step and μ is a parameter of the model. This objective function consists of two terms. The first term is the objective function of MF on the incremental graph $\Delta\mathcal{G}(t)$. The second term is a regularization term for minimizing the shifts of latent features of the same nodes between time steps. By minimizing the two terms simultaneously, we learn latent features of nodes both from the new data and from the latent features of existing nodes of the previous time step. We can easily see that the latent features of an existing node are updated if and only if there are new links connecting to it. With the second regularization term, we ensure that the latent space does not change much from a time step to the next. The parameter μ allows to tune the contribution of the previous model to the current model.

After having calculated latent factors of nodes at the time step t, we have a low-dimensional representation of the data points (nodes) at this time step. It can be used for any standard machine learning task on the nodes. For our prediction problem, we are based on the hypothesis that data collected up to t are informative for the target variable in $t + 1$ on the social nodes. With this hypothesis, we can use these factors (low dimensional representation of the data up to t) of the social nodes to train a classifier and then use the classifier to predict in the next time step.

In terms of optimization, we can use standard algorithms (e.g gradient-based) to minimize Q in Equation 3 for batch learning or Q_{inc} in Equation 4 for incremental learning. In this work, we adapt the *Alternating Least Squared (ALS)* algorithm [15]. The basic idea of this algorithm is to solve the least square problem with respect to the latent features of one node at a time (when fixing those of the others) until convergence. The complexity of the algorithm linearly depends on the number of squared terms in the objective function, which is the total number of nodes and number of links in the SAN. In other words, the learning algorithm has linear complexity with respect to the size of the data. In case of incremental learning, when optimizing only on recent data ($\Delta \mathcal{G}(t)$), we can gain a lot in terms of computational cost.

5 Experiments

5.1 Data Description and Experimental Setup

The dataset was collected via Twitter API[1] in the period from July to December 2012. The data concern the followers of the Sosh account on Twitter (@Sosh_fr[2]) in this period. We keep the identities of the followers of @Sosh_fr in our database. During the period, new followers of @Sosh_fr were constantly added. For each follower, we regularly get the following elements: all the tweets, all the retweets, the list of followers (from which we can build the who-follow-whom graph among these users). We also collected some elements of the profile of each user (e.g. some variables related to the global centrality of the user such as the number of followers, the number of tweets posted, etc.).

We collected the data regularly enough to be able to build 20 week-based snapshots (the first week begins on 15/07/2012) of the dataset. We have totally 30 400 users, about 9×10^5 who-follow-whom links, about 36×10^4 tweets and 26×10^4 retweets (on average 18×10^3 tweets and 13×10^3 retweets per week). We want to use both the social interactions (follower-followee relation, retweet) and the tweets of users. We represent each snapshots by a SAN, the SAN for the week t is built as follows:

- *The social graph.* We put a social link between two users if they are linked (one follows the other) or if they have retweeted a common tweet in the

---

[1] https://dev.twitter.com/docs/api

[2] Sosh is a French mobile brand, developed in France by the French operator Orange since 6 October 2011. Sosh is on Twitter at http://twitter.com/Sosh_fr.

period t. We can see that this is a sum of follower graph and *co-retweet* graph (we aggregate these two graphs to get a denser graph).
- *The attributes.* We consider each word in the tweet(s) of users as an attribute. We put an attribute link between a user and a word if the user posted a tweet containing the word in the period. The link is weighted by how many times this occurred.

We are interested in predicting who will talk about the brand Sosh (i.e mention @Sosh_fr, write the word "sosh" in their tweets) in the week $t + 1$ using data up to (including) the week t. Among the followers of @Sosh_fr, the ones who will talk about "sosh" are often customers of Sosh or just people interested in the brand who could become the future customers of the brand. At each time step (week), the prediction problem is a classification problem where positive labels correspond to who talk about "sosh" in the next week. Figure 1 presents the number of users and number of positive labels in each week. The portion of positive labels in each week is relatively small (less than 1%).

We apply our method (*incremental LFM*) for this prediction problem. For each time step, we calculate latent factors of all nodes by minimizing the objective function defined in Equation 4. At the time step t, we use the latent factors to learn an SVM classifier with positive labels in the next time step $t + 1$. At $t + 1$, we use the model learned in the previous time step to predict positive labels in the next time step $t + 2$. We use *Area Under ROC Curve* (AUC) [3] to measure the prediction performance. AUC is a rank measure which allows to measure prediction performance across all possible cut-off thresholds. Roughly speaking, it is the probability that a classifier will rank a randomly chosen positive instance higher than a randomly chosen negative one. The advantage of using AUC is that we do not need to fix a cut-off threshold for each method.

The number of latent factors D is set to 20. In the objective function we use the *normalized Laplacian regularization* term as we see that it achieved better performance than normal graph regularization. We fixed the number of iterations in our ALS optimization problem to 20 since we observed no improvement of performance beyond 20 iterations. The regularization parameter λ is set to 50; λ is set to 100 and μ is set to 100. The influences of these parameters are studied in Subsection 5.5.

5.2 Baselines

At each time step (week) t we apply the following baseline techniques to compare with our incremental method:

Trivial solution 1 Since a user can talk about "sosh" more than once, it is interesting to know if this is a repeated action: if a user have talked about "sosh", how likely will she/he talk about it again. This is the idea of this first baseline: who ever had a positive label (at least once) in the past will have a positive label the next week $t + 1$.

Trivial solution 2 In Twitter, users have different levels of usage. There are users who write a lot of tweets, have lots of friends or followers, etc. These

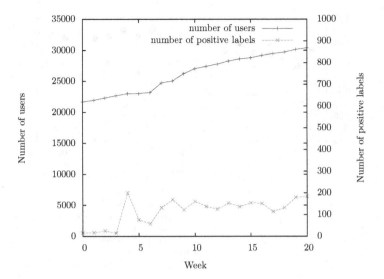

Fig. 1. Number of users and number of users having positive labels in each time step

users are more likely to talk about "sosh". From this observation, we build a prediction method that predicts the label of a user based on a score that measures how active she/he is in Twitter. We tried different measures of "activeness" of users, but we see that the number of tweets posted in the past is the best measure to differentiate between active and non-active users.

Neighbor-based method This method use the neighborhood of each user in the social graph (described in Section 3). This method does not require a learning step, the label of a node in $t+1$ is inferred from that of its neighbors in the previous time step t (Equation 1).

Social dimension This method uses the social graph. At each time step we extract the social dimensions (described in Section 3) and then use these dimensions to train an SVM classifier with positive labels of the next time step (same procedure as with the latent factors in our method). The number of dimensions is set to 10. We do not see any improvement of performance setting this value bigger than 10.

SVM on attributes We use supervised classification with the attributes. At the time step t, we train an SVM classifier from all known values of attributes up to t and positive labels in $t+1$. Because the attributes are words (in the tweets), this means that we use the "bag of words" produced by each user up to t.

Social dimensions + attributes This is a combination of the *social dimensions* and the attributes. We use supervised classification (SVM) with the social dimension and the attributes.

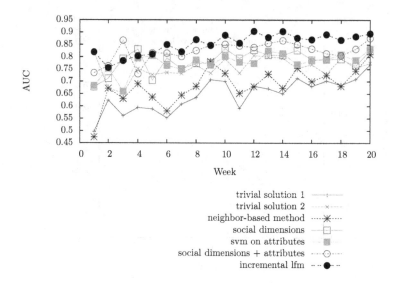

Fig. 2. Performances of different methods

5.3 Performance

Figure 2 presents the performances of all learning techniques. First of all, we can see that both the social graph and the attributes (bag of words for each user) are informative. The methods using these data (non-trivial) are often better than the trivial method. The neighbor-based method is not well adapted to this dataset : it gives even worse performance than the trivial solution 2. Combining these two sources gives even better performances (our method and Social Dimension+attributes).

Except for some perturbations in the beginning, our method (incremental LFM) achieves the highest AUC in all time steps. We conclude that, by exploiting both the social graph and the attributes we can enhance significantly the prediction performance and our incremental method based on LFM achieves relatively good performance in comparison with the best batch-learning technique.

5.4 Gain in Computational Time

Figure 3 shows computational times of the incremental LFM method and by the best baseline method - a combination of Social Dimension and attributes. For our incremental method, computational time at each time step consists of learning latent factors (optimization) and training an SVM classifier. For the other methods, computational time consists of spectral clustering of the social graph and training an SVM classifier. We measure only the learning time (i.e optimization). To be fair, the two methods are implemented and executed on the same machine (Linux 64bit, CPU 8x2.1GHz). The figure shows significant

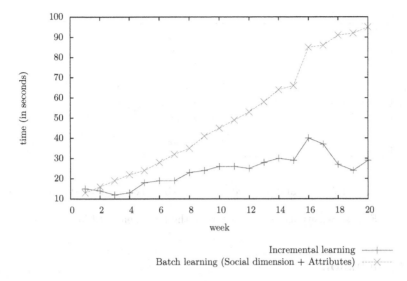

Fig. 3. Learning time

gain in time using incremental learning. This is an illustration of our theoretical analyses in previous sections: learning with aggregated data becomes more and more expensive as new data are added; incremental learning only requires time to deal with new data.

5.5   Sensitivity to Parameters

We examine the sensibility of 3 important parameters of our incremental LFM method : α, μ and the number of latent factors D. We average the performance (AUC) of all time steps to get a global performance for each parameter configuration. As shown in Figure 4a, too small or too large values of the parameter α hurt the performance. Larger α means that the social interactions have more contribution to the prediction model. When $\alpha = 0$, no interaction information is used. Maximum AUC is achieved around $\alpha = 100$. The effect of with the parameter μ is shown in Figure 4b. This parameter controls the contribution of the prediction model learned in the previous time step to the current model. $\mu = 0$ corresponds to the case where we learn latent factors only from recent data and latent factors learned in previous time step are not used. We see that when μ increases, the performance increases and achieves its maximum at $\mu = 100$. About the number of latent factors D, we observe a small influence of this parameter on the performance. We see that small values of D are adequate because we can not improve significantly the performance setting it bigger than 10.

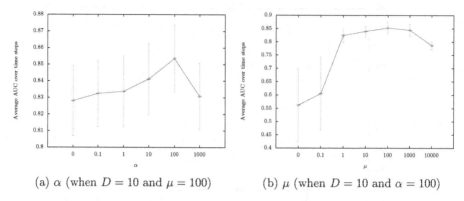

(a) α (when $D = 10$ and $\mu = 100$)   (b) μ (when $D = 10$ and $\alpha = 100$)

Fig. 4. Sensitivity to parameters of the incremental LFM method

6   Conclusion

We have proposed an incremental learning method based on *latent factor model* for a prediction problem with data collected from Twitter. Our strategy (adding a regularization term) for incremental learning leads to very promising experimental results in both performance and computational cost. The main limitation of our method is how to choose the right values of its parameters to achieve its best performance. In future work, we will consider automatic configuration for the parameters at each time step to improve the performance. We plan on extended tests on other datasets or synthetic data to understand deeply the nature of the data where the method is efficient and robust. We keep working on the Twitter dataset but for other prediction problems (other type of events), the most interesting problem is to predict whether a user talks positively or negatively about the brand. We also consider other possible extensions of our models to handle more complicated data structure from social media: there are more than one types of social links in the SAN, directed links between social nodes, link disappearance etc.

References

1. Aiello, L.M., Petkos, G., Martin, C., Corney, D., Papadopoulos, S., Skraba, R., Goker, A., Kompatsiaris, I., Jaimes, A.: Sensing trending topics in twitter. IEEE Transactions on Multimedia 15(6), 1268–1282 (2013)
2. Bartholomew, D.J., Knott, M., Moustaki, I.: Latent Variable Models and Factor Analysis: A Unified Approach. Wiley Series in Probability and Statistics. Wiley (2011)
3. Bradley, A.P.: The use of the area under the ROC curve in the evaluation of machine learning algorithms. Pattern Recogn. 30(7), 1145–1159 (1997)
4. Cortes, C., Vapnik, V.: Support-vector networks. Machine Learning 20(3), 273–297 (1995)

5. Getoor, L., Taskar, B.: Introduction to Statistical Relational Learning. Adaptive computation and machine learning, vol. L. MIT Press (2007)
6. Gong, N.Z., Talwalkar, A., Mackey, L.: Jointly Predicting Links and Inferring Attributes using a Social-Attribute Network (SAN). CoRR, abs/1112.3 (December 2011)
7. Gundecha, P., Media, Liu, H.: Mining Social Media: A Brief Introduction. In: Tutorials in Operations Research - New Directions in Informatics, Optimization, Logistics, and Production, pp. 1–17 (2012)
8. Joshi, P., Kulkarni, P.: Incremental Learning: Areas and Methods–A Survey. International Journal of Data Mining & Knowledge Management Process 2(5) (2012)
9. Kim, M., Newth, D., Christen, P.: Trends of news diffusion in social media based on crowd phenomena. In: Proceedings of the Companion Publication of the 23rd International Conference on World Wide Web Companion, WWW Companion 2014, pp. 753–758. International World Wide Web Conferences Steering Committee, Republic and Canton of Geneva, Switzerland (2014)
10. Li, W.J., Yeung, D.Y.: Relation regularized matrix factorization. In: IJCAI 2009, pp. 1126–1131. Morgan Kaufmann Publishers Inc., San Francisco (2009)
11. Luxburg, U.: A tutorial on spectral clustering. Statistics and Computing 17(4), 395–416 (2007)
12. Macskassy, S.A., Provost, F.: Macskassy and Foster Provost. A simple relational classifier. In: Proceedings of the Second Workshop on Multi-Relational Data Mining (MRDM 2003) at KDD 2003, pp. 64–76 (2003)
13. Tang, L., Liu, H.: Leveraging social media networks for classification. Data Min. Knowl. Discov. 23(3), 447–478 (2011)
14. Le Tran, D.K., Bothorel, C., Cheung-Mon-Chan, P.: Incremental learning with latent factor models for attribute prediction in social-attribute networks. In: EGC, pp. 77–82 (2014)
15. Zhou, Y., Wilkinson, D., Schreiber, R., Pan, R.: Large-Scale Parallel Collaborative Filtering for the Netflix Prize. In: Fleischer, R., Xu, J. (eds.) AAIM 2008. LNCS, vol. 5034, pp. 337–348. Springer, Heidelberg (2008)

Stacking Label Features for Learning Multilabel Rules

Eneldo Loza Mencía and Frederik Janssen

Technische Universität Darmstadt
Knowledge Engineering Group
{eneldo,janssen}@ke.tu-darmstadt.de

Abstract. Dependencies between the labels is commonly regarded as the crucial issue in multilabel classification. Rules provide a natural way for symbolically describing such relationships, for instance, rules with label tests in the body allow for representing directed dependencies like implications, subsumptions, or exclusions. Moreover, rules naturally allow to jointly capture both local and global label dependencies.

We present a bootstrapped stacking approach which uses a common rule learner in order to induce label-dependent rules. For this, we learn for each label a separate ruleset, but we include the remaining labels as additional attributes in the training instances. Proceeding this way, label dependencies can be made explicit in the rules. Our experiments show competitive results in terms of the standard multilabel evaluation measures. But more importantly, using these additional attributes is shown to allow to discover and consider label relations as well as to better comprehend the available multilabel datasets.

However, this approach is only a first step towards integrating the multilabel rule learning directly in the rule induction process, e.g., in typical separate-and-conquer rule learners. We present future perspectives, advantages, and arising issues in this regard.

1 Introduction

Rule learning has a very long history and is a well-known problem in the machine learning community. Over the years many different algorithms to learn a set of rules were introduced. The main advantage of rule-based classifiers are interpretable models as rules can be easily comprehended by humans. Also, the structure of a rule offers the calculation of overlapping of rules as well as *more specific* and *more general*-relations. Thus, the rule set can be easily modified as opposed to most statistical models such as SVMs or neural networks. However, most rule learning algorithms are currently limited to binary or multi-class classification.

On the other hand, many problems involve assigning more than a single class to an object. These so-called multilabel problems can often be found when text is classified into topics or tagged with keywords, but there are also many examples from other media such as the recognition of music instruments or emotions in audio recordings or the classification of scenes in images.

It is widely accepted that one major issue in learning from multilabel data is the exploitation of label dependencies. Learning algorithms may greatly benefit from considering label correlations, and we believe that rule induction algorithms provide a good

S. Džeroski et al. (Eds.): DS 2014, LNAI 8777, pp. 192–203, 2014.

base for this. Firstly, so called global dependencies between only labels can be explicitly modeled and expressed in form of rules. But also, and much more interesting, dependencies that include both label and regular features can be constituted, which we refer to as local dependencies. Secondly, such rules are directly interpretable and comprehensible for humans. Even if complex and long rules are generated, the implication between classes can be estimated more easily than with other approaches by focusing on the part of the rules that considers the classes. Hence, one is able to directly analyze the induced rule models and may greatly benefit from these explicit notations, in contrast to other types of models where the key information is not accessible directly.

We propose in this work to learn such interdependencies by providing the true label information directly to the rule learner. This is done by stacking the label features as additional input instance features. Although this is not the first work in making use of stacking in order to consider label dependencies (cf. Sec. 3.1), it is to our knowledge the first time that rule induction was used in order to make the label dependencies explicit. We show that the proposed method, though conceptually very simple, is suitable in order to reveal global as well as local label dependencies. Almost more importantly, the induced models allow for a detailed analysis of the datasets commonly used in the community for benchmarking w.r.t. the contained dependencies.

The proposed bootstrapping in the prediction phase remains open for discussion, though its performance is competitive to straight-forward approaches. But our ultimate goal is to have a complete framework for multilabel rule induction instead of employing special schemes for learning and predicting. We give some perspectives and ideas for further research in the end of the paper.

2   Multilabel Classification and Inductive Rule Learning

2.1   Multilabel Classification

Multilabel classification refers to the task of learning a function $h(\mathbf{x})$ that maps instances $\mathbf{x} = (x_1, \ldots, x_m) \in \mathscr{X}$ to label subsets or label vectors $\mathbf{y} = (y_1, \ldots, y_n) \in \{0, 1\}^n$, where $\mathscr{L} = \{\lambda_1, \ldots, \lambda_n\}$, $n = |\mathscr{L}|$ is a finite set of predefined labels and where each label attribute y_i corresponds to the absence (0) or presence (1) of label λ_i . Thus, in contrast to multiclass learning, alternatives are not assumed to be mutually exclusive, such that multiple labels may be associated with a single instance. This, and especially the resulting correlations and dependencies between the labels in $\mathscr{L}$, make the multilabel setting particularly challenging and interesting compared to the classical field of binary and multiclass classification.

From a probabilistic point of view, this is one of the main differences. In binary and multiclass problems the only observable probabilistic dependence is between the input variables, i.e., the attributes x_j, and the label variables y_i. A learning algorithm tries to learn exactly this dependence in form of a classifier function h. In fact, if a classifier provides a score or confidence for its prediction $\hat{\mathbf{y}} = h(\mathbf{x})$, this is often regarded as an approximation of $P(\mathbf{y} = \hat{\mathbf{y}} \mid \mathbf{x})$, i.e., the probability that $\hat{\mathbf{y}}$ is true given a document $\mathbf{x}$.

From the early beginning of multilabel classification, there have been attempts to exploit these types of *label correlations* [e.g. 12, 7, 17]. A middle way is followed by

Read et al. [14] and Dembczyński et al. [5] and their popular (probabilistic) classifier chains by stacking the underlying binary relevance classifiers with the predictions of the previous ones. However, only recently Dembczyński et al. provided a clarification and formalization of label dependence in multilabel classifications. Following their argumentation, one must distinguish between unconditional and conditional label dependence. Roughly speaking, the *unconditional dependence* or independence between labels does not depend on a specific given input instance (the condition) while *conditional dependence* does. We may also refer to these as *global* and *local* dependencies, since they are revealed globally or only in subspaces of the input space.

An example may illustrate this: Suppose a label space indicating topics from news articles and a subtopic *foreign affairs* of the topic *politics*. Obviously, there will be a dependency between both labels, since the presence of a subtopic implies the presence of the super topic and the probability of *foreign affairs* would be higher than average if *politics* is observed. These probabilities are *unconditional* or *global* since they do not depend on a particular document. Suppose now that a particular news article is about the *Euro crisis*. Under this condition, the *conditional* probabilities for both labels as well as the dependency between would likely increase and hence be different from the unconditional ones. However, if an article was about the *cardiovascular problems of Ötzi*, we would observe that both labels are *conditionally independent* for this instance, since the probability for one label would very likely not depend on the presence of the other label (both being very low).

The predominant approach in multilabel classification is *binary relevance* (BR) learning [cf. e.g. 16]. It tackles a multilabel problem by learning one classifier for each label, using all objects of this label as positive examples and all other objects as negative examples. There exists hence a strong connection to concept learning, which is dedicated to infer a model or description of a target concept from specific examples of it [see, e.g., 4]. When several target concepts are possible or given for the same set of instances, we formally have a multilabel problem. The problem of this approach is that each label is considered independently of each other, and as we have seen by the example given before, this can lead to loss of useful information for classification.

A possible simple solution to generate rules that may consider several labels in the head is to use the label powerset (LP) transformation [cf. 16], which decomposes the initial problem into a multiclass problem with $\{\mathbf{y} \mid (\mathbf{x}, \mathbf{y}) \in \text{training set}\} \subseteq \{0, 1\}^n$ as possible classes. This problem can then be processed with common rule induction algorithms, which will thus produce rules with several labels in the head.

This approach is potentially able to consider conditional dependencies, namely the case of label co-occurrences. The main drawback is that the number of classifiers that have to be learned grows exponentially. Another obvious disadvantage is that we can only predict label relations and combinations which were seen in the training data.

2.2   Inductive Rule Learning

As the goal of this paper is to make label dependencies explicit by using rules, we will also shorty introduce inductive rule learning. This is one of the oldest and therefore best researched fields in machine learning. Many algorithms were proposed over the years, *Ripper* [3] being one of the most popular and used ones. In this work, we used this algorithm, but the proposed method does also naturally work with other rule learning algorithms. *Ripper* is a so-called separate-and-conquer (SeCo) algorithm [6], i.e., it proceeds by learning a good rule on the data, then adds the rule to the ruleset, removes all examples covered by this rule, and searches the next one as long as (positive) examples are left in the dataset. In order to prevent overfitting, the two constraints that all examples have to be covered (*completeness*) and that negative examples must not be covered (*consistency*) can be relaxed so that some positive examples may remain uncovered and/or some negative examples may be covered by the set of rules. SeCo usually only works for binary datasets. Hence, a natural way of addressing multilabel problems is to consider each label separately (cf. BR), resulting in a model consisting of separate rulesets for each label.

2.3 Different Forms of Multilabel Rules

A rule learner has a set of rules (*ruleset*) as result. These rules are of the form

$$head \leftarrow body$$

where the body consists of a number of *conditions* (attribute-value tests) and, in the regular case, the head has only one single condition of the form $y_i = 0$ or 1 (in our case). We refer to this type of rules as *single-label head* rules in contrast to *multi-label head* rules, which contain several label assignments in their head and can thus conveniently express label co-occurrences. Commonly, the conditions in the body are on attributes from the instance space. However, in order to reflect label dependencies (e.g., implications, subsumptions, or exclusions), we would need to have labels on both sides of the rule. Hence, if a rule may contain conditions on the labels, we refer to it as label-dependent rules (also referred to as *contextual* rules [4]), and *label-independent* if this is not the case. Global dependencies are hence best reflected by *full label-dependent* bodies, whereas local dependencies can be described by *partially label-dependent* rules with mixed attributes in the body.

In summary: We start from **label-independent single-label** rules. Label dependencies can already be captured by **label-independent multi-label** rules. The next section describes a straight-forward approach for obtaining such rules. Future extensions are proposed in Sec. 6. This particular work focuses on learning **label-dependent single-label** rules (Sec. 3), which, as shown, are well suited for modeling and expressing label dependencies. The full expressiveness is though obtained by **label-dependent multi-label** rules, which we leave for further research (Sec. 6).

3 Learning Label-Dependent Rules

We present in the next subsection a straight-forward, yet effective approach in order to learn label-dependent rules which allows to discover valuable information in data.

3.1 Stacking of Label Features

The recently very popular classifier chains [14] were found to be an effective approach for exploiting conditional label dependencies. Classifier chains (CC) make use of stacking the previous BR classifiers' predictions in order to implement the chain rule $P(y_1,\ldots,y_n) = P(y_n \mid y_1,\ldots,y_{n-1})$ in probability theory, since they learn the binary classifiers h_i with training examples of the form $(x_1,\ldots,y_1,\ldots,y_{i-1})$ [cf. 5]. One drawback of CC is the (randomly chosen) predetermined, fixed order of the classifiers (and hence the labels) in the chain, which makes it impossible to learn dependencies in the contrary direction. This was already recognized by D. Malerba and Esposito [4] in 1997, who built up a very similar system in order to learn multiple dependent concepts. In this case, the chain on the labels was determined beforehand by a statistical analysis of the label dependencies. Still, using a rule learner for solving the resulting binary problems would only allow to induce rules between two labels in one direction.

Thus, we propose to use a full stacking approach in order to overcome the main disadvantage of CC, i.e., the fixed order. Like in binary relevance, we learn one theory for each label, but we expand our training instances by the label information of the other labels, i.e., the training examples vectors for learning label y_i are of the type $(x_1,\ldots,y_1,\ldots,y_{i-1},y_{i+1},\ldots,y_n)$ for an instance $\mathbf{x}$. The result of using this as training data is exactly what we are seeking for, namely label-dependent single-label rules. The amount of label-features in the body additionally allows us to determine the type of dependency. We refer to this technique as *stacked binary relevance* (SBR) in contrast to plain, unstacked BR.

This is very similar to the approaches of Godbole and Sarawagi [8], Guo and Gu [9], and very recently, Montañés et al. [13]. They all have in common that they are using label presence information (either directly from the training data, or from the outputs of underlying BR classifiers) as (either sole or additional) features in order to learn an ensemble of binary relevance classifiers on top. The closest related approaches to our proposition are the conditionally dependency networks (CDN) [9] and the dependent binary relevance (DBR) models [13]. Both learn their models as indicated before but with one major difference: Since they are concerned with estimating probability distributions (especially joint distribution), they both use logistic regression as their base classifier, which is particularly adequate for estimating probabilities. This type of models are obviously much harder to comprehend than rules, especially for higher number of input features. Therefore, the label dependencies would remain hidden somewhere in the model, even though they may have been taken into account and accurate classifiers may have been obtained. To make the dependencies explicit and at the same time keep a high prediction quality, we propose to use rule-based models. One additional difference between the approaches is how the prediction is conducted, which is discussed next.

3.2 Prediction by Bootstrapping

For the prediction we propose to use a bootstrapping approach in the sense that we apply our models iteratively on our own previous predictions until the predictions are stable or any other stopping criterium is met. More formally, we use the learned models h'_i to produce a prediction $\hat{\mathbf{y}}_j = (\hat{y}_{j,1}, \hat{y}_{j,2}, \ldots)$ where $\hat{y}_{j,i} = h'_i(\mathbf{x}, \hat{\mathbf{y}}_{j-1})$ is based on the predictions in the previous iteration $j-1$.

One obvious issue with this approach is the initialization of $\hat{y}_0$. A possible option, also proposed by DBR, is to use the predictions of a BR ensemble, i.e., $\hat{y}_{0,i} = h_i(\mathbf{x})$. We also evaluate the option of initializing with *unknown* label information, i.e., $\hat{\mathbf{y}}_0 =$

$(?, ?, \ldots)$, and to benefit from the natural support of symbolic approaches for such attribute states (*missing, don't care*, etc.). On the other hand, this approach only works if the rule learner found enough rulesets with label-independent rules so that the bootstrapping can proceed, which is in fact somehow contradictory to the objective of detecting as much dependencies as possible. In the future, we also plan to use random initialization. Together with enough iterations of *Gibbs sampling*, this was shown to be very effective for CDN.

We also may make use of an additional capability of rule learners, namely to abstain from classifying if no appropriate rule was found (instead of predicting the default rule) so that the label attribute may be filled up in consequent iterations.

4   Evaluation

An overview of the used datasets[1] is given in Tab. 1. They are from different domains and have varying properties. Details of the data are given in the analysis when needed. As rule learner, we use the JRip implementation of Ripper [3] with default parameters, except for the pruning, which is turned off or on depending on the experiment.

We use micro-averaged precision and recall to evaluate our results, i.e., we compute a two-class confusion matrix for each label ($y_i = 1$ vs. $y_i = 0$) and eventually aggregate the results by (component-wise) summing up all n matrices into one global confusion matrix (cf. [16]). Recall and precision is computed based on this global matrix in the usual way, F1 denotes the unweighted harmonic mean between precision and recall. In addition, we measure the subset accuracy, which is the percentage $\frac{1}{m} \sum_{i=1}^{m} [[\mathbf{y}_i = \hat{\mathbf{y}}_i]]$ of the m test instances for which the labelsets were exactly correctly predicted ($[[z]]$ returns 1 if z is true, otherwise 0). The measures, as well as other statistics, are averaged over the ten-fold cross validation results, which we use for all our experiments.

4.1   Model and Data Analysis

Tab. 2 shows the properties of the rulesets generated by using plain BR and stacked BR decomposition with JRip. As we will see in the following, these statistics not only help to analyze the algorithm, but even more importantly, they are of great use for analyzing and understanding the datasets at hand. Though it is commonly assumed that there exist label dependencies between the labels in multilabel datasets, and many works deal with exploiting such dependencies, this assumption is most often not explicitly examined. To our knowledge, this is the first work providing a systematic analysis of the label dependencies contained in seven of the most popular benchmarks.

Column (5) shows the percentage of conditions on labels w.r.t. to all conditions in the model. We see that there is a great divergence between the datasets. E.g., the models for GENBASE do not use label features at all, i.e., their rules' bodies are completely label-independent. This is a strong indicator that we have completely independent labels in this dataset, or, at least, very weak dependencies. This is remarkable, since this breaks

---

[1] We refer to the MULAN repository for details and sources:
http://mulan.sf.net/datasets.html

Table 1. Statistics of the used datasets: name of the dataset, domain of the input instances, number of instances, number of nominal/binary and numeric features, total number of unique labels, average number of labels per instance (cardinality), average percentage of relevant labels (label density), number of distinct labelsets in the data

name	domain	instances	nominal	numeric	labels	cardinality	density	distinct
EMOTIONS	music	593	0	72	6	1.87	0.311	27
SCENE	image	2407	0	294	6	1.07	0.179	15
YEAST	biology	2417	0	103	14	4.24	0.303	198
GENBASE	biology	662	1186	0	27	1.25	0.046	32
MEDICAL	text	978	1449	0	45	1.25	0.028	94
ENRON	text	1702	1001	0	53	3.38	0.064	753
CAL500	music	502	0	68	174	26.0	0.150	502

the main assumption, mentioned before, and yet this dataset may have often been used in the literature to show the ability of a certain algorithm to exploit label dependencies. In this case though, learning each label independently is already sufficient and exploiting (possibly non-existing) label dependencies clearly will not yield better performance. A look into columns (1)-(4), the prediction quality (Tab. 3) and eventually into the models, reveals that the presence of one single short amino acid chain (instance feature) is often enough to correctly predict a particular functional family (label).

For (5) it is also remarkable that pruning substantially increases the percentage of used label features. Pruning tries to remove conditions and rules which work good on a training set, but do not generalize well on a separate validation set. Hence, this increase indicates that label features are more useful for obtaining more general models than the original instance features. However, the increase does not come hand in hand with a decrease in the size of the models comparing BR and stacked BR, as can be seen by the average size of the rulesets (columns (1) and (2)) and rules ((3) and (4)), which does not reveal any trend.

While (5) may serve as an indicator of general dependency between labels, columns (6) and (7) allow to further differentiate. E.g., 20.8% of fully label-dependent rulesets for YEAST, i.e., rulesets with rules only having conditions on label features, show that (at least) 20.8% of the labels in YEAST are *unconditionally* dependent on other labels. On the other hand, by leaving out the 6.2% of labels which are independent, we can derive that (at most) 73.0% of the labels are *conditionally* dependent on other labels. Note that (6) should be considered as a lower bound, since the rate substantially suffers from the high number of instance features due to a kind of *instance feature flooding*: The probability of selecting an instance feature in the refinement step of a rule instead of an equally good label feature increases with growing number of instance features. However, the same effect cannot be observed for (7).

The datasets with the highest observed degree of label dependency are YEAST and CAL500. For CAL500, this may be explained by the categorizations of songs into emotions, which often come hand in hand or completely contradict, like *Angry-Agressive* against *Carefree-Lighthearted*.

Examples of learned rulesets for YEAST are given in Fig. 1. In this particular case, we see a much more compact and less complex ruleset for *Class4* for the stacked model

Table 2. Statistics. From left to right, for BR model: (1) avg. # rules per label ruleset, (2) avg. # conditions per rule. For stacked model: (3) avg. # rules per label ruleset, (4) avg. # conditions per rule, (5) percentage of conditions with label feature tests, (6) perc. of label rulesets depending *only* on other labels, (7) perc. of label rulesets depending *only* on instance features.

dataset	pruning	(1)	(2)	(3)	(4)	(5)	(6)	(7)
EMOTIONS	yes	3.26	2.78	2.74	3.09	35.0%	18.0%	0.0%
EMOTIONS	no	11.50	4.02	11.02	4.18	17.6%	0.0%	0.0%
SCENE	yes	6.72	4.27	5.44	4.44	16.0%	0.0%	18.0%
SCENE	no	13.58	5.40	11.10	5.09	10.2%	0.0%	2.0%
YEAST	yes	2.47	3.72	3.78	2.56	63.0%	20.8%	6.2%
YEAST	no	7.20	5.95	10.58	3.78	31.3%	0.0%	0.0%
GENBASE	yes	0.90	1.05	0.90	1.05	0.0%	0.0%	100.0%
GENBASE	no	0.99	1.29	0.99	1.29	0.0%	0.0%	100.0%
MEDICAL	yes	1.08	1.72	1.07	1.81	17.4%	0.0%	79.3%
MEDICAL	no	2.46	3.47	2.00	3.17	13.6%	0.0%	73.6%
ENRON	yes	1.54	3.38	1.89	3.37	35.9%	3.3%	35.0%
ENRON	no	5.82	4.97	6.94	4.68	25.1%	0.0%	11.0%
CAL500	yes	0.45	2.23	1.37	2.07	60.7%	29.0%	23.8%
CAL500	no	6.03	3.88	6.82	3.51	29.7%	1.2%	1.7%

than for the independently learned BR classifier. The ruleset also seems more appropriate for a domain expert to understand coherences between proteins (instance features) and protein functions (labels).

Fig. 1 also shows the models for the diagnosis *Cough* in the MEDICAL task. This dataset is concerned with the assignment of international diseases codes (ICD) to real, free text radiological reports. Interestingly, the stacked model reads very well, and the found relationship seems to be even comprehensible by non-experts: If the patient does not have *Pneumonia*, a *Pulmonary_collapse* or *Asthma* and "cough"s or is "coughing", he just has a *Cough*. Otherwise, he may also have a "mild" *Asthma*, in which case he is also considered to have a *Cough*.

In ENRON, which is concerned with the categorization of emails during the Enron scandal, the model is less comprehensible, as it is also for the BR model. However, the relation between *Personal* and *Joke* can clearly be explained from the hierarchical

Approach	YEAST	MEDICAL	ENRON
BR	*Class4* ← x23 > 0.08, x49 < -0.09	Cough ← "cough", "lobe"	Joke ← "mail", "fw",
	Class4 ← x68 < 0.05, x33 > 0.00, x24 > 0.00,	Cough ← "cough", "atelectasis"	"didn"
	x66 > 0.00, x88 > -0.06	Cough ← "cough", opacity	
	Class4 ← x3 < -0.03, x71 > 0.03, x91 > -0.01	Cough ← "cough", airways	
	Class4 ← x68 < 0.03, x83 > -0.00,	Cough ← "cough", $\overline{\text{"pneumonia"}}$, "2"	
	x44> 0.029, x93 < 0.01	Cough ← "coughing"	
	Class4 ← x96 < -0.03, x10 > 0.01, x78< -0.07	Cough ← "cough", "early"	
Stacked	*Class4* ← *Class3*, *Class2*	Cough ← "cough", *Pneumonia*,	Joke ← *Personal*,
BR	*Class4* ← *Class5*, $\overline{Class6}$	$\overline{Pulmonary\_collapse}$, *Asthma*	"day", "mail"
	Class4 ← *Class3*, *Class1*, x22 > -0.02	Cough ← "coughing"	
		Cough ← *Asthma*, "mild"	

Fig. 1. Example rulesets for one exemplary label, respectively, learned by the normal and the stacked BR approach. Attribute names in italic denote label attributes, attributes with an overline denote negated conditions.

structure on the topics. This also shows the potential of using rule learning in multilabel classification for reconstructing underlying hierarchies.

4.2 Prediction Performance

Tab. 3 shows the predictive performance of the different approaches. We compare BR, LP, Stacked BR with BR initialization and abstaining ($SBR_{BR/?}$) or predicting the default label ($SBR_{BR/d}$), respectively, in the case of the default rule firing, and lastly SBR with empty initialization and abstaining ($SBR_{?/?}$). For all approaches, we used the pruning version of JRip. Due to the space limit, we only report the results after the 10th bootstrapping iteration in the case of SBR.[2]

As expected, LP is the best approach w.r.t. subset accuracy. Somehow surprisingly, BR and both first SBRs obtain very similar avg. ranks, although the stacking of the label features is considered to particularly address the correct prediction of labelsets [5, 9, 13]. $SBR_{?/?}$ clearly suffers from the cold start problem when many label dependencies were encountered, best seen by the high precision but very low recall and subset accuracy obtained. BR is best for precision, but is always worse than $SBR_{BR/?}$ and $SBR_{BR/d}$ on recall,[3] which in general find the better trade-off between recall and precision, beating all other approaches on F1. Recall that BR's predictions are inputs for $SBR_{BR/?}$ and $SBR_{BR/d}$. Apparently, the additional iterations applying the stacked models allow labels which were initially missed to be found due to the label context.

5 Related Work

Many rule-based approaches to multilabel learning rely on association rules as those can have many conditions in the head. However, as the goal is classification, usually Classification Association Rules (CARs) are used, instead of regular association rules that would also find relations between instance-features. E.g., in Ávila et al. [2] a genetic algorithm is used to induce single-label association rules. A multilabel prediction is then built by using a combination of all covering rules of the BR rule sets. A good distribution of the labels is also ensures by using a token-based re-calculation of the fitness value of each rule. Li et al. [10] learn single-label association rules as well. For prediction, exactly those labels are set that have a probability greater than 0.5 in the covering rules.

A different idea is to introduce multi-label instead of single-label rules. Those are able to directly classify a multi-label-instance without the need to combine single-label rules [1]. Interestingly, the proposed rules also allow for postponing the classification by offering a "*don't care*"-value. The classification is then done by using a weighted voting scheme as many multilabel rules may cover the example.

Another multilabel rule algorithm is *MMAC* [15]. Here a multi-class, multilabel associative classification approach is used by not only generating from all frequent itemsets the rules that pass the confidence threshold but also include the second best rules and so on. Multilabel rules are then generated from these association rules by the frequent itemsets where covered instances are removed then. Rules with same conditions are then merged which enables a total ranking of all labels for each test instance.

---

[2] We found that more iterations consistently decrease subset acc. and recall, but increases precision and F1. However, the average absolute difference was consistently below 1%.

[3] Except of course for GENBASE, where all plain and stacked BR models are equal.

Table 3. Experimental performance on the seven datasets. The small number after the result indicates the rank of the particular approach. The last block shows the average over these ranks.

Approach	Subset Acc.	Precision	Recall	F1	Subset Acc.	Precision	Recall	F1
	SCENE				EMOTIONS			
BR	46.24% 2	68.82% 2	60.94% 4	64.55% 3	23.60% 3	65.54% 2	57.23% 3	60.97% 3
LP	58.33% 1	63.61% 4	61.09% 3	62.32% 4	20.56% 4	56.47% 5	55.66% 4	56.01% 4
$SBR_{BR/?}$	46.11% 3	65.56% 3	65.68% 2	65.58% 2	24.28% 2	64.02% 3	62.73% 2	63.24% 2
$SBR_{BR/d}$	45.32% 4	58.31% 5	77.79% 1	66.63% 1	24.96% 1	57.46% 4	75.54% 1	65.24% 1
$SBR_{?/?}$	29.13% 5	75.83% 1	33.28% 5	46.08% 5	9.09% 5	70.03% 1	21.97% 5	32.54% 5
	GENBASE				MEDICAL			
BR	96.83% 2.5	98.95% 2.5	98.42% 2.5	98.68% 2.5	66.96% 2	80.26% 2	84.29% 3	82.19% 1
LP	95.77% 5	97.30% 5	94.78% 5	95.99% 5	68.20% 1	80.18% 3	73.97% 4	76.93% 4
$SBR_{BR/?}$	96.83% 2.5	98.95% 2.5	98.42% 2.5	98.68% 2.5	66.86% 3	79.38% 4	84.78% 2	81.96% 2
$SBR_{BR/d}$	96.83% 2.5	98.95% 2.5	98.42% 2.5	98.68% 2.5	66.25% 4	78.21% 5	86.01% 1	81.89% 3
$SBR_{?/?}$	96.83% 2.5	98.95% 2.5	98.42% 2.5	98.68% 2.5	28.93% 5	82.39% 1	36.16% 5	50.13% 5
	ENRON				CAL500			
BR	9.17% 3	62.75% 1	49.09% 3	55.03% 2	0.00% 3	52.73% 1	24.88% 4	33.76% 2
LP	11.51% 1	41.06% 5	15.11% 4	22.08% 4	0.00% 3	31.90% 4	31.80% 1	31.84% 4
$SBR_{BR/?}$	9.17% 4	57.96% 2	55.09% 2	56.40% 1	0.00% 3	47.61% 2	30.90% 2	37.42% 1
$SBR_{BR/d}$	9.87% 2	43.13% 4	59.06% 1	49.71% 3	0.00% 3	44.76% 3	30.43% 3	36.20% 2
$SBR_{?/?}$	0.06% 5	53.10% 3	7.50% 5	13.08% 5	0.00% 3	26.48% 5	0.26% 5	0.51% 5
	YEAST				Average rank			
BR	9.18% 4	68.47% 1	55.33% 4	61.19% 3	2.79 3	1.64 1	3.36 3	2.50 3
LP	16.92% 1	60.04% 4	57.10% 3	58.52% 4	2.29 1	4.29 5	3.43 4	4.14 4
$SBR_{BR/?}$	10.18% 2.5	66.88% 2	57.63% 2	61.90% 2	2.86 4	2.64 3	2.07 2	1.79 1
$SBR_{BR/d}$	10.18% 2.5	58.31% 5	66.21% 1	61.98% 1	2.71 2	4.07 4	1.50 1	1.93 2
$SBR_{?/?}$	0.25% 5	65.35% 3	1.31% 5	2.56% 5	4.36 5	2.36 2	4.64 5	4.64 5

Other approaches are from the inductive logic programming field. Here, some also allow for having label features in the rule bodies, but due to the different nature disclosed by relational rules, these methods are not in the scope of this paper. In summary, label dependencies are not tackled explicitly though they might be taken into account by algorithm-specific properties. Please consider [11] for a more extensive discussion.

6 Future Challenges

All presented approaches for learning multilabel models, BR, LP and SBR decomposition, have one aspect in common, namely that they transform the original problem into several subproblems, which are then solved independently. This might be appropriate or even advantageous for certain use cases, for instance when the objective is to obtain isolated theories representing each label (cf. concept learning), or w.r.t. efficiency. But often it is more desirable to obtain one global theory comprehensively explaining a particular multilabel dataset. The induction of one global model also allows a better control over the objective loss, an important issue in multilabel classification due to the variety of existing measures, resulting directly from the diversity of the real life scenarios.

Regarding the introduced stacked BR approach which we used for learning label-dependent single-label rules, we propose to integrate the stacking of label features directly into the SeCo induction process. The idea is to start with unset label features, consequently only label-independent rules will be learnt in the beginning. However, the covered rules are not separated, but labeled accordingly and readded to the training set. Hence, we would get rid of the cold start and deadlock problem and no bootstrapping or sampling would be necessary.

Multiple labels in the head allow for representing co-occurrence relationships. In addition, only label-dependent multi-label rules allow to express all types of possible dependencies. The solution using LP can learn multilabel head rules, but with the mentioned shortcomings (Sec. 2.1). Therefore, we propose the following modifications.

In order to obtain one single global theory, we learn so-called multiclass decision lists, which allow to use different heads in consecutive rules of the decision list. If we limit ourselves to labelsets seen in the training data, this corresponds to using LP transformation with a multiclass decision list learner. However, the evaluation for each possible labelset can be very expensive ($\mathcal{O}(2^n)$ in the worst case). The following greedy approach may solve this. It starts by evaluating the condition candidates w.r.t. to each label independently in order to determine the best covered label. Having selected the best covered label for the given rule body, we can only stay the same or get worse if we now add an additional label to our head, since the number of covered examples remain the same and the number of covered *positives*, for which the head applies, cannot increase. Hence, depending on the heuristic used, we can safely prune great part of the label combinations by exploiting the anti-monotonicity of the heuristic.

Challenges to both proposed extensions, and to the self-evident combination of both, concern the rule learning process itself: The right selection of the heuristic was already a complex issue in traditional rule induction and has to be reviewed for multilabel learning. Furthermore, using unordered and multiclass decision lists gain new relevance, too.

We plan to use our method in order to analyze the datasets, and further benchmark datasets commonly used in the literature, in more detail. Regarding prediction quality, we expect to improve our performance by adopting the extensions presented in Sec. 3. An extended empirical study with additional state-of-the-art algorithms would reveal any development and allow further comparisons.

7 Conclusions

In this work, we introduced a simple yet effective approach to making label dependencies explicit with the means of rules. The proposed stacking approach is able to induce rules with labels as conditions in the bodies of the rules. In our analyses on seven multilabel datasets, the resulting models turned out to be indeed very useful in order to discover interesting aspects a normal rule learner is unable to uncover. For instance, we found out that the GENBASE dataset exhibits only very weak label dependencies, if any at all, despite the fact that it is frequently used for evaluating multilabel algorithms. In contrast to other approaches, the proposed method naturally allows for discovering and expressing local as well as global label dependencies.

The second part of the evaluation showed that our approach works particularly well for trading-off recall and precision, obtaining the best result w.r.t. F-measure. For subset accuracy, it is beaten by LP, which is particularly tailored towards this measure. However, the introduced technique of bootstrapping predictions still requires the initial input of a plain BR. Therefore, we presented two different but combinable directions for learning global theories as future challenges in the field of multilabel rule learning.

References

[1] Allamanis, M., Tzima, F.A., Mitkas, P.A.: Effective Rule-Based Multi-label Classification with Learning Classifier Systems. In: Tomassini, M., Antonioni, A., Daolio, F., Buesser, P. (eds.) ICANNGA 2013. LNCS, vol. 7824, pp. 466–476. Springer, Heidelberg (2013)

[2] Ávila-Jiménez, J.L., Gibaja, E., Ventura, S.: Evolving Multi-label Classification Rules with Gene Expression Programming: A Preliminary Study. In: Corchado, E., Graña Romay, M., Manhaes Savio, A. (eds.) HAIS 2010, Part II. LNCS, vol. 6077, pp. 9–16. Springer, Heidelberg (2010)

[3] Cohen, W.W.: Fast Effective Rule Induction. In: Proceedings of the 12th International Conference on Machine Learning (ICML 1995), pp. 115–123 (1995)

[4] Malerba, D., Semeraro, G., Esposito, F.: A multistrategy approach to learning multiple dependent concepts. In: Machine Learning and Statistics: The Interface, ch. 4, pp. 87–106 (1997)

[5] Dembczyński, K., Waegeman, W., Cheng, W., Hüllermeier, E.: On label dependence and loss minimization in multi-label classification. Machine Learning 88(1-2), 5–45 (2012)

[6] Fürnkranz, J.: Separate-and-Conquer Rule Learning. Artificial Intelligence Review 13(1), 3–54 (1999)

[7] Ghamrawi, N., McCallum, A.: Collective multi-label classification. In: CIKM 2005: Proceedings of the 14th ACM International Conference on Information and Knowledge Management, pp. 195–200. ACM (2005)

[8] Godbole, S., Sarawagi, S.: Discriminative methods for multi-labeled classification. In: Dai, H., Srikant, R., Zhang, C. (eds.) PAKDD 2004. LNCS (LNAI), vol. 3056, pp. 22–30. Springer, Heidelberg (2004)

[9] Guo, Y., Gu, S.: Multi-label classification using conditional dependency networks. In: Proceedings of the Twenty-Second International Joint Conference on Artificial Intelligence, IJCAI 2011, vol. 2, pp. 1300–1305. AAAI Press (2011)

[10] Li, B., Li, H., Wu, M., Li, P.: Multi-label Classification based on Association Rules with Application to Scene Classification. In: Proceedings of the 2008 the 9th International Conference for Young Computer Scientists, pp. 36–41. IEEE Computer Society (2008)

[11] Loza Mencía, E., Janssen, F.: Towards multilabel rule learning. In: Proceedings of the German Workshop on Lernen, Wissen, Adaptivität - LWA 2013, pp. 155–158 (2013)

[12] McCallum, A.K.: Multi-label text classification with a mixture model trained by EM. In: AAAI 1999 Workshop on Text Learning (1999)

[13] Montañés, E., Senge, R., Barranquero, J., Quevedo, J.R., del Coz, J.J., Hüllermeier, E.: Dependent binary relevance models for multi-label classification. Pattern Recognition 47(3), 1494–1508 (2014)

[14] Read, J., Pfahringer, B., Holmes, G., Frank, E.: Classifier chains for multi-label classification. Machine Learning 85(3), 333–359 (2011)

[15] Thabtah, F., Cowling, P., Peng, Y.: MMAC: A New Multi-Class, Multi-Label Associative Classification Approach. In: Proceedings of the 4th IEEE ICDM, pp. 217–224 (2004)

[16] Tsoumakas, G., Katakis, I., Vlahavas, I.P.: Mining Multi-label Data. In: Data Mining and Knowledge Discovery Handbook, pp. 667–685 (2010)

[17] Zhu, S., Ji, X., Xu, W., Gong, Y.: Multi-labelled classification using maximum entropy method. In: SIGIR 2005: Proceedings of the 28th Annual International ACM SIGIR Conference on Research and Development in Information Retrieval, pp. 274–281. ACM (2005)

Selective Forgetting for Incremental Matrix Factorization in Recommender Systems

Pawel Matuszyk and Myra Spiliopoulou

Otto-von-Guericke-University Magdeburg,
Universitätsplatz 2,
D-39106 Magdeburg, Germany
{pawel.matuszyk,myra}@iti.cs.uni-magdeburg.de

Abstract. Recommender Systems are used to build models of users' preferences. Those models should reflect current state of the preferences at any timepoint. The preferences, however, are not static. They are subject to concept drift or even shift, as it is known from e.g. stream mining. They undergo permanent changes as the taste of users and perception of items change over time. Therefore, it is crucial to select the actual data for training models and to forget the outdated ones.

The problem of selective forgetting in recommender systems has not been addressed so far. Therefore, we propose two forgetting techniques for incremental matrix factorization and incorporate them into a stream recommender. We use a stream-based algorithm that adapts continuously to changes, so that forgetting techniques have an immediate effect on recommendations. We introduce a new evaluation protocol for recommender systems in a streaming environment and show that forgetting of outdated data increases the quality of recommendations substantially.

Keywords: Forgetting Techniques, Recommender Systems, Matrix Factorization, Sliding Window, Collaborative Filtering.

1 Introduction

Data sparsity in recommender systems is a known and thoroughly investigated problem. A huge number of users and items together with limited capabilities of one user to rate items result in a huge data space that is to a great extent empty. However, the opposite problem to the data sparsity has not been studied extensively yet. In this work we investigate, whether recommender systems suffer from too much information about selected users. Although, the most algorithms for recommender systems try to tackle the problem of an extreme data sparsity, we show that it is beneficial to forget some information and not consider it for training models any more. Seemingly, forgetting information exacerbates the problem of having not enough data. We show, however, that much of the old information does not reflect the current preferences of users and training models upon this information decreases the quality of recommendations.

Reasons for the information about users being outdated are manifold. Users' preferences are not static - they change over time. New items emerge frequently,

S. Džeroski et al. (Eds.): DS 2014, LNAI 8777, pp. 204–215, 2014.

depending on the application scenario, e.g. news in the internet. Also the perception of existing items changes due to external factors such as advertisement, marketing campaigns and events related to the items. The environment of a recommender system is dynamic. A recommender system that does not take those changes into account and does not adapt to them deteriorates in quality. Retraining of a model does not help, if the new model is again based on the outdated information. Consequently, the information that a recommender is trained upon has to be selected carefully and the outdated information should be forgotten.

Since a recommender should be able to adapt constantly to changes of the environment, ideally in the real time, in our work we use an incremental, stream-based recommender. It does not learn upon a batch of ratings, but it considers them as a stream, as it is known from e.g. stream mining. Incremental methods have the advantage of learning continuously as new ratings in the stream arrive and, therefore, are always up to date with the current data. The batch-based methods on the other hand use a predefined batch of ratings to train the model and are, after arrival of new ratings, constantly out of date. Our method that uses matrix factorization still requires a retraining of latent item factors. However, the latent user factors are kept up to date constantly between the retraining phases. Also, since the general perception of items changes slower than preferences of a single user, the retraining is not needed as frequently as in the case of batch learners. A further essential advantage of incremental methods is that they can adapt immediately as changes occur. Because an incremental recommender learns upon ratings as soon as they arrive, it can react to changes immediately. Hence, it can capture short term changes, whereas a batch learner has to wait for the next retraining phase to adapt to changes.

Gradual changes in users' preferences and changes in item perception speak in favour of forgetting the outdated information in recommender systems. This type of changes can be related to concept drift in stream mining. It describes slow and gradual changes. There is also a second type of changes called concept shift. These changes are sudden, abrupt and unpredictable. In recommender systems those changes can be related e.g. to situations, where multiple persons share an online account. If we consider an online shop scenario, a recommender would experience a concept shift, when the owner of an account buys items for a different person (e.g. presents). When recommending movies a person can be influenced by preferences of other people, which can be a short-lived, single phenomenon, but it also can be a permanent change. In both cases a successful recommender system should adapt to those changes. This can be achieved by forgetting the old outdated information and learning a model based on information that reflects the current user preferences more accurately. In summary the contribution of our work is threefold: 1) We propose two selective forgetting techniques for incremental matrix factorization. 2) We define a new evaluation protocol for stream-based recommender systems. 3) We show that forgetting selected ratings increases the quality of recommendations.

This paper is structured as follows. In section 2 we discuss related work we used in our method, stressing the differences to existing approaches. Section 3 explains our forgetting mechanisms. The experimental settings and evaluation protocol are described in Section 4. Our results are explained in Section 5. Finally, in section 6, we conclude our work and discuss open issues.

2   Related Work

Recommender systems gained in popularity in recent years. The most widely used category of recommender systems are collaborative filtering (CF) methods. An intuitive, item-based approach in CF has been published in 2001 [6]. Despite its simplicity this method based on neighbourhoods of items has shown to have a strong predictive power. In contrast to content-based recommenders, CF works only with user feedback and without any additional information about users or items. Those advantages as well as the ability to cope with extremely sparse data made CF a highly interesting category of algorithms among practitioners and researchers. Consequently, many extensions of those methods have been developed. A comprehensive survey on those methods can be found in [1].

In 2012 Vinagre and Jorge noticed the need for forgetting mechanisms in recommender systems and proposed forgetting techniques for neighbourhood-based methods [9]. They introduced two forgetting techniques: sliding window and fading factors, which are also often used in stream mining. They also considered a recommender system as a stream-based algorithm and used those two techniques to define which information was used for computing a similarity matrix. According to the sliding window technique only a predefined number of the most recent user sessions was used for calculating the similarity matrix making sure that only the newest user feedback is considered for training a model. Their second technique, fading factors, assigns lower weight to old data than to new ones and, thereby, diminishes the importance of potentially outdated information. In our method we also use the sliding window technique, there are, however, three fundamental differences to Vinagre and Jorge: 1) Our method has been designed for explicit feedback e.g. ratings, whereas the method in [9] was designed for positive-only feedback. 2) We propose forgetting strategies for matrix factorization algorithms as opposed to neighbourhood-based methods in [9]. 3) Vinagre and Jorge apply forgetting on a stream of sessions of *all users*. Our forgetting techniques are *user-specific* i.e. we consider ratings of one user as a stream and apply a sliding window *selectively* on it. Vinagre and Jorge have shown that non-incremental algorithms using forgetting have lower computational requirements without a significant reduction of the predictive power, when compared to the same kind of algorithms without forgetting.

Despite the popularity of the neighbourhood-based methods, the state-of-the-art algorithms for recommenders are matrix factorization algorithms. They became popular partially due to the Netflix competition, where they showed a superior predictive performance, competitive computational complexity and high extensibility. Koren et. al proposed a matrix factorization method based on gradient descent [3], [4], where the decomposition of the original rating matrix is computed iteratively by reducing prediction error on known ratings. In the method called "TimeSVD++" Koren et al. incorporated time aspects accounting for e.g. changes in user preferences. Their method, however, does not encompass any forgetting strategy i.e. it always uses all available ratings no matter, if they are still representative for users' preferences. Additionally, some of the changes of the environment of a recommender cannot be captured by time factors proposed by Koren et. al. To this category of changes belong the abrupt, non-predictable changes termed before as concept shift. Furthermore, the method by Koren et. al is not incremental, therefore it cannot adapt to changes in real time.

An iterative matrix factorization method has been developed by Takács et. al in [8]. They termed the method biased regularized incremental simultaneous matrix factorization (BRISMF). The basic variant of this method is also batch-based. Takács et. al, however, proposed an incremental variant of the algorithm that also uses stochastic gradient descent. In this variant the model can be adapted incrementally as new ratings arrive. The incremental updates are carried out by fixating the latent item factors and performing further iterations of the gradient descent on the user latent factors. This method still requires an initial training and an eventual retraining of the item factors, but the latent user factors remain always up to date. In our work we use the BRISMF algorithm and extend it by forgetting techniques.

3 Method

Our method encompasses forgetting techniques for incremental matrix factorization. We incorporated forgetting into the algorithm BRISMF by Takács et. al [8]. The method is general and can be applied to any matrix factorization algorithm based on stochastic gradient descent analogously. BRISMF is a batch learning algorithm, the authors, however, proposed an incremental extension for retraining user features (cf. Algorithm 2 in [8]). We adopted this extension to create a forgetting, stream-based recommender. Our recommender system still requires an initial training, which is the first of its two phases.

3.1 Two Phases of Our Method

Phase I - Initial Training creates latent user and items features using the basic BRISMF algorithm in its unchanged form [8]. It is a pre-phase for the actual stream-based training. In that phase the rating matrix R is decomposed into a product of two matrices $R \approx PQ$, where P is a latent matrix containing user features and Q contains latent item vectors. For calculating the decomposition stochastic gradient descent (SGD) is used, which requires setting some parameters that we introduce in the following together with the respective notation.

As an input SGD takes a training rating matrix R and iterates over ratings $r_{u,i}$ for all users u and items i. SGD performs multiple runs called epochs. We estimate the optimal number of epochs in the initial training phase and use it later in the second phase. The results of the initial phase are the matrices P

and Q. As p_u we denote hereafter the latent user vector from the matrix P. Analogously, q_i is a latent item vector from the matrix Q. Those latent matrices serve as input to our next phase. The vectors p_u and q_i are of dimensionality k, which is set exogenously. In each iteration of SGD within one epoch the latent features are adjusted by a value depending on the learning rate η according to the following formulas [8]:

$$p_{u,k} \leftarrow p_{u,k} + \eta \cdot (predictionError \cdot q_{i,k} - \lambda \cdot p_{u,k}) \tag{1}$$

$$q_{i,k} \leftarrow q_{i,k} + \eta \cdot (predictionError \cdot p_{u,k} - \lambda \cdot q_{i,k}) \tag{2}$$

To avoid overfitting long latent vectors are penalized by a regularization term controlled by the variable λ. As $\vec{r}_{u*}$ we denote a vector of all ratings provided by the user u. For further information on the initial algorithm we refer to [8].

Despite the incremental nature of SGD, this phase, as also the most of matrix factorization algorithms, is a batch algorithm, since it uses a whole training set at once and the evaluation is performed after the learning on the entire training set has been finished. In our second phase evaluation and learning take place incrementally same as e.g. in stream mining.

Phase II - Stream-Based Learning. After the initial training our algorithm changes into a streaming mode, which is its main mode. From this time point it adapts incrementally to new users' feedback and to potential concept drift or shift. Also the selective forgetting techniques are applied in this mode, where they can affect the recommendations immediately. Differently from batch learning, evaluation takes place iteratively before the learning of a new data instance, as it is known from stream mining under the name "prequential evaluation" [2]. We explain our evaluation settings more detailed in Section 4.

In Algorithm 1 is pseudo-code of our method, which is an extension and modification of the algorithm presented in [8]. This code is executed at arrival of a new rating, or after a predefined number n of ratings. A high value of n results in a higher performance in terms of computation time, but also in a slower adaptation to changes. A low n means that the model is updated frequently, but the computation time is higher. For our experiments we always use $n = 1$.

The inputs of the algorithm are results of the initial phase and parameters that we also defined in the previous subsection. When a new rating $r_{u,i}$ arrives, the algorithm first makes a prediction $\hat{r}_{u,i}$ for the rating, using the item and user latent vectors trained so far. The deviation between $\hat{r}_{u,i}$ and $r_{u,i}$ is then used to update an evaluation measure (cf. line 4 in Algorithm 1). It is crucial to perform an evaluation of the rating prediction first, before the algorithm uses the rating for updating the model. Otherwise the separation of the training and test datasets would be violated. In line 6 the new rating is added to the list of ratings provided by the user u. From this list we *remove* the outdated ratings using one of our forgetting strategies (cf. line 7). The forgetting strategies are described in Section 3.2. In the line 9 SGD starts on the newly arrived rating.

It uses the optimal number of epochs estimated in the initial training. Contrary to the initial phase, here only user latent features are updated. For updating the user features the SGD iterates over all ratings of the corresponding user that *remained* after a forgetting technique has been applied. For the update of each dimension k the formula in line 16 is used.

Algorithm 1 Incremental Learning with Forgetting

Input: $r_{u,i}, R, P, Q, \eta, k, \lambda$
1. $\vec{p_u} \leftarrow$ getLatentUserVector(P, u)
2. $\vec{q_i} \leftarrow$ getLatentItemVector(Q, i)
3. $\hat{r}_{u,i} = \vec{p_u} \cdot \vec{q_i}$ //predict a rating for $r_{u,i}$
4. evaluatePrequentially$(\hat{r}_{u,i}, r_{u,i})$ //update evaluation measures
5. $\vec{r}_{u*} \leftarrow$ getUserRatings(R, u)
6. $(\vec{r}_{u*})$.addRating$(r_{u,i})$
7. **applyForgetting**$(\vec{r}_{u*})$ **//old ratings removed**
8. $epoch = 0$
9. **while** $epoch < optimalNumberOfEpochs$ **do**
10. $epoch{+}{+};$ //for all retained ratings
11.   **for all** $r_{u,i}$ in $\vec{r}_{u*}$ **do**
12. $\vec{p_u} \leftarrow$ getLatentUserVector(P, u)
13. $\vec{q_i} \leftarrow$ getLatentItemVector(Q, i)
14. $predictionError = r_{u,i} - \vec{p_u} \cdot \vec{q_i}$
15.   **for all** latent dimensions $k \neq 1$ in $\vec{p_u}$ **do**
16. $p_{u,k} \leftarrow p_{u,k} + \eta \cdot (predictionError \cdot q_{i,k} - \lambda \cdot p_{u,k})$
17.   **end for**
18.   **end for**
19. **end while**

Our variant of the incremental BRISMF method has the same complexity as the original, incremental BRISMF. In terms of computation time, it performs even better, since the number of ratings that the SGD has to iterate over is lower due to our forgetting technique. The memory consumption of our method is, however, higher, since the forgetting is based on a sliding window (cf. Section 3.2) that has to be kept in the main memory.

3.2 Forgetting Techniques

Our two forgetting techniques are based on a sliding window over data instances i.e. in our case over ratings. Ratings that enter the window are incorporated into a model. Since the window has a fixed size, some data instances have to leave it, when new ones are incorporated. Ratings that leave the window are forgotten and their impact is removed from the model. The idea of sliding window has been used in numerous stream mining algorithms, especially in a stream-based classification e.g. in Hoeffding Trees. In stream mining the sliding window is, however, defined over the entire stream. This approach has also been chosen

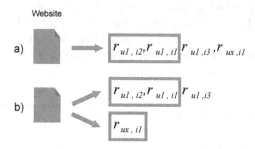

Fig. 1. Conventional definition of a sliding window a) vs. a *user-specific* window b). In case a) information on some users is forgotten entirely and no recommendations are possible (e.g. for user u_x). In case b) only users with too much information are affected (e.g. u_2). Ratings of new users, such as u_x, are retained.

Algorithm 2 applyForgetting($r_{u,*}$) - Instance-based Forgetting

---

Input: $r_{u,*}$ a list of ratings by user u sorted w.r.t. time, w - window size
1. **while** $|r_{u,*}| > w$ **do**
2.   removeFirstElement($r_{u,*}$)
3. **end while**

---

by Vinagre and Jorge in [9]. Our approach is user-specific i.e. a virtual sliding window is defined for each user separately. Figure 1 illustrates this difference.

On the left side of the figure there is a website that generates streams of ratings by different users. The upper part a) of the figure represents a conventional definition of a sliding window (blue frame) over an entire stream. In this case all ratings are considered as one stream. In our example with a window of size 2 this means that in case a) the model contains the ratings $r_{u1,i2}$ and $r_{u1,i1}$. All remaining ratings that left the window have been removed from the model. This also means that all ratings by the user u_x have been forgotten. Consequently, due to the cold start problem, no recommendations for that user can be created. Case b) represents our approach. Here each user has his/her own window. In this case all ratings of the user u_x are retained. Only users, who provided more ratings than the window can fit, are affected by the forgetting (e.g. u_1). User with very little information are retained entirely. Due to the user-specific forgetting the cold start problem is not exacerbated. The size of the window can be defined in multiple ways. We propose two implementations of the *applyForgetting()* function from Algorithm 1, but further definitions are also possible.

Instance-Based Forgetting. The pseudo code in Algorithm 2 represents a simple forgetting function based on the window size w. In Algorithm 1 new ratings are added into the list of user's ratings $r_{u,*}$. If due to that the window grows above the predefined size, the oldest rating is removed as many times as needed to reduce it back to the size w.

Time-Based Forgetting. In certain application scenarios it is reasonable to define current preferences with respect to time. For instance, we can assume that after a few years preferences of a user have changed. In very volatile applications a time span of one user session might be reasonable. Algorithm 3 implements a forgetting function that considers a threshold a for the age of user's feedback. In this implementation the complexity of forgetting is less than $O(w)$, where w is size of the window, since it does not require a scan over the entire window.

Algorithm 3 applyForgetting($r_{u,*}$) - Time-based Forgetting

---

Input: $r_{u,*}$ a list of ratings by user u sorted w.r.t. time, a - age threshold
1. forgettingApplied $\leftarrow$ true
2. **while** forgettingApplied $==$ true **do**
3.   *oldestElement* $\leftarrow$ getFirstElement($r_{u,*}$) //the oldest rating
4.   **if** *age(oldestElement)* $> a$ **then**
5.   removeFirstElement($r_{u,*}$)
6.   forgettingApplied $\leftarrow$ true
7.   **else**
8.   forgettingApplied $\leftarrow$ false
9.   **end if**
10. **end while**

---

4 Evaluation Setting

We propose a new evaluation protocol for recommender systems in a streaming environment. Since our method requires an initial training, the environment of our recommender is not entirely a streaming environment. The evaluation protocol should take the change from the batch mode (for initial training) into streaming mode (the actual method) into account.

4.1 Evaluation Protocol

Figure 2 visualizes two modes of our method and how a dataset is split between them. The initial training starts in a batch mode, which corresponds to the part 1) in the Figure (batch train). For this part we use 30% of the dataset. The ratios are example values we used in our experiments, but they can be adjusted to the idiosyncrasies of different datasets. The gradient descent used in the initial training iterates over instances of this dataset to adjust latent features. The adjustments made in one epoch of SGD are then evaluated on the batch test dataset (part 2). After evaluation of one epoch the algorithm decides, if further epochs are needed. After the initial phase is finished the latent features serve as input for the streaming mode.

For the stream based evaluation we use the setting proposed by Gama et al. called prequential evaluation [2]. In this setting ratings arrive sequentially in a

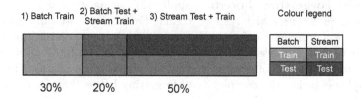

Fig. 2. Visualization of two modes of our method and split between the training and test datasets. The split ratios are example values.

stream. To keep the separation of a test and training dataset every rating is first predicted and the prediction is evaluated before it is used for training. This setting corresponds to part 3) of our Figure. Two different colours symbolize that this part is used both for training and evaluation. This also applies to part 2) of the figure. Since the latent features have been trained on part 1) and the streaming mode starts in part 3) this would mean a temporal gap in the training set. Since temporal aspects play a big role in forgetting we should avoid it. Therefore, we also train the latent features incrementally on part 2). Since this part has been used for evaluation of the batch mode already, we do not evaluate the incremental model on it. The incremental evaluation starts on part 3).

The incremental setting also poses an additional problem. In a stream new users can occur, for whom no latent features in the batch mode have been trained. In our experiments we excluded those users. The problem of inclusion of new users into a model is subject to our future work.

4.2 Evaluation Measure - *slidingRMSE*

A popular evaluation measure is the root mean squared error (RMSE), which is based on the deviation between a predicted and real rating [7]:

$$RMSE = \sqrt{\frac{1}{|T|} \sum_{(u,i)\in T} (r_{u,i} - \widehat{r}_{u,i})^2} \tag{3}$$

where T is a test set. This evaluation measure was developed for batch algorithms. It is a static measure that does not allow to investigate, how the performance of a model changes over time. We propose *slidingRMSE* - a modified version of RMSE that is more appropriate for evaluating stream recommenders. The formula for calculating *slidingRMSE* is the same as for RMSE, but the test set T is different. *slidingRMSE* is not calculated over the entire test set, but only over a sliding window of the last n instances. Prediction error of ratings that enter the sliding window are added to the squared sum of prediction errors and the ones that leave it are subtracted. The size of the window n is independent from the window size for forgetting techniques. A small n allows to capture short-lived effects, but it also reveals a high variance. A high value of n reduces the variance, but it also makes short-lived phenomena not visible. For our experiments we use $n = 500$. *slidingRMSE* can be calculated at any timepoint in a stream, therefore, is is possible to evaluate how RMSE changes over time.

Since we are interested in measuring how the forgetting techniques affect the prediction accuracy, we measure the performance of an algorithm with and without forgetting, so that the difference can be explained only by application of our forgetting techniques. Forgetting is applied only on a subset of users, who have sufficiently many ratings. Consequently, all other users are treated equally by both variants of the algorithm. Thus, we measure *slidingRMSE* only on those users, who were treated differently by the forgetting and non-forgetting variants.

5   Experiments

We performed our experiments on four real datasets: Movielens 1M[1], Movielens 100k, Netflix (a random sample of 1000 users) and Epinions (extended) [5]. The choice of datasets was limited by the requirement to have timestamped data. In all experiments we used our modified version of the BRISMF algorithm [8] with and without forgetting. Since BRISMF requires parameters to be set, on each dataset we performed a grid search over the parameter space to find approximately optimal parameters. In Fig. 3 we present the results of the best parameters found by the grid search. As an evaluation measure we used *slidingRMSE* (lower values are better). The left part of Fig. 3 represents the *slidingRMSE* over time. The red curves represent our method with forgetting technique denoted in the legend and the blue ones the method without forgetting. "Last20" stands for an instance-based forgetting, when only 20 last ratings of a user are retained. The best results were achieved constantly by the instance-based forgetting. Time-based forgetting also performed better than no forgetting on all datasets. However, we do not present its results due to space constraints.

The box plots on the right side are centred around the median of *slidingRMSE*. They visualize the distribution of *slidingRMSE*.Please, consider that box plots are normally used for visualizing independent observations, this is, however, not the case here. From Fig. 3 we see that our method with forgetting dominates the non-forgetting strategy on all datasets at nearly all timepoints. In Table 1 we present numeric, averaged values of *slidingRMSE* for each dataset.

Table 1. Average values of *slidingRMSE* for each dataset (lower values are better). Our forgetting strategy outperforms the non-forgetting strategy on all datasets.

Dataset	ML1M	ML100k	Epinions	Netflix
avg. slidingRMSE - Forgetting	**0.9151**	**1.0077**	**0.6627**	**0.9138**
avg. slidingRMSE - NO Forgetting	1.1059	1.0364	0.8991	1.0162

---

[1] http://www.movielens.org/

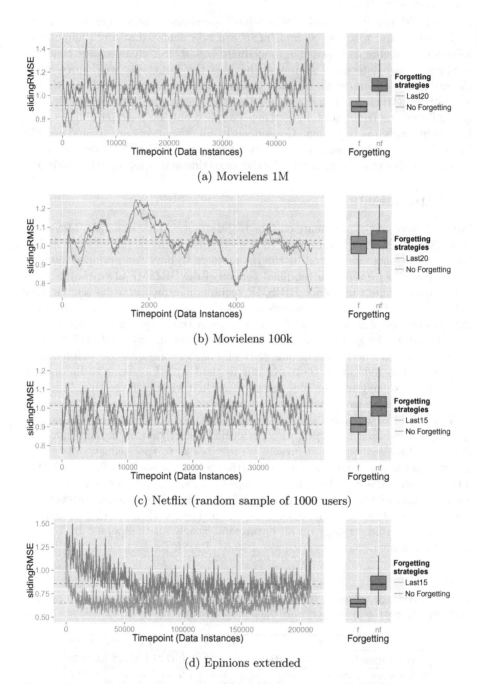

(a) Movielens 1M

(b) Movielens 100k

(c) Netflix (random sample of 1000 users)

(d) Epinions extended

Fig. 3. *SlidingRMSE* on four real datasets with and without forgetting (lower values are better). Application of forgetting techniques yields an improvement on all datasets at nearly all timepoints.

6 Conclusions

In this work we investigated, whether selective forgetting techniques for matrix factorization improve the quality of recommendations. We proposed two techniques, an instance-based and time-based forgetting, and incorporated them into a modified version of the BRISMF algorithm. In contrast to existing work, our approach is based on a user-specific sliding window and not on a window defined over an entire stream. This has an advantage of selectively forgetting information about users, who provided enough feedback.

We designed a new evaluation protocol for stream-based recommenders that also takes the initial training and temporal aspects into account. We introduced an evaluation measure, *slidingRMSE*, that is more appropriate for evaluating recommender systems over time and capturing also short-lived phenomena. In experiments on real datasets we have shown that a method that uses our forgetting techniques, outperforms the non-forgetting strategy on all datasets at nearly all timepoints. This also proves that user preferences and perception of items change over time. We have shown that it is beneficial to forget the outdated user feedback despite the extreme data sparsity known in recommenders.

In our future work we plan to develop more sophisticated forgetting strategies for recommender systems. Our immediate next step is also a research on a performant inclusion of new users into an existing, incremental model.

References

1. Desrosiers, C., Karypis, G.: A Comprehensive Survey of Neighborhood-based Recommendation Methods. In: Ricci, F., Rokach, L., Shapira, B., Kantor, P.B. (eds.) Recommender Systems Handbook, pp. 107–144. Springer US
2. Gama, J., Sebastião, R., Rodrigues, P.P.: Issues in evaluation of stream learning algorithms. In: KDD (2009)
3. Koren, Y.: Collaborative filtering with temporal dynamics. In: KDD (2009)
4. Koren, Y., Bell, R., Volinsky, C.: Matrix Factorization Techniques for Recommender Systems. Computer 42(8), 30–37 (2009)
5. Massa, P., Avesani, P.: Trust-aware bootstrapping of recommender systems. In: ECAI Workshop on Recommender Systems, pp. 29–33. Citeseer (2006)
6. Sarwar, B., Karypis, G., Konstan, J., Riedl, J.: Item-based collaborative filtering recommendation algorithms. In: WWW 2001 (2001)
7. Shani, G., Gunawardana, A.: Evaluating Recommendation Systems. In: Ricci, F., Rokach, L., Shapira, B., Kantor, P.B. (eds.) Recommender Systems Handbook
8. Takács, G., Pilászy, I., Németh, B., Tikk, D.: Scalable Collaborative Filtering Approaches for Large Recommender Systems. J. Mach. Learn. Res. 10 (2009)
9. Vinagre, J., Jorge, A.M.: Forgetting mechanisms for scalable collaborative filtering. Journal of the Brazilian Computer Society 18(4), 271–282 (2012)

Providing Concise Database Covers Instantly by Recursive Tile Sampling

Sandy Moens[1,2], Mario Boley[2,3], and Bart Goethals[1]

[1] University of Antwerp, Belgium
firstname.lastname@uantwerpen.be
[2] University of Bonn, Germany
firstname.lastname@uni-bonn.de
[3] Fraunhofer IAIS, Germany
firstname.lastname@iais.fgh.de

Abstract. Known pattern discovery algorithms for finding tilings (covers of 0/1-databases consisting of 1-rectangles) cannot be integrated in instant and interactive KD tools, because they do not satisfy at least one of two key requirements: a) to provide results within a short response time of only a few seconds and b) to return a concise set of patterns with only a few elements that nevertheless covers a large fraction of the input database. In this paper we present a novel randomized algorithm that works well under these requirements. It is based on the recursive application of a simple tile sample procedure that can be implemented efficiently using rejection sampling. While, as we analyse, the theoretical solution distribution can be weak in the worst case, the approach performs very well in practice and outperforms previous sampling as well as deterministic algorithms.

Keywords: Instant Pattern Mining, Sampling Closed Itemsets, Tiling Databases.

1 Introduction

Recently, data mining tools for interactive exploration of data have attracted increased research attention [7,9,11,19]. For such an interactive exploration process, a tight coupling between user and system is desired [4,17], i.e., the user should be allowed to pose and refine queries at any moment in time and the system should respond to these queries instantly [16]. This allows to transfer subjective knowledge and interestingness better as opposed to a batch setting with high computational overhead [4,17].

When finding collections of patterns that cover large fractions of the input data, existing techniques often fail to deliver the requirement of instant results. The reason is that they iteratively find the best pattern that covers the remaining data [10,13], which involves an NP-hard problem [10]. Another approach employed in the literature is first enumerating a large collection of patterns and then selecting distinct patterns that optimize the quality [20]. The large bottleneck for such procedures is the enumeration of many patterns.

S. Džeroski et al. (Eds.): DS 2014, LNAI 8777, pp. 216–227, 2014.
© Springer International Publishing Switzerland 2014

In this paper, we address the issue of finding patterns that have good descriptive capabilities, i.e., they individually describe a substantial amount of data, and that can together be used to describe the data. Such collections are helpful in exploratory data mining processes to get a quick overview of the prominent structures in the data [20]. Then, iteratively a user can drill down to specific parts of the data to explore even further. To this end, we propose a randomized procedure to quickly find small collections of patterns consisting of large tiles. Our method is based on a recursive sampling scheme that selects individual cells in a conditional database. The sampling process is based on a heuristic computation reflecting the potential of a cell for being part of a large pattern.

In summary, the contributions of our work are:

- We introduce a sampling method for finding database covers in binary data with near instants results in Section 3. Our sampling method heuristically optimizes the area of individual tiles using a recursive extension process.
- We introduce a new measure for evaluating pattern collections specifically in interactive systems which ensures the total representativeness of a pattern collection while guaranteeing the individual quality of patterns.
- We evaluate our novel sampling method with respect to the proposed measure in a real-time setting, in which algorithms are given only a short time budget of one second to produce results. We compare to state-of-the-art techniques and show that our method outperforms these techniques.

2   Preliminaries

In this paper we consider binary databases, as in itemset mining and formal concept analysis. A formal context is a triple $(O, A, \mathcal{R})$ with a set of **objects** O, a set of **attributes** A and a binary **relation** or database defined between the objects and attributes $\mathcal{R} \subseteq O \times A$. We use o_k and a_l as mappings to individual objects and attributes from the sets. Two Galois operators are defined as $O[X] = \{o \in O : \forall a \in X, (o, a) \in \mathcal{R}\}$ and $A[Y] = \{a \in A : \forall o \in Y, (o, a) \in \mathcal{R}\}$. $O[X]$ is also called the **cover** $cov(X)$. Applying both Galois operators sequentially, yields two closures operators, $\tilde{o}[.] = O[A[.]]$ and $\tilde{a}[.] = A[O[.]]$.

The binary relation is essentially a binary matrix $\{0, 1\}^{|O| \times |A|}$ such that a region consisting of only 1's is called a **tile** $\mathcal{T} = (X, Y)$ [10]. A tile can be adopted directly to transactional databases as an itemset X and its corresponding cover Y, such that $\forall a \in X, \forall o \in Y : (o, a) \in \mathcal{R}$. A tile is said to support a **cell** $(k, l) \in \mathcal{R}$ if $o_k \in Y$ and $a_l \in X$. In this work we are interested specifically in formal concepts (also maximal tiles or closed itemsets), such that $X = A[Y]$ and $Y = O[X]$. The interestingness of a tile is defined by its **area** in the data $area(\mathcal{T}) = |X| \cdot |Y|$. Note that the area of a tile is neither monotonic nor anti-monotonic. Hence, typical enumeration strategies [21] can not be used directly.

Geerts et al. [10] developed an algorithm for mining all tiles with at least a given area by adopting Eclat [21]. Using an upper bound on the maximum area of tiles that can still be generated, they prune single attributes during the mining process. Given a set of attributes X and a test attribute a, they count the number of objects $o \in O[X]$ such that $(o, a) \in \mathcal{R}$ and $|A[\{o\}]| \geq \ell$. Denote this count as $count_{\geq \ell}(X, a)$. The upper bound on the area of a tile with $|X| = s$ is given by $s \cdot count_{\geq s}$. The total **upper bound** over all sizes and possible tiles is obtained by taking the maximum:

$$UB_{X \cup \{a\}} = \arg \max_{\ell \in |X|+1,\ldots} (\ell \cdot count_{\geq \ell}(X, a)). \tag{1}$$

A database **tiling** is a collection of possibly overlapping tiles [10]. The problem statement we consider, is finding a tiling covering as much of the data as possible with only few patterns, in short time budgets. Therefore, each pattern should individually have large area. We stress that our setting is new [7,9,11] and existing measures do not satisfy these requirements. Given a collection of tiles $\mathcal{F}$, then the quality combines total collection and individual pattern qualities:

$$qual(\mathcal{F}) = \frac{cov(\mathcal{F})}{|\mathcal{R}|} \cdot \frac{1}{|\mathcal{F}|} \sum_{\mathcal{T} \in \mathcal{F}} \frac{area(\mathcal{T})}{|\mathcal{R}|}, \tag{2}$$

with $cov(\mathcal{F})$ the total number of cells covered. If known, for comparison purposes the normalization for area can be replaced by the area of the largest tile in the data $\mathcal{T}_{larg}$. Note that this measure indirectly favors small pattern collections.

3   Biased Sampling of Large Tiles

In this section we introduce a sampling procedure on individual cells from a conditional database. Our method heuristically optimizes the area of tiles.

3.1   Sampling Individual Cells

The upper bound from Equation (1) is good but intensive to compute. We propose to use a less intensive bound to guide the search for large area patterns. Consider a cell $(k, l) \in \mathcal{R}$, corresponding to object o_k and attribute a_l. If $(o_k, a_l) \in \mathcal{R}$, the maximum area of a tile having this cell is given by the **maximum upper bound** $MB(k, l) = |O[\{a_l\}]| \cdot |A[\{o_k\}]|$, where the first part is called the **column marginal** M^A, and the second part the **row marginal** M^O.

Proposition 1. *Given a cell (k, l) in a dataset, $MB(k, l)$ is never less than the true area of a tile containing the cell.*

Proof. Suppose $\mathcal{T} = (X, Y)$ contains cell (k, l) but has area greater than $MB(k, l)$. Then $\mathcal{T}$ either (1) contains object $o_i \notin O[\{a_l\}]$, or (2) contains attribute $a_j \notin A[\{o_k\}]$. Since the cover of $\mathcal{T}$ is an intersection relation over the covers of the attributes, (1) is not possible. Suppose (2) holds, then o_k will never be part of the cover of $\mathcal{T}$, which is in contrast with the original assumption.   □

We can use MB as probabilities for sampling individual cells from a binary database by using them as probabilities for sampling a single cell from the data:

$$\Pr(k, l) = \frac{MB(k, l)}{Z} = \frac{|O[\{a_l\}]| \cdot |A[\{o_k\}]|}{Z}. \tag{3}$$

Algorithm 1 RTS($\mathcal{R}, S$)

Require : relation $\mathcal{R}$, current state S
Return : collection of large tiles

1: $(i, j) \sim \Pr(k, l, S)$
2: $S \leftarrow (S^A \cup \{a_j\}, S^O \cup \{o_i\})$
3: $Tiles \leftarrow \{\mathcal{T}_V, \mathcal{T}_H\}$ // Closures of S
4: **if** $A(S^O) \setminus S^A \neq \phi$ and $O(S^A) \setminus S^O \neq \phi$ **then**
5:     $Tiles \leftarrow Tiles \cup RTS(\mathcal{R}, S)$
6: **end if**
7: **return** $Tiles$

The heuristic behind this sampling scheme is that a cell which is part of tiles with large area, also has a larger sampling probability. Using the Equation 3 we sample individual cells having higher chance of being in a large area tile.

3.2   Recursive Tile Sampling

Large tiles can be sampled by repeatedly sampling individual cells that consitute a tile. As such, instead of sampling cells independently, we restrict the sampling of new cells, to those that definitely form a tile. We do so by considering only the conditional database constructed by the previous samples.

We define a **current state** S by a set of current attributes S^A and a set of current objects S^O (in fact a current state is an intermediate tile). The sampling probabilities from Equation 3 can be updated to incorporate the previous knowledge reflected by the current state. The new probabilities become

$$\Pr(k, l, S) = \frac{|O[\{a_l\}] \cap O[S^A]| \cdot |A[\{o_k\}] \cap A[S^O]|}{Z}, \tag{4}$$

with Z a normalization constant over remaining cells. The intersections construct the conditional database $\mathcal{R}^S$. As we condition on the current state we know that an object in $O[S^A]$ or an attribute in $A[S^O]$ contains only ones. We therefore remove these cells from the conditional database.

Pseudo code for **Recursive Tile Sampling**, RTS, is given in Algorithm 1. The sampling procedure extends a current state in two directions simultaneously (see Figure 1). When either no attributes or no objects remain, the algorithm stops. Therefore, the current state S itself, never represents a rectangular tile. As such, each time when constructing the conditional database we report a pair of tiles defined by the Galois and closure operators on S: the tile containing all remaining attributes is $\mathcal{T}_H = (A[S^O], \tilde{o}[S^O])$ and the tile containing all remaining objects is $\mathcal{T}_V = (\tilde{a}[S^A], O[S^A])$. $\mathcal{T}_H$ is also called a horizontal extension tile and $\tilde{o}[S^O] \supseteq S^O$. Likewise, $\mathcal{T}_V$ is also called a vertical extension tile and $\tilde{a}[S^A] \supseteq S^A$. In the end we return all tiles that are found during the recursive steps.

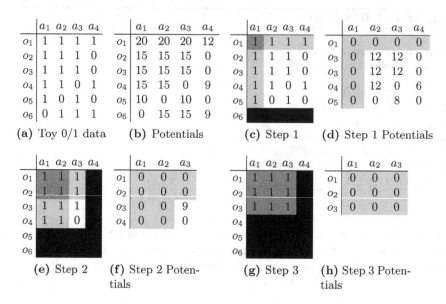

Fig. 1. Running example of RTS: dark gray = S, light gray = cells without sampling probability, black = cells that are not extensions of S

Example 1. Given the toy dataset from Figure 1a, the initial unnormalized sampling probabilities (or potentials) are given in Figure 1b. First cell (1,1) is sampled and object o_6 is removed because $(o_6, a_1) \notin \mathcal{R}$. Two extension tiles are formed by applying the closure operators: $\mathcal{T}_{H_1} = (\{a_1, a_2, a_3, a_4\}, \{o_1\})$ and $\mathcal{T}_{V_1} = (\{a_1\}, \{o_1, o_2, o_3, o_4, o_5\})$. The recursive step is applied to the database excluding attributes $a_i \notin A[\{o_1\}]$ and objects $o_j \notin O[\{a_1\}]$. In step 2, cell (2,2) is sampled and $\mathcal{T}_{H_2} = (\{a_1, a_2, a_3\}, \{o_1, o_2, o_3\})$ and $\mathcal{T}_{V_2} = (\{a_1, a_2\}, \{o_1, o_2, o_3, o_4\})$ are reported. In the last step, (3,3) is sampled and the last extension tiles are found: $\mathcal{T}_{H_3} = \mathcal{T}_{V_3} = (\{a_1, a_2, a_3\}, \{o_1, o_2, o_3, \})$. The process then stops as no recursion can be applied.

3.3 Efficient Sampling of Individual Cells

The distribution $f(k, l, S)$, $f(x)$ for simplicity, defined by Equation 4 is

$$f(x) = \begin{cases} MB(k, l, S)/Z & \text{if } (k, l) \in \mathcal{R}^S, t_k \notin S^O, i_l \notin S^A \\ 0 & \text{otherwise.} \end{cases} \quad (5)$$

We show how to sample efficiently from $f(x)$ without completely materializing it. We can use rejection sampling with non-uniform distributions for this purpose. Rejection sampling is a general method for sampling from a probability distribution f by sampling from an auxiliary distribution g that acts as an envelope. It uses the fact that we can sample uniformly from the density of the area under the curve $cg(x)$ [15], $c > 1$. Samples from cg that are generated outside of the density region of $f' = \beta f$ are then rejected, $\beta > 1$. Formally, we have $f'(x) < cg(x)$ and acceptance probability $\alpha \leq f'(x)/cg(x)$, with $\alpha \sim Unif[0,1]$.

To use rejection sampling we first set f' to the unnormalized version of the true distribution f, using the same condition $(i,j) \in \mathcal{R}$ and $\beta = Z$. As auxiliary distribution we choose $g(x) = (|O[\{a_l\}] \cap O[S^A]| \cdot |A[\{o_k\}] \cap A[S^O]|)/Z$, without the condition on the presence of (k,l). Setting $c = Z$ we obtain two rules

$$f'(x) = \begin{cases} cg(x) & \text{if } (k,l) \in \mathcal{R} \\ 0 \leq cg(x) & \text{if } (k,l) \notin \mathcal{R}, \end{cases} \tag{6}$$

such that by construction the acceptance probability boils down to accepting if the cell is present, rejecting if not.

What is left is obtaining samples from g. Since this distribution is based on the independence of rows and columns we sample a_l with probability $\Pr_A(l, S) = |O[\{a_l\}] \cap O[S^A]|/Z$ and o_k with probability $\Pr_O(k, S) = |A[\{o_k\}] \cap A[S^O]|/Z$ independently. This results in exact samples from g.

3.4 Incorporating Knowledge

Our sampling method can integrate prior knowledge by assigning weights to cells instead of marginal counts. Suppose a weight function $w : (k,l) \to [0,1]$, then we have marginals $M^O(o_k) = \sum_{a_i \in A[\{o_k\}]} w(k,i)$ and $M^A(a_l) = \sum_{o_j \in O[\{a_l\}]} w(j,l)$. and potentials $\Pr(k,l) = (\sum_{a_i \in A[\{o_k\}]} w(k,i) \cdot \sum_{o_j \in O[\{a_l\}]} w(j,l))/Z$, with Z a normalization constant to obtain a probability distribution. For the weights we propose the use of multiplicative weights [12] based on the number of times a cell has already been covered by previous knowledge, i.e., $w(k,l) = \gamma^m$, with m the number of times a cell has been covered and γ a discounting factor.

The main bottleneck of this weighting scheme is the computation of individual marginals. When explicitly storing the previous marginal, we can efficiently update them. Suppose for each of the attributes and objects we keep the current marginals in memory. Given that new knowledge is provided in the form of a tile $\mathcal{T}_n = (X, Y)$, then only marginals of attributes and objects supporting cells described by $\mathcal{T}_n$ have to be updated. Using the following scheme over the individual cells of the tile we obtain fast incorporation of knowledge:

$$\forall a_l \in X, \forall o_k \in Y$$
$$\to$$
$$M^A(a_l) = M^A(a_l) - w_{old}(k,l) + w_{new}(k,l)$$
$$M^O(o_k) = M^O(o_k) - w_{old}(k,l) + w_{new}(k,l).$$

This results in updates in $\mathcal{O}(2|X||Y|)$ time rather than an update for the complete database in $\mathcal{O}(2|A||O|)$ time. The factor 2 comes from the fact that we have to update the margins for attributes as well as for objects.

3.5 Worst Case Analysis

Individual Cell Sampling. We analyze the worst case performance of our sampling probabilities wrt $area^2$, which is the distribution that is closest to the distribution simulated by our sampling method. We define the worst case scenario as a dataset where the sampling probabilities do not reflect the $area^2$ of sampled closed tiles. In our framework this happens when $\{0,1\}^{n \times n}$ contains all ones except on the diagonal. Then, the number of closed tiles equals $2^n - 2$ and the largest tile in the region has area $\lceil n/2 \rceil \lfloor n/2 \rfloor$. The total sampling potential is $n(n-1)^3$, because each row and column has a count of $(n-1)$ and there are $n(n-1)$ non-zero entries. The true $area^2$ mass induced by all closed tiles equals $\sum_{k=1,\cdots,n-1} \binom{n}{k}(k(n-k-1))^2 \geq n(n-1)^3$ and holds for all $n \geq 2$. Moreover for $n \geq 4$ it holds that the second parts is strictly smaller and we have a constant undersampling of this density region.

Rejection Sampling. We first analyze the general sampling complexity and then the worst-case time complexity when many samples are rejected.

Using Section 3.3 we sample independently one column and one row. The materialization of the marginal distributions has $\mathcal{O}(|A| + |O|)$ time and space complexity. (A naïve direct approach for sampling relies on full materialization of the matrix and therefore has time and space complexity $\mathcal{O}(|A||O|)$) For the time complexity we also have to take into account the time for sampling one element. This can be achieved in logarithmic time using a binary seach. The total sampling time with rejection sampling becomes $\mathcal{O}((|A| + \log|A|) + (|O| + \log|O|))$ compared to $\mathcal{O}(|A||O|) + (\log|A| \log|O|)$ for the direct approach.

This does not yet conclude our time analysis as we did not yet take into account the number of times we have to resample due to rejections. We use as basis the worst-case scenario for rejection sampling, which is the setting where the binary representation of the data is an identity matrix I_n of size $n \times n$. The marginal probabilities equal $1/n$ and the probability of sampling a single 1-valued cell equals $1/n^2$. Since the data contains exactly n ones, the total probability of sampling a valid cell with rejection sampling, and thus accepting the sample, equals $n \cdot 1/n^2 = 1/n$. Setting $|\mathcal{R}|$ to n we obtain a total time sampling complexity for rejection sampling equal to $\mathcal{O}((|A| + |O|) + (\log|A| + \log|O|) \cdot |\mathcal{R}|)$. Note that the first term is the time for materializing the distribution and has to be done just once. The last part is the time for sampling the distribution, which has to be repeated in worst case $|\mathcal{R}|$ times.

4 Experiments

We experimented with our sampling method on several real world datasets shown in Table 1 together with their main characteristics. Pumsb, Connect and Acc are publicly available from the FIMI Repository [2]. Adult, Cens-Inc, CovType and Pokhand are made available at the UCI Machine Learning Repository [1].

For the experiments our interests are two-fold:
- How well does RTS work in real instant conditions?
- How does the quality of patterns evolve over time?

Table 1. Characteristics for different datasets, quality of pattern collections and the number of patterns in the collection obtained in 1 second

$\mathcal{R}$	$|O|$	$|A|$	RTS quality	$|\mathcal{F}|$	$CDPS_{area*fq^2}$ quality	$|\mathcal{F}|$	Asso quality	$|\mathcal{F}|$	Tileminer$_{first}$ quality	$|\mathcal{F}|$	Tileminer$_{unif}$ quality	$|\mathcal{F}|$
Adult	48,842	97	**0.64**	25.0	0.40	24.8	0.58	7.0	0.55	25.0	0.61	25.0
Pumsb	49,046	2.113	**0.31**	25.0	0.00	1.10	0.00	0.0	0.00	0.0	0.00	0.0
Connect	67,557	129	**0.47**	25.0	0.09	25.0	0.00	0.0	0.00	25.0	0.00	25.0
Cens-Inc	199,523	519	**0.35**	12.7	0.03	25.0	0.00	0.0	0.04	25.0	0.04	25.0
Acc	340,183	468	**0.18**	10.5	0.00	0.3	0.00	0.0	0.00	0.0	0.00	0.0
CovType	581,012	5.858	0.28	3.0	0.33	24.5	0.00	0.0	**0.45**	14.0	**0.45**	14.0
Pokhand	1,000,000	95	**0.09**	9.2	0.02	5.1	0.00	0.0	0.00	0.0	0.00	0.0

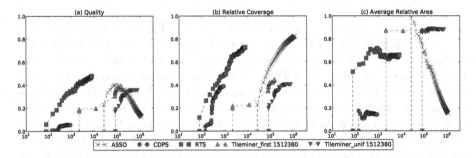

Fig. 2. Quality of 100 patterns over time in millisecond (log scale) for Cens-Inc

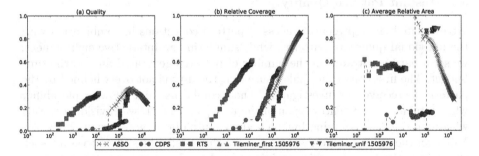

Fig. 3. Quality of 100 patterns over time in millisecond (log scale) for Acc

The first topic is related to intensive interactive environments: a user is exploring data and is assisted by several pattern mining algorithms. The user hereby expects instant condense results. The second topic is related also to an interactive environment with more relaxed conditions. For instance, instead of having instant results, a user is willing to wait up to 1 minute. Due to space limitations, we do not show our evaluation for the incorporation of knowledge.

For the experiments we compared our method to three other techniques for finding tiles. As deterministic baseline we used a boolean matrix factorization algorithm called Asso [13], which greedily optimizes the coverage of k basis

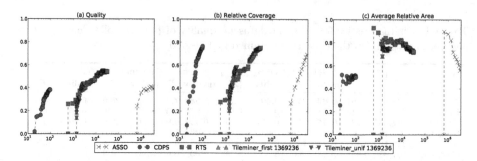

Fig. 4. Quality of 100 patterns over time in millisecond (log scale) for CovType

vectors. Asso is implemented in C++. We used a Tileminer implementation that finds all tiles given a minimum area, which can be seen as a baseline for finding large area tiles. Tileminer is also written in C++. We use Controlled Direct Pattern Sampling [8] (CDPS), written in Java, as alternative sampling method and as baseline for a fast anytime sampling technique. CDPS provides i.i.d. samples in rapid succession from a user customizable distribution over the complete pattern space. At last, RTS[1] is implemented in Java. In our experiments we reported a single pattern for each recursive run, which is the largest area pattern produced. The tests themselves are executed on our local server running Ubuntu 12.04. The machine contains 32 Intel Xeon CPUs and 32 GB of RAM.

4.1   Instant Pattern Quality

We evaluated the representativeness of pattern collections in combination with the individual quality of patterns, while simulating an interactive environment: we set a hard constraint on the number of patterns generated and on the time budget. The first constraint makes sure that the user is not overwhelmed by the patterns he receives for investigation. The second constraint allows explorability while retaining the attention span of the user [16]. In fact, we envision a setting where a user clicks a mining button and wants to obtain good results instantly. Moreover the mining can only start when the user tells the system to start, such that knowledge from a previous mining round can be taken into account.

We used a hard time budget of 1 second for all datasets and reported the top 25 results obtained. Throughout the experiments we did not take into account the loading times because in interactive environments the data is already in memory. For Asso we did not allow to find fault-tolerant covers and set the penalty to 100. Since Asso is greedy, the first 25 results are the top 25. For Tileminer we made a sweep over the parameter space, varying the area threshold from 5% to 95% of $area(\mathcal{T}_{larg})$ (area of largest tile) with a 5% step increase. Moreover, we used two settings: Tileminer$_{first}$, uses the first k patterns obtained and Tileminer$_{unif}$, selects k patterns uniformly at random from all tiles found in one second. For CDPS we used several area distribution settings [8]. However, in the results

---

[1] Implementation can be found at http://adrem.ua.ac.be/rts

we show only CDPS$_{area*fq^2}$ which is the setting that performed best overall. We selected the first 25 non duplicates. For the samplers we made 10 runs and averaged over the results. Each collection was scored wrt Equation (2). For comparison, we normalized the *areas* over $area(\mathcal{T}_{larg})$ rather than $|\mathcal{R}|$.

The qualities of pattern collections are shown in Table 1. It shows per dataset and for each of the algorithms the quality and the number of patterns obtained when running for at most 1 second and selecting 25 top patterns. The experiments show that RTS performs good on our quality measure and even outperforms other algorithms on 6 out of 7 datasets. Only for the sparse dataset CovType it is not able to produce enough patterns to beat the other methods. Moreover, it is the only method that produces at least one pattern within one second. Aside from quality this also is very important in our experiment because even while retrieving just one pattern the knowledge of the user is influenced! The user can then incorporate this knowledge when exploring the data further. It is clear that RTS outperforms all other methods on this experiments.

4.2   Pattern Quality over Time

We evaluate the quality using larger time budgets and show the evolving quality over time. In this scenario all algorithms are given at most one hour to produce 100 patterns. For Tileminer we used the same parameter selection procedure as in Section 4.1. We selected the setting giving the highest overall total coverage for the first 100 patterns and generated a random selection of 100 patterns for that complete collection of patterns. Evolving scores for quality, relative coverage and average relative area are shown in Figures 2, 3 and 4 for a selection of datasets. The graphs show elapsed time in milliseconds in log scale on the x-axis and the respective qualities, with *areas* normalized by $area(\mathcal{T}_{larg})$ rather than $|\mathcal{R}|$.

The results show similar behaviour except again on CovType, a sparse dataset with a low number of high area patterns. Generally, the experiments show that Asso is good at providing large covers while maximizing new information. However, the mining process often takes long and due to optimization of uncovered parts, the area of tiles reduces drastically after the first pattern. Tileminer is good at providing multiple large area patterns but lacks the ability to cover the data with the patterns found. This is due to the enumeration strategy. CDPS is not able to outperform other methods due to i.i.d sampling of $area * fq^2$, which does not optimize enough the area. It is however one of the fastest methods. RTS is the only method that is quick and robust in terms of quality. For all datasets it is at least an order of magnitude faster than Asso with only slightly lower total coverage. It is possible to obtain better coverage by also incorporating previous samples (Section 3.4). This, however, negatively influences the score.

5   Related Work

We describe some related research to this paper. We cite two types of research material, work based on finding database covers and work based on sampling.

Geerts et al. [10] adopted Eclat to mine non fault-tolerant (FT) large area tiles, i.e., tiles with no false positives (0 values) in the result. They also describe how to produce tilings, by recursively finding the largest tile in the data excluding previous tiles. Vreeken et al. [18] optimize the Minimum Description Length in pattern collections. As such, they maximize the number of times patterns are used when encoding the data, rather than maximizing the total coverage.

In the FT setting multiple algorithms exist that try to find database covers. These methods allow false positives in their patterns. Miettinen et al. [13] use boolean matrix factorization in Asso, to find products of matrices that reconstruct the complete data with a low number of errors. Asso can be adopted to mine non FT patterns in binary data. Xiang et al. [20] use the concept of hyperrectangles, which is similar to hierarchical tiles. They find hyperrectangles by combining frequent itemsets with a given budget of false positives.

Sampling the output space is relatively new in pattern set mining. Existing algorithms enumerate a large fraction of the pattern space and filter a selection of patterns a posteriori. Not many techniques exist that sample the output space directly. Al Hasan and Zaki [3] coined the term output space sampling on graphs. They used a simple random walk on the partially ordered graph (POG), by allowing only extensions of the current POG. Moens and Goethals [14] used the same technique to sample maximal itemsets. They use an objective functions to prune the search space and to propose new transitions for the random walk.

Boley et al. [6] introduced two-step sampling procedures for the discovery of patterns following a target distribution. In step one a single data object is materialized and in step two a pattern is sampled from the object. Boley et al. [8] extended this into a general framework, where a user specifies a full distribution in terms of frequency factors over specific parts of the data. This framework can not optimize enough the area of patterns to find good covers. Boley [5] uses Metropolis-Hastings to uniformly sample closed itemsets, however he does not incorporate the area of patterns for biasing the results towards large area tiles.

6   Conclusion and Future Work

Interactive KD tools demanding short response times and concise pattern collections are becoming increasingly popular. Existing techniques for finding large covers in 0/1 databases fail in at least one of the two requirements and can therefore not be integrated properly in such frameworks. We presented a novel technique for sampling large database covers given a real-time situation of interactive data mining with very short time budgets. We showed how our method can be implemented efficiently using rejection sampling. Moreover, we showed that our technique outperforms existing techniques for finding large database covers given very short time. For larger time budgets we showed that our method obtains comparable results to greedy optimizations, yet, much faster.

Interesting future work on this topic relates to tiles that are enumerated in one recursive step. In the current implementation we select only the largest tile in the collection. Another technique is to maintain an evolving list of top-k tiles using for instance reservoir sampling. Other reseach directions for this technique are the FT setting and probabilistic databases.

References

1. Uci machine learning repository, http://archive.ics.uci.edu/ml/
2. Frequent itemset mining dataset repository (2004), http://fimi.ua.ac.be/data
3. Al Hasan, M., Zaki, M.J.: Output space sampling for graph patterns. In: Proc. VLDB Endow, pp. 730–741 (2009)
4. Blumenstock, A., Hipp, J., Kempe, S., Lanquillon, C., Wirth, R.: Interactivity closes the gap. In: Proc. of the KDD Workshop on Data Min. for Business Applications, Philadelphia, USA (2006)
5. Boley, M.: The Efficient Discovery of Interesting Closed Pattern Collections. PhD thesis (2011)
6. Boley, M., Lucchese, C., Paurat, D., Gärtner, T.: Direct local pattern sampling by efficient two–step random procedures. In: Proc. ACM SIGKDD (2011)
7. Boley, M., Mampaey, M., Kang, B., Tokmakov, P., Wrobel, S.: One click mining: Interactive local pattern discovery through implicit preference and performance learning. In: IDEA 2013 Workshop in Proc. ACM SIGKDD, pp. 27–35. ACM (2013)
8. Boley, M., Moens, S., Gärtner, T.: Linear space direct pattern sampling using coupling from the past. In: Proc. ACM SIGKDD, pp. 69–77. ACM (2012)
9. Dzyuba, V., van Leeuwen, M.: Interactive discovery of interesting subgroup sets., pp. 150–161 (2013)
10. Geerts, F., Goethals, B., Mielikäinen, T.: Tiling databases. In: Suzuki, E., Arikawa, S. (eds.) DS 2004. LNCS (LNAI), vol. 3245, pp. 278–289. Springer, Heidelberg (2004)
11. Goethals, B., Moens, S., Vreeken, J.: Mime: a framework for interactive visual pattern mining. In: Proc. ACM SIGKDD, pp. 757–760. ACM (2011)
12. Lavrač, N., Kavšek, B., Flach, P., Todorovski, L.: Subgroup discovery with cn2-sd. J. Mach. Learn. Res, 153–188 (2004)
13. Miettinen, P., Mielikäinen, T., Gionis, A., Das, G., Mannila, H.: The discrete basis problem. IEEE Trans. on Knowl. and Data Eng., 1348–1362 (2008)
14. Moens, S., Goethals, B.: Randomly sampling maximal itemsets. In: IDEA 2013 Workshop in Proc. ACM SIGKDD (2013)
15. Neal, R.M.: Slice sampling. In: Ann. Statist., pp. 705–767 (2003)
16. Ng, R.T., Lakshmanan, L.V., Han, J., Pang, A.: Exploratory mining and pruning optimizations of constrained association rules. ACM SIGMOD Record, 13–24 (1998)
17. van Leeuwen, M.: Interactive data exploration using pattern mining. In: Holzinger, A., Jurisica, I. (eds.) Interactive Knowledge Discovery and Data Mining. LNCS, vol. 8401, pp. 169–182. Springer, Heidelberg (2014)
18. Vreeken, J., van Leeuwen, M., Siebes, A.: Krimp: Mining itemsets that compress. Data Min. Knowl. Discov., 169–214 (2011)
19. Škrabal, R., Šimůnek, M., Vojíř, S., Hazucha, A., Marek, T., Chudán, D., Kliegr, T.: Association rule mining following the web search paradigm. In: Flach, P.A., De Bie, T., Cristianini, N. (eds.) ECML PKDD 2012, Part II. LNCS, vol. 7524, pp. 808–811. Springer, Heidelberg (2012)
20. Xiang, Y., Jin, R., Fuhry, D., Dragan, F.F.: Summarizing transactional databases with overlapped hyperrectangles. Data Min. Knowl. Discov, 215–251 (2011)
21. Zaki, M.J., Parthasarathy, S., Ogihara, M., Li, W.: Parallel algorithms for discovery of association rules. Data Min. Knowl. Discov., 343–373 (1997)

Resampling-Based Framework for Estimating Node Centrality of Large Social Network

Kouzou Ohara[1], Kazumi Saito[2], Masahiro Kimura[3], and Hiroshi Motoda[4]

[1] Department of Integrated Information Technology, Aoyama Gakuin University, Japan
ohara@it.aoyama.ac.jp
[2] School of Administration and Informatics, University of Shizuoka, Japan
k-saito@u-shizuoka-ken.ac.jp
[3] Department of Electronics and Informatics, Ryukoku University, Japan
kimura@rins.ryukoku.ac.jp
[4] Institute of Scientific and Industrial Research, Osaka University, Japan
School of Computing and Information Systems, University of Tasmania, Australia
motoda@ar.sanken.osaka-u.ac.jp

Abstract. We address a problem of efficiently estimating value of a centrality measure for a node in a large social network only using a partial network generated by sampling nodes from the entire network. To this end, we propose a resampling-based framework to estimate the approximation error defined as the difference between the true and the estimated values of the centrality. We experimentally evaluate the fundamental performance of the proposed framework using the closeness and betweenness centralities on three real world networks, and show that it allows us to estimate the approximation error more tightly and more precisely with the confidence level of 95% even for a small partial network compared with the standard error traditionally used, and that we could potentially identify top nodes and possibly rank them in a given centrality measure with high confidence level only from a small partial network.

Keywords: Error estimation, resampling, node centrality, social network analysis.

1 Introduction

Recently, Social Media such as Facebook, Digg, Twitter, Weblog, Wiki, etc. becomes increasingly popular on a worldwide scale, and allows us to construct large-scale social networks in cyberspace. An article that is posted on social media can rapidly and widely spread through such networks and can be shared by a large number of people. Since such information can substantially affect our thought and decision making, a large number of studies have been made by researchers in many different disciplines such as sociology, psychology, economy, and computer science [8,4] to analyze various aspects of social networks and information diffusion on them.

In the domain of social network analysis, several measures called centrality have been proposed so far [7,5,1,3,13]. They characterize nodes in a network based on its structure, and give an insight into network performance. For example, a centrality provides us with the information of how important each node is through node ranking

S. Džeroski et al. (Eds.): DS 2014, LNAI 8777, pp. 228–239, 2014.

derived directly from the centrality. It also provides us with topological features of a network. For example, scale free property is derived from the degree distribution. As a social network in World Wide Web easily grows in size, it is becoming pressingly important that we are able to efficiently compute values of a centrality to analyze such a large social network. However, if a centrality measure is based not only on local structure around a target node, e.g. its neighboring nodes, but also on global structure of a network, e.g. paths between arbitrary node pairs, its computation becomes harder as the size of the network increases. Thus, it is crucial to reduce the computational cost of such centralities for large social networks. Typical examples are the closeness and the betweenness centralities which we consider in this paper (explained later).

It is worth noting that such a centrality is usually defined as a summarized value of more primitive ones that are derived from node pairs in a network. For example, the closeness centrality is defined as the average of the shortest path lengths from a target node to each of the remaining nodes in a network. Considering this fact, it is inevitable to employ a sampling-based approach as a possible solution of this kind of problem on scalability. It is obvious that using only a limited number of nodes randomly sampled from a large social network can reduce the computational cost. However, the resulting value is an approximation of its true value, and thus it becomes important to accurately estimate the approximation error. It is well known from the statistical view point that the margin of error (difference between sample mean and population mean) is $\pm 2 \times \sigma / \sqrt{N}$ with the confidence level of 95%, where σ and N are the standard deviation of a population and the number of samples, respectively. However, this traditional boundary does not necessarily give us a tight approximation error.

In this paper, we propose a framework that provides us with a tighter error estimate of how close the approximation is to the true value. The basic idea is that we consider all possible partial networks of a fixed size that are generated by resampling nodes according to a given coverage ratio, and then estimate the approximation error, referred to as *resampling error*, using centrality values derived from those partial networks. We test our framework using two well-known centrality measures, the closeness and the betweenness centralities, both of which require to use the global structure of a network for computing the value of each node. Extensive experiments were performed on three real world social networks varying the sample coverage for each centrality measure. We empirically confirmed that the proposed framework is more promising than the traditional error bound in that it enables us to give a tighter approximation error with a higher confidence level than the traditional one under a given sampling ratio. The framework we proposed is not specific to computation of node centralities for social network analysis. It is very generic and is applicable to any other estimation problems that require aggregation of many (but a finite number of) primitive computations.

The paper is organized as follows. Section 2 gives the formal definitions of both the resampling-based framework that we propose and the traditional bound of approximation error. Section 3 explains the closeness and the betweenness centralities we used to evaluate our framework and presents how to estimate their approximation error. Section 4 reports experimental results for these centralities on three real world networks. Section 5 concludes this paper and addresses the future work.

2 Related Work

As mentioned above, it is crucial to employ a sampling-based approach when analyzing a large social network. Many kinds of sampling methods have been investigated and proposed so far [6,11,10]. Non-uniform sampling techniques give higher probabilities to be selected to specific nodes such as high-degree ones. Similarly, results by traversal/walk-based sampling are biased towards high-degree nodes. In our problem setting the goal is to accurately estimate centralities of an original network and thus uniform sampling that selects nodes of a given network uniformly at random is essential because biased samples might skew centrality values derived from a resulting network.

This motivates us to propose the framework that ensures the accuracy of the approximations of centrality values under uniform sampling. Although we use a simple random sampling here, our framework can adopt a more sophisticated technique such as MH-sampling [6] in so far as it falls under uniform sampling. In this sense, our framework can be regarded as a meta-level method that is applicable to any uniform sampling technique.

3 Resampling-Based Estimation Framework

For a given set of objects S whose number of elements is $L = |S|$, and a function f which calculates some associated value of each object, we first consider a general problem of estimating the average value μ of the set of entire values $\{f(s) \mid s \in S\}$ only from its arbitrary subset of partial values $\{f(t) \mid t \in T \subset S\}$. For a subset T whose number of elements is $N = |T|$, we denote its partial average value by $\mu(T) = (1/N)\sum_{t \in T} f(t)$. Below, we formally derive an expected estimation error $RE(N)$ which is the difference between the average value μ and the partial average value $\mu(T)$, with respect to the number of elements N. Hereafter, the estimated error based on $RE(N)$ is referred to as resampling error.

Now, let $\mathcal{T} \subset 2^S$ be a family of subsets of S whose number of elements is N, that is, $|T| = N$ for $T \in \mathcal{T}$. Then, we obtain the following estimation formula for the expected error:

$$RE(N) = \sqrt{\langle(\mu - \mu(T))^2\rangle_{T \in \mathcal{T}}}$$

$$= \sqrt{\binom{L}{N}^{-1} \sum_{T \in \mathcal{T}} \left(\mu - \frac{1}{N}\sum_{t \in T} f(t)\right)^2} = \sqrt{\binom{L}{N}^{-1} \frac{1}{N^2} \sum_{T \in \mathcal{T}} \left(\sum_{t \in T}(f(t) - \mu)\right)^2}$$

$$= \sqrt{\binom{L}{N}^{-1} \frac{1}{N^2} \left(\binom{L-1}{N-1}\sum_{s \in S}(f(s) - \mu)^2 + \binom{L-2}{N-2}\sum_{s \in S}\sum_{t \in S, t \neq s}(f(s) - \mu)(f(t) - \mu)\right)}$$

$$= \sqrt{\binom{L}{N}^{-1} \frac{1}{N^2} \left(\left(\binom{L-1}{N-1} - \binom{L-2}{N-2}\right)\sum_{s \in S}(f(s) - \mu)^2 + \binom{L-2}{N-2}\left(\sum_{s \in S}(f(s) - \mu)\right)^2\right)}$$

$$= C(N)\sigma. \tag{1}$$

Here the factor $C(N)$ and the standard deviation σ are given as follows:

$$C(N) = \sqrt{\frac{L-N}{(L-1)N}}, \quad \sigma = \sqrt{\frac{1}{L}\sum_{s \in S}(f(s) - \mu)^2}.$$

In this paper we consider a huge social network consisting of millions of nodes as a collection of a large number of objects, and propose a framework in which we use the partial average value as an approximate solution with an adequate confidence level using the above estimation formula, Equation (1). More specifically, we claim that for a given subset T whose number of elements is N, and its partial average value $\mu(T)$, the probability that $|\mu(T) - \mu|$ is larger than $2 \times RE(N)$, is less than 5%. This is because the estimation error of Equation (1) is regarded as the standard deviation with respect to the number of elements N. Hereafter this framework is referred to as the resampling estimation framework.

In order to confirm the effectiveness of the proposed resampling estimation framework, we also consider a standard approach based on the i.i.d. (independently identical distribution) assumption for comparison purpose. More specifically, for a given subset T whose number of elements is N, we assume that each element $t \in T$ is independently selected according to some distribution $p(t)$ such as an empirical distribution $p(t) = 1/L$. Then, by expressing elements of T as $T = \{t_1, \cdots, t_N\}$, we obtain the following estimation formula for the expected error:

$$SE(N) = \sqrt{\langle(\mu - \mu(T))^2\rangle}$$

$$= \sqrt{\sum_{t_1 \in S} \cdots \sum_{t_N \in S}\left(\mu - \frac{1}{N}\sum_{n=1}^{N}f(t_n)\right)^2 \prod_{n=1}^{N}p(t_n)} = \sqrt{\frac{1}{N^2}\sum_{t_1 \in S} \cdots \sum_{t_N \in S}\left(\sum_{n=1}^{N}(f(t_n) - \mu)\right)^2 \prod_{n=1}^{N}p(t_n)}$$

$$= \sqrt{\frac{1}{N^2}\sum_{t_1 \in S} \cdots \sum_{t_N \in S}\left(\sum_{n=1}^{N}(f(t_n) - \mu)^2 + \sum_{n=1}^{N}\sum_{m=1,m\neq n}^{N}(f(t_n) - \mu)(f(t_m) - \mu)\right)\prod_{n=1}^{N}p(t_n)}$$

$$= D(N)\sigma, \tag{2}$$

where $D(N) = 1/\sqrt{N}$ and σ is the standard deviation. Hereafter, the estimated error based on $SE(N)$ is referred to as standard error. The difference between Equations (1) and (2) is only their coefficients, $C(N)$ and $D(N)$. We can easily see that $C(N) \leq D(N)$, $C(L) = 0$ and $D(L) \neq 0$. For more details, we empirically compare these resampling error $RE(N)$ and standard error $SE(N)$ through experiments on node centrality calculation of social networks as described below. Note that the standard deviation σ is needed in both Equations (1) and (2). We are assuming that $|S|$ is too large to compute σ. Otherwise, sampling is not needed. We can use, instead of σ, the standard deviation σ' that is derived from a subset S' ($\subset S$) such that $|S'| = L'$ is small enough to compute σ' within a reasonable time.

4   Application to Node Centrality Estimation

We investigate our proposed resampling framework on node centrality estimation of a social network represented by a directed graph $G = (V, E)$, where V and E ($\subset V \times V$)

are the sets of all the nodes and the links in the network, respectively. When there is a link (u, v) from node u to node v, u is called a *parent node* of v and v is called a *child node* of u. For any node $v \in V$, let $A(u)$ and $B(v)$ respectively denote the set of all child nodes of u and the set of all parent nodes of v in G, i.e., $A(u) = \{v \in V; (u, v) \in E\}$ and $B(v) = \{u \in V; (u, v) \in E\}$.

4.1 Closeness Centrality Estimation

The closeness $cls_G(u)$ of a node u on a graph G is defined as

$$cls_G(u) = \frac{1}{(|V| - 1)} \sum_{v \in V, v \neq u} \frac{1}{spl_G(u, v)}, \tag{3}$$

where $spl_G(u, v)$ stands for the shortest path length from u to v in G. Namely, the closeness of a node u becomes high when a large number of nodes are reachable from u within relatively short path lengths. Here note that we set $spl_G(u, v) = \infty$ when node v is not reachable from node u on G. Thus, in order to naturally cope with this infinite path length, we employ the inverse of the harmonic average as shown in Equation (3).

The burning algorithm [12] is a standard technique for computing $cls_G(u)$ of each node $u \in V$. More specifically, after initializing a node subset X_0 to $X_0 \leftarrow \{u\}$, and path length d to $d \leftarrow 0$, this algorithm repeatedly calculates a set X_{d+1} of newly reachable nodes from X_d and set $d \leftarrow d + 1$ unless X_d is empty. Here, newly reachable nodes from X_{d-1} is defined by $X_d = (\bigcup_{v \in X_{d-1}} A(v)) \setminus (\bigcup_{c < d} X_c)$. Then the shortest path length of node $v \in X_d$ from u is obtained as $spl_G(u, v) = d$. Here recall that $spl_G(u, v) = \infty$ if v is not reachable from u. Since the computational complexity of computing $cls_G(u)$ for each node $u \in V$ become $O(|E|)$, it takes a large amount of computation time for a huge social networks consisting of millions of nodes.

Now, we present a method for computing $cls_G(u)$ of each node $u \in V$ under our resampling estimation framework. The method first constructs the reverse network of $G = (V, E)$ by reversing the direction of each link from (u, v) to (v, u). Namely, the reverse network is defined by $H = (V, F)$ and $F = \{(v, u) \mid (u, v) \in E\}$. Then, by using the burning algorithm starting from node v over the reverse network, we can calculate each shortest path length from v to u as $spl_H(v, u)$. Clearly, $spl_H(v, u)$ is the shortest path length from node u to v, i.e., $spl_G(u, v)$. Namely, for each node $u \in V$, by setting $S_u = V \setminus \{u\}$ and $f_u(v) = spl_H(v, u)$, we can calculate partial average value from an arbitrary subset $T \subset S_u \cup \{u\}$. Here note that, due to the nature of the burning algorithm, we can obtain such partial average value simultaneously for all nodes $u \in V$.

4.2 Betweenness Centrality Estimation

The betweenness $btw_G(u)$ of a node u on a graph G is defined as

$$btw_G(u) = \frac{1}{(|V| - 1)(|V| - 2)} \sum_{v \in V, v \neq u} \left(\sum_{w \in V, w \neq u, w \neq v} \frac{nsp_G(v, w; u)}{nsp_G(v, w)} \right), \tag{4}$$

where $nsp_G(v, w)$ is the number of the shortest paths from v to w in G and $nsp_G(v, w; u)$ is the number of the shortest paths from v to w in G that passes through node u. Namely, the betweenness of a node u becomes high when a large number of shortest paths between two nodes pass through node u. Here note that although $cls_G(u)$ and $cls_H(u)$ is not generally equal, since any node pair (v, w) is examined in Equation (4) we can easily see that $btw_G(u) = btw_H(u)$.

The Brandes algorithm [2] is a standard technique for computing $btw_G(u)$ of each node $u \in V$. The algorithm utilizes a series of node subsets $(X_0, \cdots, X_D)$ produced by the burning algorithm described in Section 4.1 starting from node $v \in V$, where D stands for the maximum burning step. Then, after setting $nsp_G(v, w) \leftarrow 1$ for $w \in X_1$, the algorithm in turn computes $nsp_G(v, w) \leftarrow \sum_{x \in B(w) \cap X_{d-1}} nsp_G(v, x)$ for $w \in X_d$ from $d = 2$ to D. Next, we define the following betweenness $btw_G(u; v)$ of node u, which restricts its starting node to v,

$$btw_G(u; v) = \sum_{w \in V, w \neq u, w \neq v} \frac{nsp_G(v, w; u)}{nsp_G(v, w)}. \tag{5}$$

Then, after setting $btw_G(u; v) \leftarrow 0$ for $u \in X_D$, the algorithm in turn computes $btw_G(u; v)$ $\leftarrow \sum_{x \in A(u) \cap X_{d+1}} (nsp_G(v, u)/nsp_G(v, x))(1 + btw_G(u; x))$ for $u \in X_d$ from $d = D - 1$ to 2. Finally, by computing and summing $btw_G(u; v)$ by changing the starting node v, we can obtain the betweenness $btw_G(u)$ of each node $u \in V$. Again, the computational complexity of computing $btw_G(u)$ for each node $u \in V$ become $O(|E|)$.

Now, we present a method based on the Brandes algorithm for computing $btw_G(u)$ of each node $u \in V$ under our resampling estimation framework. Namely, for each node $u \in V$, by setting $S_u = V \setminus \{u\}$ and $f_u(v) = btw_G(u; v)/(|V| - 2)$, we can calculate partial average value from an arbitrary subset $T \subset S_u \cup \{u\}$. Again note that, due to the nature of the Brandes algorithm, we can obtain such partial average value simultaneously for all nodes $u \in V$.

5 Experiments

5.1 Datasets

To experimentally evaluate the methods proposed in the previous sections, we employed three datasets of real networks, where all networks are represented as directed graphs. The first one is a reader network extracted from a Japanese blog service site "Ameba"[1], in which each blog can have a list of reader links. A reader link is directional and a link is constructed from blog u to blog v if blog v registers blog u as her favorite one. We crawled the lists of $117, 374$ blogs of "Ameba" in June 2006, and extracted a large connected network that has $56, 604$ nodes and $734, 737$ directed links. We refer to this network as the Ameblo network. The second one is a network extracted from "@cosme"[2], a Japanese word-of-mouth communication site for cosmetics, in which each user page can have *fan links*. A fan link (u, v) means that user v registers user u as her favorite user. We traced up to ten steps in the fan-link network from a randomly chosen user in December 2009, and extracted a large connected network consisting

---

[1] http://www.ameba.jp/
[2] http://www.cosme.net/

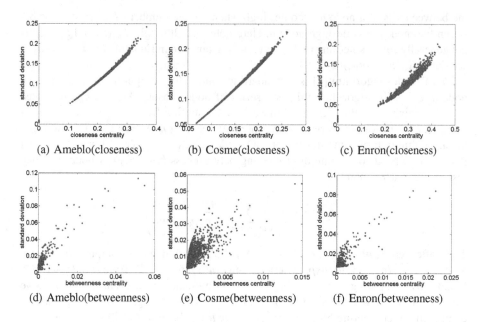

Fig. 1. Results for "centrality value vs. standard deviation"

of 45,024 nodes and 351,299 directed links. We refer to this directed network as the Cosme network. The last one is a network derived from the Enron Email Dataset [9], in which an email address that appears in the dataset as either a sender or a recipient is regarded as a node and two email addresses u and v are linked by a directional link (u, v) if u sent an email to v. We refer to this directed network as the Enron network, which has 19,603 nodes and 210,950 links. These three networks are not very huge, *i.e.*, networks with millions of nodes. We dare chose them to investigate the basic performance from various angles.

5.2 Statistical Analysis

For each of the three real networks, $G = (V, E)$, we first computed the value of the closeness centrality $cls_G(u)$ and betweenness centrality $btw_G(u)$ of each node $u \in V$ by means of the algorithms presented in Sections 4.1 and 4.2, respectively. In addition, we investigated their standard deviations given by

$$\sigma_{cls}(u) = \sqrt{\frac{1}{|V| - 1} \sum_{v \in V, v \neq u} \left(\frac{1}{spl_G(u, v)} - cls_G(u)\right)^2}$$

for the closeness centrality, and

$$\sigma_{btw}(u) = \sqrt{\frac{1}{|V| - 1} \sum_{v \in V, v \neq u} \left(\frac{btw_G(u; v)}{|V| - 2} - btw_G(u)\right)^2}$$

for the betweenness centrality. Figures 1(a) to 1(c) plot the pair $(cls_G(u), \sigma_{cls}(u))$ for the Ameblo, Cosme, and Enron networks, and Figs. 1(d) to 1(f) plot the pair $(btw_G(u), \sigma_{btw}(u))$ for the same three networks. In each figure, the horizontal and vertical axes indicate the values of corresponding centrality, $cls_G(u)$ or $btw_G(u)$, and its standard deviation, $\sigma_{cls}(u)$ or $\sigma_{btw}(u)$, respectively.

We can observe that there exists positive correlation between the centrality value of each node and its standard deviation. This tendency can be found more clearly in the results for the closeness centrality compared to the results for the betweenness centrality in which nodes are scattered over a larger area. It is noted that, for every network, higher-ranked nodes in each centrality measure are distinguishable from each other because of their distinctive values of the centrality, while it looks hard to do the same for lower-ranked nodes. This implies that there is a possibility that we can detect a cluster of such high ranked nodes or estimate their ranking with a high confidence level only using a smaller partial network if we can secure a tight approximation error.

5.3   Results

In this section, we evaluated the fundamental performance of the resampling error $RE(N)$, *i.e.*, how tightly and accurately it estimates the approximation error, using the closeness and betweenness centralities on the three networks. To this end, we considered a problem of estimating $\mu_G(u)$, the true value of a centrality measure for node u in network $G(V, E)$ using its partial network G' generated by sampling N nodes from V, and empirically investigated whether or not the estimation $\mu_{G'}(u)$, the partial average derived from G', falls within the range of $\mu_G(u) \pm 2 \times RE(N)$. Here, $\mu_G(u)$ stands for either $cls_G(u)$ or $btw_G(u)$. In addition, we considered the range of $\mu_G(u) \pm 2 \times SE(N)$ for comparison.

Figures 2 and 3 show the results for the closeness and betweenness centralities, respectively. In this experiment, we considered the top and second nodes in each network that respectively have the largest and second largest true values of the corresponding centrality in Fig. 1. In each figure, the horizontal axis "coverage" means the ratio of the number of sampled nodes N to the total number of nodes L, *i.e.*, N/L, in each network, while the vertical axis means the value of the centrality, and how the estimated value fluctuates as a function of the coverage is depicted. We conducted five independent trials for each of these two nodes in each network, and plotted estimated values $\mu_{G'}(u)$ for a given coverage N/L with green jagged lines. The red horizontal center line in each figure presents the true value of the centrality $\mu_G(u)$ for node u, while the red broken and blue chain lines show the ranges of $\mu_G(u) \pm 2 \times RE(N)$ and $\mu_G(u) \pm 2 \times SE(N)$, respectively.

From these results, we can confirm that the boundary determined by $RE(N)$ estimates the approximation error more tightly and converges to 0.0 as the coverage approaches 1.0, while the boundary by $SE(N)$ is looser and does not converge to 0.0 even if the coverage becomes 1.0. Furthermore, in most cases, the estimated value falls within the range of $\mu_G(u) \pm 2 \times RE(N)$ for every network regardless of the centrality used. From these results, we can say that the resampling error $RE(N)$ provides us with a better error bound with the confidence level of 95% compared to the standard error $SE(N)$. Besides, it is found that in Fig. 2(d) the value of the upper-bound of the range given

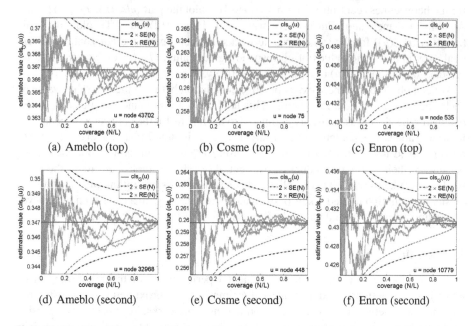

Fig. 2. Fluctuation of the estimated value of the closeness centrality as a function of the coverage for the top and second nodes that respectively have the largest and second largest true values of the centrality in the Ameblo, Cosme, and Enron networks

by $RE(N)$ is approximately 0.3505 when the coverage is 0.2, and it is smaller than the corresponding value of the lower-bound of the range given by $RE(N)$ in Fig. 2(a), which is approximately 0.3628. These observations enable us to decide that in the Ameblo network the value of the closeness centrality of node 43702 (the top node) is higher than the value of node 32968 (the second node) with the confidence level of 95% only from the results obtained under the coverage of 0.2 because their error bounds derived from $RE(N)$ for the confidence level of 95% do not overlap each other. The same holds for the results of the Ameblo network in Fig. 3 although the coverage must be slightly larger in this case. This may not necessarily generalize to other networks, but it suggests that we could potentially detect top-K nodes and possibly their ranking in a given centrality measure with such a high confidence level even under a small coverage.

Next, we quantitatively confirmed the accuracy of the proposed resampling error in Fig. 4, in which it is shown how the difference $\delta(N)$ between the true approximation error and the estimated error fluctuates as a function of coverage in the same fashion as in Figs. 2 and 3. Here, we computed RMSE (Root Mean Squared Error) by conducting $R = 1,000$ independent trials for each value of N, which is defined as follows:

$$\varepsilon_{RMSE}(N) = \sqrt{\frac{1}{R}\sum_{r=1}^{R}(\mu_{G',r}(u) - \mu_G(u))^2},$$

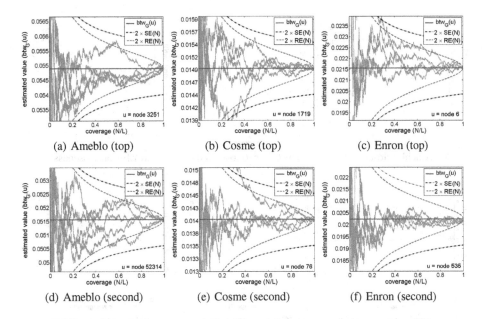

Fig. 3. Fluctuation of the estimated value of the betweenness centrality as a function of the coverage for the top and second nodes that respectively have the largest and second largest true values of the centrality in the Ameblo, Cosme, and Enron networks

where $\mu_{G',r}(u)$ denotes the estimated value of the centrality of node u for partial graph G' in the r-th trial. Then, we used $\mathcal{E}_{RMSE}(N)$ as the true approximation error, and $RE(N)$ and $SE(N)$ as the estimated error.

Namely, in Fig. 4, the difference $\delta(N)$ is defined as $RE(N) - \mathcal{E}_{RMSE}(N)$ for the resampling error (the red curves), and $SE(N) - \mathcal{E}_{RMSE}(N)$ for the standard error (the blue broken curves). Here, we only show the results for the top node of each network in both centralities and omit the results for the second node because the tendency observed for the second node was quite similar to the one for the top nodes.

From these results, we can observe that the difference fluctuates when the value of coverage is less than 0.2 in both cases of $RE(N)$ and $SE(N)$, but for a larger coverage it becomes remarkably stable and almost equal to 0.0 in the case of $RE(N)$, while it increases as the value of coverage becomes larger in the case of $SE(N)$. This tendency is common to every network regardless of the centrality used. These results show that the proposed resampling error can precisely estimate the approximation error from the true values of a centrality measure if the coverage is larger than a certain threshold, say 0.2, while the standard error tends to overestimate the true approximation error.

Consequently, we can say that the resampling error we proposed is more promising than the standard error in this kind of estimation problem, and can give a tighter and more precise estimate of the approximation error with high confidence level than the standard error does.

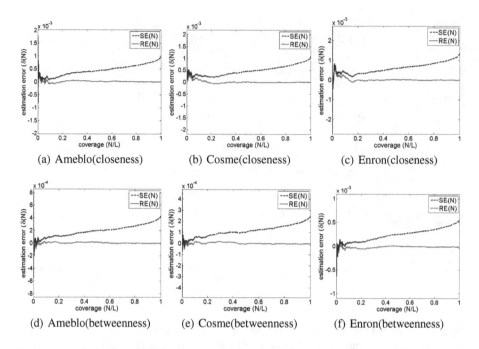

Fig. 4. Fluctuation of the difference between the true and the estimated approximation errors as a function of the coverage for the top node that has the largest true value of the centrality in the Ameblo, Cosme, and Enron networks.

6   Conclusion

In this paper, we addressed a problem of estimating the value of a centrality measure for a node in a social network. Centrality measure plays an important role in social network analysis since it characterizes nodes in a network and its values indicate the importance of nodes in some respects. Thus, it is crucial to efficiently calculate the value of a centrality measure for each node, but its computation could be intractable for those centrality measures that require use of a global network structure for their computation when the network becomes very large. It is inevitable to take a sampling-based approach to deal with the scalability problem, in which we approximate the true value of a centrality only from a partial network that can be generated by sampling nodes from the whole network. What is important is that we ensure the accuracy of the approximations without knowing the truth. To this end, we proposed a resampling-based framework to estimate the approximation error of the estimated values of a centrality measure for each node. We have conducted extensive experiments on three real world networks varying the coverage ratio of nodes to be sampled, and evaluated the proposed framework by comparing it with the standard error known in statistics using two typical centrality measures, the closeness and betweenness centralities. We empirically confirmed that the proposed framework enables us to estimate the approximation error more tightly and more precisely with the confidence level of 95% even for a partial network whose coverage is small, say 0.2, than using the standard error estimate. Furthermore, the experimental results suggest that we could potentially estimate top-K nodes for a small K, say 10, and possibly their ranking in a given centrality measure with high confidence level only from a small partial network. It is noted that the framework we proposed is not specific to computation of centrality measures. Indeed, it is very generic and applicable to any other estimation problems that require aggregation of many (but a finite number of) primitive computations. We believe that the conclusion obtained in this paper can generalize but we have yet to test out the proposed framework in a broader setting and also in different domains, too.

Acknowledgments. This work was partly supported by Asian Office of Aerospace Research and Development, Air Force Office of Scientific Research under Grant No. AOARD-13-4042, and JSPS Grant-in-Aid for Scientific Research (C) (No. 26330261).

References

1. Bonacichi, P.: Power and centrality: A family of measures. Amer. J. Sociol. 92, 1170–1182 (1987)
2. Brandes, U.: A faster algorithm for betweenness centrality. Journal of Mathematical Sociology 25, 163–177 (2001)
3. Brin, S., Page, L.: The anatomy of a large-scale hypertextual web search engine. Computer Networks and ISDN Systems 30, 107–117 (1998)
4. Chen, W., Lakshmanan, L., Castillo, C.: Information and influence propagation in social networks. Synthesis Lectures on Data Management 5(4), 1–177 (2013)
5. Freeman, L.: Centrality in social networks: Conceptual clarification. Social Networks 1, 215–239 (1979)
6. Henzinger, M.R., Heydon, A., Mitzenmacher, M., Najork, M.: On near-uniform url sampling. The International Journal of Computer and Telecommunications Networking 33(1-6), 295–308 (2000)
7. Katz, L.: A new status index derived from sociometric analysis. Sociometry 18, 39–43 (1953)
8. Kleinberg, J.: The convergence of social and technological networks. Communications of ACM 51(11), 66–72 (2008)
9. Klimt, B., Yang, Y.: The enron corpus: A new dataset for email classification research. In: Boulicaut, J.-F., Esposito, F., Giannotti, F., Pedreschi, D. (eds.) ECML 2004. LNCS (LNAI), vol. 3201, pp. 217–226. Springer, Heidelberg (2004)
10. Kurant, M., Markopoulou, A., Thiran, P.: Towards unbiased bfs sampling. IEEE Journal on Selected Areas in Communications 29(9), 1799–1809 (2011)
11. Leskovec, J., Faloutsos, C.: Sampling from large graphs. In: Proceedings of the 12th ACM SIGKDD International Conference on Knowledge Discovery and Data Mining (KDD 2006), pp. 631–636 (2006)
12. Newman, M.E.J.: Scientific collaboration networks. ii. Shortest paths, weighted networks, and centrality. Physical Review E 64, 016132 (2001)
13. Zhuge, H., Zhang, J.: Topological centrality and its e-science applications. Journal of the American Society of Information Science and Technology 61, 1824–1841 (2010)

Detecting Maximum k-Plex
with Iterative Proper ℓ-Plex Search

Yoshiaki Okubo, Masanobu Matsudaira, and Makoto Haraguchi

Graduate School of Information Science and Technology
Hokkaido University
N-14 W-9, Sapporo 060-0814, Japan
{yoshiaki,mh}@ist.hokudai.ac.jp

Abstract. In this paper, we are concerned with the notion of k-plex , a relaxation model of clique, where degree of relaxation is controlled by the parameter k. Particularly, we present an efficient algorithm for detecting a *maximum k-plex* in a given simple undirected graph. Existing algorithms for extracting a maximum k-plex do not work well for larger k-values because the number of k-plexes exponentially grows as k becomes larger. In order to design an efficient algorithm for the problem, we introduce a notion of *properness* of k-plex . Our algorithm tries to *iteratively* find a maximum proper ℓ-plex, decreasing the value of ℓ from k to 1. At each iteration stage, the maximum size of proper ℓ-plex found so far can work as an effective lower bound which makes our branch-and-bound pruning more powerful. Our experimental results for several benchmark graphs show that our algorithm can detect maximum k-plexes much faster than SPLEX, the existing most efficient algorithm.

Keywords: proper k-plex , maximum k-plex , clique relaxation model.

1 Introduction

Given a graph or a network G, finding *dense subgraphs* in G is one of the important challenging tasks for many domains. For example, detecting *communities* as dense subgraphs is a fundamental problem in *Social Network Analysis* [12].

In order to formalize meaningful and useful dense subgraphs, various *relaxation models of cliques* have been proposed and investigated (e.g. [17,3,13]). Some of those including *k-clique, k-club* and *k-plex* have been surveyed in [11]. In this paper, we are especially concerned with one of the relaxation models, *k-plex* [13], and try to tackle a problem of finding a k-plex with the *maximum cardinality* also known as *Maximum k-Plex Problem* [1].

Given a graph G, a set of vertices S is called a k-plex in G if each vertex in S is *not adjacent* to *at most* k vertices in S [1]. Since the number of non-neighbors of each vertex is restricted to be at most $k - 1$, a k-plex is regarded as a *familiarity-based relaxation model* of clique, where familiarity is an important

---

[1] Since we assume a simple undirected graph, each vertex in G is not adjacent to itself.

S. Džeroski et al. (Eds.): DS 2014, LNAI 8777, pp. 240–251, 2014.
© Springer International Publishing Switzerland 2014

property required for meaningful communities in Social Network Analysis. It is, however, obvious that such a relaxation model would be also useful in graph-based data mining applications such as *Systems Biology*.

Although the problem of finding a k-plex with the maximum cardinality has been theoretically proved *NP-complete* [1], several practical and exact algorithms for the problem have already been proposed due to its wide applicability. For example, BC is a system based on a $0/1$ *integer linear programming formulation* and IPBC is a variant of BC [1]. OsterPlex proposed in [8] is an adapted version of an algorithm for detecting a clique with the maximum cardinality [10]. Moreover, SPLEX is an algorithm based on *bounded-degree vertex deletion* process [9]. As the authors have ever known, SPLEX is the most efficient system for the problem.

Those existing systems can actually work well in the range of smaller k (that is, degree of clique relaxation is relatively small). However, if the value of k becomes larger, they often fail to detect solutions with reasonable computation times. We therefore try to design an algorithm based on a different idea which can exactly and efficiently solve the problem even for larger values of k.

Concretely speaking, following the success in finding a maximum clique [15], we present in this paper a *depth-first branch-and-bound* algorithm for detecting a maximum k-plex . In this type of algorithms, it is very important to detect a tentative solution close to the optimal one *as early as possible* so that we can enjoy powerful branch-and-bound prunings. In order to realize it in our search process, we introduce a notion of *properness* of k-plex .

Based on properness, we can divide the family of k-plexes in G into k disjoint classes, $\mathcal{P}_1, \ldots, \mathcal{P}_k$, where the class $\mathcal{P}_\ell$ is the set of proper ℓ-plexes. Therefore, by finding a maximum ℓ-plex in each $\mathcal{P}_\ell$, we can detect an optimal k-plex with the maximum size in G. It should be emphasized here that in general, k-plexes with larger size in G tend to be contained in $\mathcal{P}_k$. It is, thus, very probable that a maximum proper k-plex in $\mathcal{P}_k$ is in fact an optimal k-plex . Therefore, changing the value of ℓ from k to 1, we *iteratively* try to find a maximum proper ℓ-plex in $\mathcal{P}_\ell$. At each iteration, since a maximum ℓ-plex found to be maximum in previous iterations works as a lower bound of the optimal size, we can expect our branch-and-bound pruning becomes more powerful. In addition, at each iteration, we can concentrate on finding only proper ℓ-plex with a particular ℓ. This provides us some pruning rules which are very effective in reducing our search space.

Compared with the most efficient existing algorithm, our experimental results for several benchmark graphs show rationality and practicality of our approach.

2 Preliminaries

In this paper, we are concerned with *simple undirected graphs*.

A graph is denoted by $G = (V, E)$, where V is a set of *vertices* and $E \subseteq V \times V$ a set of *edges*. For any vertices $x, y \in V$, if $(x, y) \in E$, x is said to be *adjacent* to y and vice versa. For a vertex $x \in V$, the set of vertices adjacent to x is denoted by $N_G(x)$, that is, $N_G(x) = \{v \in V | (x, v) \in E\}$, where $|N_G(x)|$ is called the *degree* of x. The degree of x is often referred to as $deg_G(x)$. If it is clear from the context, they are just denoted by $N(x)$ and $deg(x)$, respectively.

Given a graph $G = (V, E)$, for a vertex $v \in V$ and a set of vertices $X \subseteq V$, the number of vertices in X not adjacent to v is denoted by $\#miss(v, X)$, that is, $\#miss(v, X) = |\{u \in X \mid (u, v) \notin E\}| = |X \setminus N(v)|$. Note here that if $v \in X$, then $\#miss(v, X) \geq 1$ because each vertex has no edge connecting to itself.

For a graph $G = (V, E)$ and a subset $X \subseteq V$, a graph $G[X]$ defined by $G[X] = (X, E \cap (X \times X))$ is called a *subgraph* of G and is said to be *induced by* X. If each pair of vertices in the subgraph are connected, then it is called a *clique* in G. A clique $G[X]$ is simply referred to as the set of vertices X.

For a graph $G = (V, E)$, let us consider a sequence of vertices in V, $P = (v_0, v_1, \ldots, v_\ell)$. If $(v_i, v_{i+1}) \in E$ for any i such that $1 \leq i < \ell$, P is called a *path* from v_0 to v_ℓ and its length is defined by ℓ. For any pair of vertices $u, v \in V$, the *distance* between u and v in G is defined as the minimum length of path from u and v in G. It is often denoted by $dist_G(u, v)$.

3   Maximum k-Plex Problem

Given a graph (network), cliques can be regarded as a perfect structure of communities in the graph. However, it seems to be difficult to find them in real-world except trivial ones like triangles. The notion of k-plex has been introduced as a relaxation model of cliques [13].

Definition 1 (k-Plex). Let $G = (V, E)$ be a graph and $G[S]$ a subgraph of G induced by $S \subseteq V$. If for each $v \in S$, $\#miss(v, S) \leq k$, then S is called a k-plex in G.   ∎

As is similar to the case of cliques, a k-plex is just referred to as the set of vertices by which it is induced.

A k-plex is regarded as a pseudo-clique in the sense that it can be obtained by deleting *at most* $k - 1$ edges from each vertex in a clique. Thus, a clique is a special case of k-plex with $k = 1$.

From the definition, it is easy to see that a k-plex has the following antimonotone property.

Observation 1. Let S be a k-plex in G. Then for any S' such that $S' \subseteq S$, S' is also a k-plex in G.   ∎

As is easily imagined, since the number of k-plexes found in a graph is very huge in general, finding a k-plex with the *maximum-cardinality* is an interesting challenge to be tackled [1]. It is formalized as *Maximum k-Plex Problem*.

Definition 2 (Maximum k-Plex Problem). Given a graph $G = (V, E)$, *Maximum k-Plex Problem* is to identify a set of vertices $X \subseteq V$ such that

1. X is a k-plex .
2. There exists no k-plex X' in G such that $|X'| > |X|$.   ∎

Several exact algorithms for the problem have already been proposed (e.g. [1,8,9]). As the authors have ever known, SPLEX is the most efficient exact algorithm for finding a maximum k-plex in a given graph [9]. For many benchmark graphs publicly available, the algorithm has succeeded in finding maximum k-plexes with reasonable computation times for small k-values.

In the following sections, we present our depth-first branch-and-bound algorithm for the problem and verify its effectiveness.

4 Algorithm for Finding Maximum k-Plex with Iterative Proper ℓ-Plex Search

As has been mentioned, existing algorithms for Maximum k-Plex Problem can work only with relatively smaller k-values. We present in this section an efficient algorithm which can identify a maximum k-plex for larger k-values.

In case of $k = 1$, it has been reported that a *depth-first branch-and-bound algorithm* is highly successful in finding the target, that is, a maximum clique [15]. Since it is expected the search strategy would be still effective for k larger than 1, our algorithm is also designed with a depth-first branch-and-bound strategy. Before going to details of our algorithm, we discuss useless k-plexes to be excluded from our consideration.

4.1 Useless k-Plexes

As we have seen, the definition of k-plex is very simple. According to the definition, k-plexes with larger size (cardinality) are regarded as pseudo-cliques. From the following observation, on the other hand, we can find a number of k-plexes with smaller size difficult to be pseudo-cliques.

Observation 2. For a graph $G = (V, E)$, any (non-empty) set of vertices $X \subseteq V$ such that $|X| \leq k$ is a k-plex in G. ∎

That is, every vertex set with at most k vertices is always a k-plex . As a critical case, even any independent set with at most k vertices must be a k-plex . In this sense, taking such k-plex into our consideration would be quite nonsense. Therefore, we exclude all k-plexes with at most k vertices as *trivial* ones [2].

In addition to k-plexes with size at most k, we exclude every k-plex which is *not connected* (that is, consists of two or more connected components).

As has been mentioned above, our algorithm takes a depth-first branch-and-bound search strategy. The most important point in this type of algorithms is to find tentative solutions close to the optimal one *as early as possible* so that the branch-and-bound pruning can work well. In order to realize it in our search, we consider *properness* for k-plexes .

---

[2] Although one might claims that even cliques and dense subgraphs with at most k vertices are excluded, their size would be too small to be actually meaningful ones as long as we assume the relaxation parameter k.

4.2   Partitioning Family of k-Plexes Based on Properness

From the definition of k-plex , we can easily observe the following property.

Observation 3. Given a graph G, if S is a k-plex , then for any k' such that $k' \geq k$, S is also a k'-plex. ∎

Then, we introduce a notion of k-*properness* for k-plexes .

Definition 3. Let G be a graph and S a k-plex in G. If $\max_{v \in S}\{\#miss(v, S)\} = k$, then S is called a *proper k-plex* . ∎

From the above observation, it is easy to see that a k-plex is not always a proper k-plex . In other words, the family of k-plexes in G can be partitioned into k classes. More formally speaking, let $\mathcal{P}(k)$ be the family of k-plexes in G. Based on the properness, $\mathcal{P}(k)$ can be divided into k classes as $\mathcal{P}(k) = \mathcal{P}_1 \cup \cdots \cup \mathcal{P}_k$, where $\mathcal{P}_\ell$ is the set of proper ℓ-plex in $\mathcal{P}(k)$, that is,

$$\mathcal{P}_\ell = \{S \in \mathcal{P}(k) \mid \max_{v \in S}\{\#miss(v, S)\} = \ell\}.$$

From the partition of $\mathcal{P}(k)$, it is possible to identify a maximum k-plex in G. More precisely speaking, for each ℓ such that $1 \leq \ell \leq k$, we try to detect a proper ℓ-plex, referred to as S_ℓ, in $\mathcal{P}_\ell$ with the maximum cardinality. Then, a maximum k-plex in G, S_{max}, can be extracted as

$$S_{max} = \arg_{S \in \{S_1, \ldots, S_k\}} \max\{|S|\}.$$

It is easy to see the procedure can correctly detect a maximum k-plex in G.

This simple idea provides us an iterative algorithm for detecting a maximum k-plex in G with powerful branch-and-bound prunings.

4.3   Finding Maximum k-Plex with Iterative Search of Maximum Proper ℓ-Plex

In algorithms with branch-and-bound strategy, a tentative solution found so far gives a lower bound of evaluation value which the optimal solution must have. The procedure just discussed above can be modified so that we can enjoy this powerful pruning mechanism.

More concretely speaking, assume some (total) ordering on $\{1, \ldots, k\}$, say $\ell_1 \prec \cdots \prec \ell_k$. At i-th iteration stage, we try to find a maximum proper ℓ_i-plex, S_{ℓ_i}, which must be larger than $\tilde{S}$ already found to be a (tentatively) maximum one at the previous stages. Thus, in our search at i-th stage, we can safely prune every search path which can never generate a proper ℓ_i-plex with cardinality larger than $|\tilde{S}|$. If we can find a maximum proper ℓ_i-plex S_{ℓ_i} larger than $\tilde{S}$, S_{ℓ_i} is used as the updated $\tilde{S}$ at the following iteration stage.

Efficiency of this iterative procedure is strongly affected by the ordering $\ell_1 \prec \cdots \prec \ell_k$ on $\{1, \ldots, k\}$. As has been mentioned above, we strongly desire to detect a larger $\tilde{S}$ as early as possible in order to make our branch-and-bound pruning more powerful. As a reasonable ordering, we propose to begin our iteration in descending order of $\{1, \ldots, k\}$, that is, we first try to find a maximum proper k-plex, then try to find a maximum proper $(k-1)$-plex, and so forth. Rationality and practicality of this ordering are supported by an empirical fact that cardinality of ℓ-plex tends to become larger as the value of ℓ becomes larger. It is, therefore, quite probable that a maximum proper k-plex found at the first iteration stage is actually a maximum k-plex in G.

4.4   Finding Maximum Proper ℓ-plex

Given a graph $G = (V, E)$ and a value of k, we present here our algorithm for finding a maximum proper ℓ-plex for a given ℓ such that $1 \leq \ell \leq k$.

Assume that we have a tentative maximum k-plex already found at some previous iteration stage. Its size is given by a global variable $maxsize$.

Basic Search Strategy

We assume a total ordering on V, $\prec$. In what follows, we suppose that a set of vertices $X \subseteq V$ is always arranged in the ordering of $\prec$, where the last vertex in X is referred to as $tail(X)$. Particularly, if $X = \emptyset$, then $tail(X) = \perp$, where $\perp \prec v$ for any vertex $v \in V$. Moreover, for $v \in V$ and $X \subseteq V$, the set of vertices in X preceding v is referred to as $X_{\prec v}$ and those in X succeeding to v as $X_{\succ v}$.

We now introduce the notion of *primary vertex* in a proper ℓ-plex.

Definition 4 (Primary Vertex). For an ℓ-plex S, the minimum vertex v in S such that $\#miss(v, S) = \ell$ is called the *primary vertex* in S.   ∎

Let $\mathcal{P}_\ell$ be the family of proper ℓ-plex in G and $\mathcal{P}_\ell(v)$ the family of proper ℓ-plex with the primary vertex v. Then it is easy to see that $\mathcal{P}_\ell = \bigcup_{v \in V} \mathcal{P}_\ell(v)$. That is, for each vertex $v \in V$, if we can identify a maximum proper ℓ-plex with the primary vertex v, a maximum proper ℓ-plex in G can be exactly identified.

We can search for a maximum proper ℓ-plex with the primary vertex v in a depth-first manner. Roughly speaking, a k-plex X containing the vertex v is expanded with a vertex x resulting in a larger k-plex, $X \cup \{x\}$, which possibly becomes a proper ℓ-plex. Starting with the initial singleton k-plex with v, this expansion process is recursively performed in depth-first manner until no k-plex remains to be examined.

To provide a more precise description, we define a set of vertices, called *candidate vertices*, which may be added to a k-plex in our expansion process.

Definition 5 (Candidate Vertices for k-Plex). Let S be a k-plex in $G = (V, E)$. For a vertex $v \in V \setminus S$, if $S \cup \{v\}$ is still a k-plex, then v is said to *expand* S and is called a *candidate vertex* for S. The set of candidate vertices for S is denoted by $cand(S)$, that is,

$$cand(S) = \{v \in V \mid S \cup \{v\} \text{ is a } k\text{-plex}\}.$$ ∎

In our search for finding a maximum proper ℓ-plex with the primary vertex v, it is sufficient to examine every k-plex containing v. Such a k-plex can always be represented in the form of $\{v\} \cup X$, where $X \subseteq V \setminus \{v\}$. Therefore, initializing X to $\emptyset$, we try to expand X by adding a vertex $x \in cand(\{v\} \cup X)$ in a depth-first manner. The resultant k-plex , $\{v\} \cup X \cup \{x\}$, is further expanded with a vertex in $cand(\{v\} \cup X \cup \{x\})$, and so forth.

More precisely speaking, along with a *set enumeration tree*, each k-plex $\{v\} \cup X$ is tried to expand with each vertex $x \in cand(\{v\} \cup X)$ such that $tail(X) \prec x$. Thus, we can completely examine every k-plex containing v without duplication.

Pruning Useless Search Branches

Our final target is a maximum k-plex in G. As its subtask, we try to detect a maximum proper ℓ-plex with the primary vertex v for each $v \in V$. From the assumption that we already have a lower bound on the optimal size, referred to as *maxsize*, any set of vertices from which we can never obtain a proper ℓ-plex with size smaller than or equal to *maxsize* is of no interest. Furthermore, any set of vertices from which no proper ℓ-plex with the primary vertex v can be obtained is also useless. In our search, those hopeless vertex sets can be excluded with some pruning mechanisms.

Branch-and-Bound Pruning Based on *maxsize***:** For any pair of k-plexes , S and S', such that $S \subseteq S'$, $S \cup cand(S) \supseteq S' \cup cand(S')$ holds. Assume here that in our search, a k-plex $vX = \{v\} \cup X$ is expanded to a k-plex $vX' = \{v\} \cup X'$ [3]. Then, from the observation, we have $vX' \cup cand(vX') \subseteq vX \cup cand(vX)$. Particularly, at each expansion step for vX, since we expand vX with a vertex in $cand(vX)$ succeeding to $tail(X)$, we have $vX' \cup cand(vX')_{\succ tail(X')} \subseteq vX \cup cand(vX)_{\succ tail(X)}$. That is, for any k-plex obtained by expanding vX, its size is at most $|vX \cup cand(vX)_{\succ tail(X)}|$. Therefore, if $|vX \cup cand(vX)_{\succ tail(X)}| < maxsize$ holds, we can never detect a maximum k-plex in G by expanding vX. In this case, we can safely prune any expansion of vX without loosing completeness.

Pruning Based on Unconnectivity to Primary Vertex: From the definition, for a proper ℓ-plex with the primary vertex v, $v\tilde{X} = \{v\} \cup \tilde{X}$, we must have $\#miss(v, v\tilde{X}) = \ell$. Let us assume here that we try to obtain a proper ℓ-plex with the primary vertex v by expanding a k-plex $vX = \{v\} \cup X$ with vertices in $cand(vX)_{\succ tail(X)}$. In order to make v the primary vertex, vX must be expanded by adding $\ell - \#miss(v, vX)$ vertices in $cand(vX)_{\succ tail(X)}$ which are *not adjacent* to v. That is, $cand(vX)_{\succ tail(X)}$ must contain *at least* $\ell - \#miss(v, vX)$ vertices not adjacent to v. If it is *not* the case, we can obtain no proper ℓ-plex with the primary vertex v by expanding vX. Thus we can prune any expansion of vX.

Pruning Based on Connectivity to Primary Vertex: By expanding a k-plex $vX = \{v\} \cup X$ with vertices in $cand(vX)_{\succ tail(X)}$, we now try to obtain a

---

[3] For concise expression, curly brackets of singleton set are often omitted. A set $\{a\} \cup X \cup \{b\}$ is simply abbreviated to aXb.

proper ℓ-plex with the primary vertex v and size larger than *maxsize*. In order to get such a proper ℓ-plex, we have to add *at least*

$$m = (maxsize + 1) - |vX| - (\ell - \#miss(v, vX))$$

vertices which are *adjacent* to v. In other words, $cand(vX)_{\succ tail(X)}$ is required to have at least m vertices adjacent to v. Therefore, if $cand(vX)_{\succ tail(X)}$ does not have such vertices, any proper ℓ-plex with enough size can never be detected by expanding vX. In that case, any expansion of vX is hopeless and can be safely pruned without affecting completeness.

Excluding Useless Candidate Vertices: Our basic procedure is to expand a k-plex X with a vertex $x \in cand(X)$. Therefore, if $cand(X)$ contains useless vertices, efficiency of our computation would become worse. The following property implies that we can exclude some of them from our candidate vertices.

Observation 4. Let S be a connected k-plex in $G = (V, E)$. For any pair of vertices $u, v \in S$, we have $dist_{G[S]}(u, v) \leq k$. ∎

The observation tells us that in our search for finding a maximum proper ℓ-plex with a primary vertex v, the initial set of candidate vertices can be reduced from $cand(\{v\})$ to $cand(\{v\}) \cap D_\ell(v)$, where $D_\ell(v)$ is the set of vertices in G each of which is far from v with distance no larger than ℓ, that is, $D_\ell(v) = \{x \in V \mid dist_G(v, x) \leq \ell\}$. Such a reduction of candidate vertices is very effective in improving our computational cost.

It should be emphasized here that for a larger k-plex , we can observe a more desirable property.

Observation 5. Let S be a k-plex in $G = (V, E)$ such that $|S| \geq 2k - 1$. For any pair of vertices $u, v \in S$, we have $dist_{G[S]}(u, v) \leq 2$. ∎

Based on the property, if we concentrate on finding a maximum proper ℓ-plex with the primary vertex v and size no less than $2\ell - 1$, the initial set of candidates can be further reduced to $cand(\{v\}) \cap D_2(v)$, likely to be much smaller than the original $cand(\{v\})$.

From the above, it would be reasonable to divide our computational procedure into two phases, one for finding a maximum proper ℓ-plex with size no less than $2\ell - 1$ and the other for that with size less than $2\ell - 1$. Once we succeed in finding some proper ℓ-plex in the former phase, the latter is no longer necessary.

Iterative Algorithm for Finding Maximum k-Plex
Summarizing the above discussion, we propose an iterative algorithm for finding a maximum k-plex . Its pseudo-code is presented in Figure 1. In the algorithm, decreasing the value of ℓ from k to 1, the procedure MAXIMUMPROPERKPLEX tries to find a maximum proper ℓ-plex with the help of a lower bound of the optimal size, *maxsize*, identified so far. For each vertex v, the procedure calls a sub-procedure MAXKPLEXWITHPRIMARYVERTEX to detect a maximum proper ℓ-plex with the primary vertex v and size larger than *maxsize*. The pruning mechanisms presented above are all incorporated into the algorithm.

```
procedure MAIN(G, k):
  [Input] G = (V, E): an undirected graph.
          k: an integer for k-plex such that k > 1.
  [Output]: a maximum k-plex in G.
  [Global Variables] G, S_max, k, phase, maxsize and maxdist
  begin
    phase ← 1; // Search in phase 1 (for finding larger k-plexes )
    maxsize ← 0; // Initializing tentative maximum size
    ℓ ← k;
    S_max ← ∅;
    while ℓ > 0
      MAXIMUMPROPERKPLEX(ℓ); // Finding a maximum proper ℓ-plex in phase 1
      ℓ ← ℓ - 1;
    endwhile
    if maxsize ≥ 2k - 1 then return S_max; // Output a maximum proper k-plex in G
    else phase ← 2 // Search in phase 2 (for finding smaller k-plexes )
    ℓ ← k;
    while ℓ > 0
      MAXIMUMPROPERKPLEX(ℓ); // Finding a maximum proper ℓ-plex in phase 2
      ℓ ← ℓ - 1;
    endwhile
    return S_max; // Output a maximum proper k-plex in G
  end
```

```
procedure MAXIMUMPROPERKPLEX(ℓ):
  begin
    if phase == 1 then
      maxdist ← 2; //maximum distance in Phase 1 (Observation 5.)
    else
      maxdist ← ℓ; //maximum distance in Phase 2 (Observation 4.)
    endif
    for each v ∈ V
      C_v ← {u ∈ V \ {v} | dist_G(u, v) ≤ maxdist}; // Based on Observation 4. and 5.
      MAXKPLEXWITHPRIMARYVERTEX(v, ∅, V \ {v}, C_v, ℓ);
    endfor
  end
```

```
procedure MAXKPLEXWITHPRIMARYVERTEX(v, X, Cand, C_v, ℓ);
  if Cand ∩ C_v = ∅ then
    if |vX| > maxsize and vX is a proper ℓ-plex with primary vertex v then
      maxsize ← |vX| ;
      S_max ← vX ;
    endif
    return;
  endif
  for each x ∈ Cand
    N_x ← {u ∈ V \ {x} | dist_G(x, u) ≤ maxdist};
    NextCand ← cand(vXx)_{≻x} ∩ N_x ∩ C_v;
    if |vXx ∪ NextCand| ≤ maxsize then
      continue; // Branch-and-bound pruning based on tentative maximum size
    endif
    m ← (maxasize + 1) - |vXx| - (ℓ - #miss(v, vXx));
    Adj ← NextCand ∩ N(v); // Candidate vertices adjacent to v
    if |Adj| < m then
      continue; // Pruning based on connectivity to primary vertex
    endif
    m ← ℓ - #miss(v, vXx);
    NonAdj ← NextCand \ N(v); // Candidate vertices not adjacent to v
    if |NonAdj| < m then
      continue; // Pruning based on unconnectivity to primary vertex
    endif
    MAXKPLEXWITHPRIMARYVERTEX(v, Xx, NextCand, C_v, ℓ);
  endfor
```

Fig. 1. Iterative Algorithm for Finding Maximum k-Plex

Table 1. Scale of Graphs

Name	# of Vert.	# of Edges	Density
ERDÖS-97-1	472	1314	0.01182
ERDÖS-98-1	485	1381	0.01177
ERDÖS-99-1	492	1417	0.01173
ERDÖS-97-2	5488	8972	0.00060
ERDÖS-98-2	5822	9505	0.00056
ERDÖS-99-2	6100	9939	0.00053
GEOM-0	7343	11898	0.00044
GEOM-1	7343	3939	0.00015
GEOM-2	7343	1976	0.00007
DAYS-3	13332	5616	0.00006
DAYS-4	13332	3251	0.00004
DAYS-5	13332	2179	0.00003

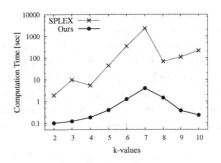

Fig. 2. Computation Time for ERDÖS-99-2

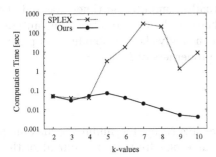

Fig. 3. Computation Time for GEOM-2

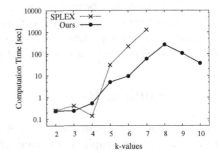

Fig. 4. Computation Time for DAYS-3

5   Experimental Results

We present in this section our experimental results [4].

In order to verify effectiveness of our system, we compare it with SPLEX[5] [9], the most efficient system for Maximum k-Plex Problem. The literature has reported that for many benchmark graphs, SPLEX can detect maximum k-plexes much faster than existing systems, BC [1], IPBC [1] and OsterPlex [8]. For several benchmark graphs used in the literature and publicly available, ERDÖS [5], GEOM [2] and DAYS [2], we observe computation times by each system. ERDÖS and GEOM are collections of collaboration networks and DAYS is that of word co-occurrence networks. Scale of those networks (graphs) are summarized in Table 1.

For each graph, we have observed computation times taken by SPLEX and our system, changing values of k from 2 to 10. Due to space limitation, three of them

---

[4] Our system has been implemented in C and executed on a PC with Intel® Core™-i7 (1.7GHz) processor and 8GB memory.

[5] http://ftp.akt.tu-berlin.de/splex/

are presented in Figure 2 – 4. We can observe similar computational behavior for the other graphs.

In the figures, a missing data point means that the system has failed to detect a solution within 1 hour. The results show that for most of the cases, our system has detected the optimal solutions much faster than SPLEX. Particularly, in range of larger k, our system clearly outperforms the existing system. In some cases, we can observe drastic improvements in computation times (e.g., several orders of magnitude faster). It is due to our iterative search of ℓ-plex for each ℓ from k to 1.

In fact, the maximum k-plexes in the benchmark graphs are proper k-plexes . That is, the tentative solutions detected by our system at the first iteration are actually optimal. Throughout the following iterations, therefore, those tentative solutions can work as *the most effective lower bounds* which provide us the most powerful branch-and-bound pruning. Actually, computation times taken at the following iterations are much less than those at the first iteration. The authors consider that our experimental results convincingly show rationality and practicality of our iterative approach based on properness of k-plex .

6   Concluding Remarks

In this paper, we presented an iterative algorithm for finding a maximum k-plex in a given graph. Decreasing the value of ℓ from k to 1, our algorithm tries to detect a maximum proper ℓ-plex. Particularly, we can strongly expect that a larger tentative solution can be found at early iteration stages, say, the first iteration in many cases. We can therefore enjoy powerful branch-and-bound prunings and detect an optimal solution very efficiently. Compared with the most efficient existing system, rationality and practicality of our approach were shown by experimental results for several benchmark graphs.

Effectiveness for larger scale graph and another types of graphs still remains to be verified. Analyzing computational complexity of our algorithm would be also important future work to be investigated. A polynomial-time solvability of finding a maximum clique has been proved under a certain condition on the maximum degree in a given graph [14]. Our computation at each iteration stage can be regarded as an extension of a depth-first algorithm based on which the complexity has been analyzed. As a future direction, therefore, it would be interesting to analyze our algorithm following [14].

Although efficient methods for enumerating maximal cliques have already been proposed [16,4], such an enumerator for *maximal k-plexes* has not been developed except [18]. An algorithm for the task based on our iterative strategy would be worth investigating. Since there are in general a huge number of maximal k-plexes with smaller size, it would be required to impose some reasonable constraint on our target to be enumerated. In this direction, a promising algorithm based on the notion of j-core [11] has been designed in [6,7] and is currently under further improvement.

References

1. Balasundaram, B., Butenko, S., Hicks, I.V.: Clique Relaxations in Social Network Analysis: The Maximum k-Plex Problem. Operations Research 59(1), 133–142 (2011), INFORMS
2. Batagelj, V., Mrvar, A.: Pajek Datasets (2006), http://vlado.fmf.uni-lj.si/pub/networks/data/
3. Brunato, M., Hoos, H.H., Battiti, R.: On Effectively Finding Maximal Quasi-cliques in Graphs. In: Maniezzo, V., Battiti, R., Watson, J.-P. (eds.) LION 2007 II. LNCS (LNAI), vol. 5313, pp. 41–55. Springer, Heidelberg (2008)
4. Eppstein, D., Strash, D.: Listing All Maximal Cliques in Large Sparse Real-World Graphs. In: Pardalos, P.M., Rebennack, S. (eds.) SEA 2011. LNCS, vol. 6630, pp. 364–375. Springer, Heidelberg (2011)
5. Grossman, J., Ion, P., Castro, R.D.: The Erdös Number Project (2007), http://www.oakland.edu/enp/
6. Matsudaira, M., Haraguchi, M., Okubo, Y., Tomita, E.: An Algorithm for Enumerating Maximal j-Cored Connected k-Plexes. In: Proc. of the 28th Annual Conf. of the Japanese Society for Artificial Intelligence, 3J3-1 (2014) (in Japanese)
7. Matsudaira, M.: A Branch-and-Bound Algorithm for Enumerating Maximal j-Cored k-Plexes, Master Thesis, Graduate School of Information Science and Technology, Hokkaido University (2014) (in Japanese)
8. McClosky, B., Hicks, I.V.: Combinatorial Algorithms for The Maximum k-Plex Problem. Journal of Combinatorial Optimization 23(1), 29–49 (2012)
9. Moser, H., Niedermeier, R., Sorge, M.: Exact Combinatorial Algorithms and Experiments for Finding Maximum k-Plexes. Journal of Combinatorial Optimization 24(3), 347–373 (2012)
10. Östergård, P.R.J.: A Fast Algorithm for the Maximum Clique Problem. Discrete Applied Mathematics 120(1-3), 197–207 (2002)
11. Pattillo, J., Youssef, N., Butenko, S.: Clique Relaxation Models in Social Network Analysis. In: Thai, M.T., Pardalos, P.M. (eds.) Handbook of Optimization in Complex Networks: Communication and Social Networks, Springer Optimization and Its Applications, vol. 58, pp. 143–162 (2012)
12. Scott, J.P., Carrington, P.J. (eds.): The SAGE Handbook of Social Network Analysis. Sage (2011)
13. Seidman, S.B., Foster, B.L.: A Graph Theoretic Generalization of the Clique Concept. Journal of Mathematical Sociology 6, 139–154 (1978)
14. Tomita, E., Nakanishi, H.: Polynomial-Time Solvability of the Maximum Clique Problem. In: Proc. of the European Computing Conference - ECC 2009 and the 3rd Int'l Conf. on Computational Intelligence - CI 2009, pp. 203–208 (2009)
15. Tomita, E., Kameda, T.: An Efficient Branch-and-Bound Algorithm for Finding a Maximum Clique with Computational Experiments. Journal of Global Optimization 37(1), 95–111 (2007)
16. Tomita, E., Tanaka, A., Takahashi, H.: The Worst-Case Time Complexity for Generating All Maximal Cliques and Computational Experiments. Theoretical Computer Science 363(1), 28–42 (2006)
17. Uno, T.: An Efficient Algorithm for Solving Pseudo Clique Enumeration Problem. Algorithmica 56, 3–16 (2010)
18. Wu, B., Pei, X.: A Parallel Algorithm for Enumerating All the Maximal k-Plexes. In: Washio, T., et al. (eds.) PAKDD 2007. LNCS (LNAI), vol. 4819, pp. 476–483. Springer, Heidelberg (2007)

Exploiting Bhattacharyya Similarity Measure to Diminish User Cold-Start Problem in Sparse Data

Bidyut Kr. Patra[1,*], Raimo Launonen[1], Ville Ollikainen[1], and Sukumar Nandi[2]

[1] VTT Technical Research Centre of Finland, P.O. Box 1000, FI-02044 VTT, Finland
[2] Indian Institute of Technology Guwahati, Guwahati, Assam, PIN-789 039, India
{ext-bidyut.patra,raimo.launonen,ville.ollikainen}@vtt.fi,
sukumar@iitg.ernet.in

Abstract. Collaborative Filtering (CF) is one of the most successful approaches for personalized product recommendations. Neighborhood based collaborative filtering is an important class of CF, which is simple and efficient product recommender system widely used in commercial domain. However, neighborhood based CF suffers from *user cold-start* problem. This problem becomes severe when neighborhood based CF is used in sparse rating data. In this paper, we propose an effective approach for similarity measure to address user cold-start problem in sparse rating dataset. Our proposed approach can find neighbors in the absence of co-rated items unlike existing measures. To show the effectiveness of this measure under cold-start scenario, we experimented with real rating datasets. Experimental results show that our approach based CF outperforms state-of-the art measures based CFs for cold-start problem.

Keywords: data sparsity, cold-start problem, Bhattacharyya measure, similarity measure, neighborhood based CF.

1 Introduction

Recommender system (RS) is a tool to cope with information overload problem. The primary task of RS is to provide personalized suggestions of products or items to individual user so that user can select desired products or items directly without surfing over long list of items (products). Many recommender systems have been developed in various domains such as e-commerce, digital library, electronic media, on-line advertising, etc. [1–3].

Collaborative Filtering (CF) is the most successful and widely used recommendation system [1, 4]. In CF, item recommendations to an user are performed by analyzing rating information of the other users or other items in the system. There are two main approaches for recommending items in CF category, *viz. neighborhood based CF* and *model based CF*.

Neighborhood based CF relies on a simple intuition that an item might be interesting to an active user if the item is appreciated by a set of similar users (neighbors) or she has appreciated similar items in the system. Generally, a similarity measure is used for finding similar items or similar users.

---

* On leave from National Institute of Technology Rourkela, Odisha, PIN- 769 008, India.

S. Džeroski et al. (Eds.): DS 2014, LNAI 8777, pp. 252–263, 2014.
© Springer International Publishing Switzerland 2014

Model-based CF algorithms learn a model from the training rating data using machine learning and other techniques [5, 4]. Subsequently, model is used for predictions. Main advantage of the model-based approach is that it does not need to access whole rating data once model is built. Few model based approaches provide more accurate results than neighborhood based CF [6, 7]. However, most of the electronic retailers such as Amazon, Netflix deployed neighborhood based recommender systems to help out their customers. This is due to the fact that neighborhood based approach is simple, intuitive and it does not have learning phase so it can provide immediate response to new user after receiving upon her feedback [8].

Despite the huge success of neighborhood based CF in industry, it suffers from *limited coverage* and *data sparsity* problems [9]. The *limited coverage* appears due to improper choice of similarity measures while finding neighbors of an active user. The problem of *data sparsity* becomes severe for a RS when it encounts a new user. Recommender system cannot provide recommendation as traditional similarity measures cannot compute neighbors of the new user. As a result, a new user may stop using RS after providing few ratings- a problem commonly known as *user cold-start*. Few methods incorporate additional information into the neighborhood based CF to tackle this problem [10, 11]. However, extracting additional information is hard from many domains such as multimedia data. Therefore, researchers address this problem using ratings information only. Few similarity measures [12–14] are introduced to deal with the cold-start problem. Main constrain of these measures is that a new user must rate on items (co-rated) on which a number of existing users rated.

To address above problem, we propose a formulation for similarity measure which can find neighbors effectively and it can compute similarity between two users if there is no co-rated item between them unlike most of the existing solutions [12–14]. We do not intend to introduce yet another similarity measure. However, we propose a generalized formula in which existing similarity measure can be incorporated to tackle data sparsity problem. The proposed formulation is simple and it can easily be deployed into the neighborhood based CF. The proposed approach utilizes global information of items and local information of ratings of active users. The Bhattacharyya measure [15] is exploited to find global information of the items (item similarity) on which the active users rated. To show the effectiveness, proposed scheme is implemented and tested on real rating datasets. Experimental results show that our measure based CF outperforms CFs using PIP [12], MJD [14] measures for cold-start problem.

The rest of the paper is structured as follows. In section 2, we discuss necessary background and related work. In Section 3, we present our similarity measure. Experimental results of proposed similarity based CF are provided in Section 4. We conclude our paper in Section 5.

2   Background and Related Work

In this section, we discuss working principle of neighborhood based CF and different similarity measures introduced to address user cold-start problem.

2.1   Neighborhood-Based CF

The neighborhood (memory) based algorithms use entire rating dataset to generate a prediction for an item or a product. Let $\mathcal{R} = [r_{ui}]^{M \times N}$ be a given rating matrix (dataset) in a CF based recommender system with M users and N items, where each entry r_{ui} represents a rating value made by an user U_u on an item i. Generally, rating values are confined within a range (say, 1-5 in MovieLens dataset). The task of neighborhood-based CF algorithm is to predict rating of the ith item using the neighborhood information of the uth user or i^{th} item. There are two types of approaches for finding prediction: *user-based* and *item-based*. In this paper, we focus on the former approach. The user-based method makes prediction using the ratings made by the neighbors of uth user on the ith item [9, 16]. For this purpose, it computes similarity between an active user (here, U_u) and U_p, $p = 1 \ldots M$, $p \neq u$. Subsequently, it selects K closest users to form neighborhood of the active user. Finally, it computes a prediction of ith item using each neighbor's rating on i^{th} item weighted by similarity between the neighbor and active user [16]. Traditional similarity measures such as pearson correlation coefficient, cosine similarity and their variants are frequently used to find neighbors of an active user.

2.2   The User Cold-Start Problem

The user cold-start problem is a situation in which a new user cannot get personalized recommendations after providing ratings on few items. The common strategy to tackle this problem is to use additional information along with the ratings of new users. Kim et al. [10] proposed collaborative tagging to address the user cold-start problem. Loh et al. [11] extract information from user's scientific publications. Park et al. [17] use filterbots and surrogate users to extract additional information of new users. The main disadvantages of these approaches are that additional information are not always reliable and complete. Therefore, research community in recommender system addresses cold-start problem using only rating information [12–14]. It is hard to find intersection between items rated by a new user and an existing user oftenly, hence system could not compute neighbors of the new user using traditional similarity measures.

   PIP [12] is the most popular (cited) measure after traditional similarity measures in RS. The PIP measure captures three important aspects (factors) namely, *Proximity*, *Impact* and *Popularity* between a pair of ratings on the same item. These factors are combined to obtain PIP value between each pair of ratings made on a co-rated item. PIP based CF outperforms correlation based CF in providing recommendations to the new users. Heung et al. [13] address the cold-start problem building a model which first predicts rating and computes prediction error on known ratings for each user. From this error information finally an error-reflected model is built. However, approach uses traditional similarity measures for initial predictions.

   Bobadilla et al. in [14] proposed a similarity measure for cold start users called MJD (Mean-Jaccard-Difference). The MJD has started gaining popularity among the research community. In MJD, six basic measures are combined. It computes five basic measures $(v^0, v^1, v^3, v^4$ and $\mu)$ using numerical values of the ratings and one basic measure $(Jaccard$ measure) using distributions (variation) of ratings. Let I^{UV} be the set of co-rated items between a pair of users U and V. Let r_{Ui} be the rating made by user U on an item i. Each rating difference measure v^k ($k = 0, 1, 3, 4$) is ratio of the number of co-rated items on which rating difference is $'k'$ ($|\{i \mid i \in I^{UV}, |r_{Ui} - r_{Vi}| = k\}|$) to the total number of co-rated items ($|I^{UV}|$). The mean difference (μ) is the function of ratings made on corated items. Neural learning technique is used to compute weight of each basic measure. They showed that the MJD based CF outperforms PIP based CF, error-reflected model based CF and traditional measures based CFs. However, all these measures use only co-rated ratings for similarity computations. Therefore, these could not be used for sparse rating data under cold-start scenario.

3 Proposed Similarity Measure

The proposed measure uses Bhattacharyya Coefficient [15], which is discussed next. Subsequently, we show how this measure can be used to alleviate cold-start problem.

3.1 Bhattacharyya Measure

The Bhattacharyya measure has been widely used in signal and image processing domains [18–20]. It measures similarity between two probability distributions. Let $p_1(x)$ and $p_2(x)$ be two density distributions over discrete domain X. Bhattacharyya Coefficient (BC) (similarity) between these densities is defined as follows (equation (1)).

$$BC(p_1, p_2) = \sum_{x \in X} \sqrt{p_1(x)\, p_2(x)} \tag{1}$$

The BC Coefficient is exploited to find similarity between a pair of items in a sparse rating data. Densities of $p_1(x)$ and $p_2(x)$ are estimated from the given rating data. Histogram formulation can be used to estimate these densities. Let $\hat{p}_i$ and $\hat{p}_j$ be the estimated discrete densities of the two items i and j obtained from rating data. The BC similarity between item i and item j is computed as

$$BC(i, j) = BC(\hat{p}_i, \hat{p}_j) = \sum_{h=1}^{m} \sqrt{(\hat{p}_{ih})\, (\hat{p}_{jh})} \tag{2}$$

where, m is the number of bins (number of distinct rating values, i.e., 5 for MovieLens) and $\hat{p}_{ih} = \frac{\#h}{\#i}$, where $\#i$ is the number of users rated the item i, $\#h$ is the number of users rated the item i with rating value $'h'$, $\sum_{h=1}^{m} \hat{p}_{ih} = \sum_{h=1}^{m} \hat{p}_{jh} = 1$.

3.2 Proposed Similarity Measure for New User in Neighborhood Based CF

In this section, we propose a generalized formula for similarity measure which can work in the absence of co-rated items. Any existing measure in RS can be incorporated in the proposed formulation. To the best of our knowledge, this is the first approach to provide a generalized formula in which any existing measure can be incorporated to tackle data sparsity problem [9] in general and cold-start problem in particular. Main idea of the proposed measure is based on the fact that a new user may rate few co-rated items but she also rates items which are similar to the items rated by a significant number of existing users in the system. We aim to utilize ratings made on co-rated items as well as ratings made on all pair of similar items. The correlation (similarity) between a pair of ratings made on a pair of similar items can be computed in the same way that an existing measure computes similarity (correlation) between ratings on co-rated items.

Let U be a new user and V be an existing user in a system. Let I_U and I_V be the two sets of items on which user U and V rated, respectively. The set of all pairs of rated items $\mathcal{I} = I_U \times I_V$ can be divided into two subsets: I_{UV} and $\mathcal{I} \setminus I_{UV}$, where I_{UV} denotes pair of co-rated item set, and $\mathcal{I} \setminus I_{UV}$ denotes relative complement set of I_{UV} in $\mathcal{I}$. The proposed approach finds each pair of similar items $(i, j) \in \mathcal{I} \setminus I_{UV}$ and computes similarity (correlation) between ratings (r_{Ui}, r_{Vj}) made on items i and j. The Bhattacharyya measure is exploited to find all pairs of similar items between U and V using Definition 1. We choose BC measure over other measures (adjusted cosine) as BC can compute similarity between a pair of items in absence of users rated both items. Mutual information cannot be used to find similar item pairs [21].

Definition 1 (Similar item) *An item j is called similar to another item i, if BC similarity between them is maximum, i.e., $BC(i, j) = 1$.*

The proposed measure is termed as B̲hattacharyya C̲oefficient in S̲parse data (BCS). Proposed generalized formula for similarity measure is given as follows.

$$BCS(U, V) = C_1 \sum_{(i,i) \in I_{UV}} sim(r_{Ui}, r_{Vi}) + C_2 \sum_{(i,j) \in \mathcal{I} \setminus I_{UV} \,\wedge\, BC(i,j)=1} sim(r_{Ui}, r_{Vj})$$

(3)

where $sim(.)$ denotes similarity between a pair of ratings, and constants C_1, C_2 are normalization factors. The function $sim(.)$ can be computed using any standard measure directly in case of co-rated items and with little modification in case of similar item pairs (second term in equation (3)).

It can be noted that BC value on a same (co-rated) item is 1, *i.e,* $BC(i, i) = 1$ (equation (2)). Therefore, co-rated item-pairs (I_{UV}) and similar item-pairs ($\{(i, j) \mid (i, j) \in \mathcal{I} \setminus I_{UV} \wedge BC(i, j) = 1\}$) can be combined into a set of relevant item pairs $\mathcal{I}_r = \{(i, j) \mid (i, j) \in \mathcal{I} \wedge BC(i, j) = 1\}$. Recent study [14, 22] combines non numerical similarity measure (such as Jaccard) with simialrity on numerical ratings ($sim(.)$). Finally, generalized formula is modified as

$$BCS(U, V) = C_f \, f(U, V) \circledast C_0 \sum_{(i,j) \in \mathcal{I} \,\wedge\, BC(i,j)=1} sim(r_{Ui}, r_{Vj})$$

(4)

where $f(U, V)$ denotes additional similarity function such as non-numerical similarity measure between two rated item sets I_U, I_V, $\circledast$ denotes binary (combination) operator such as $+$, $\times$, and C_f, C_0 are normalized factors. It can be noted that values of C_f, C_0 are assumed 1 if one uses normalized similarity measures for computing $f(.)$ and summation term in the equation (4), respectively.

Commonly used similarity measures in RS such as pearson correlation coefficient (PC), constrained PC, cosine vector similarity, adjusted cosine can be incorporated in the proposed generalized equation to work with sparse rating data. With little modification PIP can also be incorporated. These measures emphasize only numerical values

Table 1. Existing measures incorporated into proposed BCS measure

$$BCS(U,V) \quad = \quad C_f\, f(U,V) \quad \circledast \quad C_0 \sum_{\mathcal{I}_r=\{(i,j)\,\in\,\mathcal{I}\,|\,BC(i,j)=1\}} sim(r_{Ui}, r_{Vj})$$

Measure	C_f	$f(U,V)$	$\circledast$	C_0	$\sum_{\mathcal{I}_r} sim(r_{Ui}, r_{Vj})$
Pearson	1	0	+	1	$\dfrac{\sum_{(i,j)\in\mathcal{I}_r} (r_{Ui}-\bar{r}_U)(r_{Vj}-\bar{r}_V)}{\sqrt{\sum_{(i,.)\in\mathcal{I}_r} (r_{Ui}-\bar{r}_U)^2}\sqrt{\sum_{(.,j)\in\mathcal{I}_r} (r_{Vj}-\bar{r}_V)^2}}$, $\sum_{(i,.)}$ = summation over first items. $\bar{r}_U$= average rating of user U.
Constrained PC	1	0	+	1	$\dfrac{\sum_{\mathcal{I}_r} (r_{Ui}-\bar{r}_{med})(r_{Vj}-\bar{r}_{med})}{\sqrt{\sum_{(i,.)\in\mathcal{I}_r} (r_{Ui}-\bar{r}_{med})^2}\sqrt{\sum_{(.,j)\in\mathcal{I}_r} (r_{Vj}-\bar{r}_{med})^2}}$, $\bar{r}_{med}$ = median of rating scale.
Cosine	1	0	+	1	$\dfrac{\sum_{(i,j)\in\mathcal{I}_r} r_{Ui}\times r_{Vj}}{\sqrt{\sum_{(i,.)\in\mathcal{I}_r} r_{Ui}^2}\,\sqrt{\sum_{(.,j)\in\mathcal{I}_r} r_{Vj}^2}}$
JMSD	1	$\frac{|I_U\cap I_V|}{|I_U\cup I_V|}$	$\times$	1	$1 - \dfrac{\sum_{(i,j)\in\mathcal{I}_r} (r_{Ui}-r_{Vj})^2}{|\mathcal{I}_r|}$
PIP	1	0	+	1	$\sum_{(i,j)\in\mathcal{I}_r} proximity(r_{Ui},r_{Vj})*impact(.)*pop(.),$ $pop(r_{Ui},r_{Vj})=1+(\frac{r_{Ui}+r_{Vj}}{2} - \frac{(\mu_i+\mu_j)}{2})^2,$ μ_i = average ratings of item i.
MJD	w_f	$\frac{|I_U\cap I_V|}{|I_U\cup I_V|}$	+	1	$\sum_{k\in\{0,1,3,4\}} w_k\, v^k + w_\mu\, \mu,$ $v^k = \frac{|I_k|}{|\mathcal{I}_r|}, I_k=\{(i,j)\in\mathcal{I}_r \mid |r_{Ui}-r_{Vj}|=k\},$ $\mu=1-\frac{1}{|\mathcal{I}_r|}\sum_{(i,j)\in\mathcal{I}_r}(\frac{r_{Ui}-r_{Vj}}{R_{max}-R_{min}})^2,$ $R_{max}(=5), R_{min}(=1)$ are maximum and minimum values in rating scale, respectively.
					$\{w_f, w_0, w_1, w_3, w_4, w_\mu\}$ are computed using neural learning technique as described in [14].

of the ratings. Therefore, additional similarity function takes the value 0 ($f(.) = 0$) for these measures (Table 1). Numerical and non-numerical information are combined in JMSD [22] and it is incorporated in BCS. Similarly, MJD can also be incorporated in the proposed formula. Table 1 shows how existing similarity measures can be incorporated in BCS to tackle sparsity problem.

The proposed BCS measure has important properties: (i) BCS does not depend on the number of co-rated items, (ii) BCS removes the main constrain (co-rated items) of the existing measures by exploiting global item similarity from sparse rating data.

4   Experimental Evaluation

To evaluate performance of proposed similarity based CF, we implemented user-based collaborative filterings using BCS, PIP and MJD similarity measures. For the sake of experimental analysis, we incorporate MJD in the present paper as described in Table 1. We used two real datasets, namely MovieLens [1] and Netflix [2] in experiments. Brief description of these datasets is given in Table 2. To show effectiveness of our BCS

---

[1] http://www.grouplens.org
[2] http://www.netflixprize.com

Table 2. Description of the datasets used in the experiments

Dataset	Purpose	#User (M)	#Item (N)	#Rating (R)	Density index $(\kappa = \frac{R \times 100}{M \times N})$	Rating domain
MovieLens	Movie	6,040	3706	1 M	4.46	$\{1,2,3,4,5\}$
Netflix	Movie	480,189	17770	100 M	1.17	$\{1,2,3,4,5\}$

Table 3. Statistics of Sparse subsets

Dataset (original)	Data-subset	#User (M)	#Item (N)	#Rating (R)	Density index $(\kappa = \frac{R \times 100}{M \times N})$	Rating per user $(\frac{R}{M})$	Rating per item $(\frac{R}{N})$
Netflix	Net_1	8141	9318	196,656	0.25	24.2	21.1
	Net_2	8141	9318	72,184	0.10	8.8	7.4
Movie-Lens	ML_1	6040	3706	40,957	0.18	6.8	11.1
	ML_2	1000	2994	6,000	0.20	6.0	2.0

measure in sparse data, we obtained four subsets in various sparsity levels removing ratings randomly from these original datasets. The sparsity level is parametarized by the *density index* (κ), which is the percentage of all possible ratings available in a dataset. The characteristics of all these subsets is summarized in Table 3.

We used popular evaluation metrics *Mean Absolute Error (MAE)* and $F1$ metric in the experiments. The *MAE* is the average absolute errors over all valid predictions and a smaller value indicates a better accuracy. Let r_i and $\hat{r}_i$ be the actual rating and predicted rating by a CF algorithm, respectively. Let *MAX* be the number of times valid predictions performed by the CF algorithm. The MAE is calculated as follows. $MAE = \frac{1}{MAX} \sum_{i=1}^{MAX} | r_i - \hat{r}_i |$. The $F1$ metric provides qualitative performance of a recommender system. The $F1$ measure is the combination of other two important performance metrics (Precision and Recall) in information retrieval domain.

In addition to the above two metrics, we also report number of successful (valid) predictions and number of *perfect predictions* of each CF. Many neighborhood based CF cannot make a valid prediction for a target item due to absence of an user who rates the target item in its neighborhood. Therefore, number of valid predictions is an important metric of a CF. The number of *successful prediction* is the number of valid prediction made by a CF. The number of *perfect prediction* is the number of times a CF correctly predicts the actual ratings.

To evaluate the performance of these CF approaches, we selected a fixed number (say, 5 for MovieLens and 3 for Netflix subsets) of good movies (rating ≥ 4) randomly as the target items for each user. We created artificial cold-start scenarios in each sparse subset removing ratings of an user under consideration keeping the ratings of all other users unchanged.

4.1   Experiments and Results Analysis

We made detailed analysis on all subsets. In this study, we consider each user as an active user and find average number of users who rate on same items as the active user

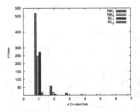

Fig. 1. Characteristics of data subsets

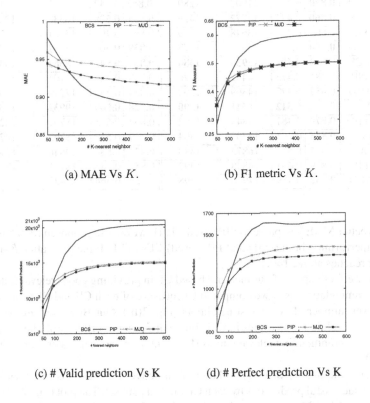

(a) MAE Vs K.   (b) F1 metric Vs K.

(c) # Valid prediction Vs K   (d) # Perfect prediction Vs K

Fig. 2. Under cold-start scenario (new user's ratings < 4) on Net_1 subset

rates. In Net_1 subset, each active user has $570, 57, 13$ neighbors (users) who respectively, share one, two and three co-rated items (Fig. 1). The Net_2 has an average of 247 users who share only one co-rated item with an active user. The ML_1 subset has an average of less than 275 users who share only one (1) co-rated item with each (active) user. The ML_2 has an average of less than 20 users who share only one co-rated item (Fig. 1).

We executed each CF on Net_1 subset and two important performance metrics such as MAE and F1 measures are reported in Fig. 2. It is observed that MJD based CF

Table 4. Experimental results with ML_1 subset

# K	Method	Rating < 4				Ratings < 6			
		MAE	$F1$ metric	# Valid predictions	#perfect predictions	MAE	$F1$ metric	# Valid prediction	#perfect prediction
40	BCS	**0.873**	**0.320**	**8325**	**982**	**0.849**	**0.324**	**8434**	**894**
	PIP	0.897	0.312	8067	890	0.885	0.316	8390	747
	MJD	0.879	0.317	8146	970	**0.849**	0.322	8310	870
80	BCS	0.873	**0.430**	**12700**	**1350**	0.841	**0.444**	**13333**	**1221**
	PIP	0.896	0.413	11895	1167	0.870	0.433	13012	1032
	MJD	**0.870**	0.418	11907	1289	0.842	0.442	13121	1212
120	BCS	**0.860**	**0.477**	**14772**	**1459**	0.839	**0.508**	**16530**	**1340**
	PIP	0.882	0.454	13648	1319	0.863	0.491	15901	1175
	MJD	**0.860**	0.458	13562	1378	**0.839**	0.501	15997	1291
160	BCS	0.862	**0.501**	15958	**1486**	0.834	**0.545**	18562	**1334**
	PIP	0.881	0.472	14462	1370	0.859	0.524	17701	1218
	MJD	**0.857**	0.477	14369	1390	**0.834**	0.533	17714	1306
200	BCS	0.858	**0.512**	**16541**	**1500**	0.828	**0.569**	**19932**	**1335**
	PIP	0.879	0.481	14848	1392	0.854	0.545	18761	1224
	MJD	**0.854**	0.485	14763	1404	0.829	0.553	18762	1274
240	BCS	0.857	**0.519**	**16873**	**1498**	**0.820**	**0.584**	**20840**	**1298**
	PIP	0.879	0.481	14848	1392	0.854	0.545	18761	1224
	MJD	**0.852**	0.489	14936	1398	0.821	0.564	19423	1248

provides better MAE compared to PIP based CF. However, our proposed BCS based CF starts outperforming both (MJD and PIP based) CFs in MAE measure after K-nearest neighbor reaches above 150.

To show effectiveness of our measure based CF in providing good (relevant) items in their recommendation lists, we computed F1 measure of each CF and F1 measures are plotted over number of K nearest neighbors (Fig. 2(b)). Our BCS measure based CF substantially well in F1 measures in wide range of K values (100-600). The CF_{BCS} is the best for recommending relevant items (F1 > 0.60) to active users on subset Net_1 at $K = 600$.

We tested the quality of neighborhood of an active user in terms of providing the number of successful predictions by each CF on Net_1 subset. The plot (Fig. 2(c)) shows that BCS based CF provides maximum number of successful predictions $(20, 363)$, which is close to the number of requested prediction to the system. The closest competitor is the PIP based CF, which makes as much as $15, 095$ successful predictions at $K = 600$. Number of perfect predictions by each CF is also reported in Fig. 2(d). The BCS based CF performs significantly well with large range of K values ($[100, 600]$).

Detailed results are reported in Table 4 after executed all three CFs on ML_1 subset under two different cold-start situations. It is found that our BCS measure based CF can make more successful predictions than MJD based CF and PIP based CF when new user has less than four and six ratings, respectively (Table 4). Number of successful predictions of each CF increases with the size of neighborhood (K). However, CF_{BCS} provides more valid predictions than the other two approaches. This shows that BCS is

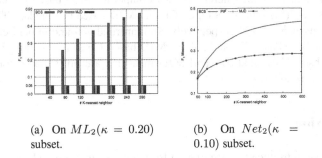

(a) On $ML_2(\kappa = 0.20)$ subset.

(b) On $Net_2(\kappa = 0.10)$ subset.

Fig. 3. F1 metric under cold-start scenario (new user's # ratings < 4)

more suitable for sparse data to draw effective neighbors for new users. Proposed BCS based CF outperforms PIP based CF and MJD based CF in providing number of perfect predictions. Proposed BCS based CF makes highest number of perfect predictions (1500) at $K = 200$ (Table 4).

The BCS based CF has better MAE value compared to PIP based CF over wide range of K values (Table 4). The BCS measure based CF provides MAE values, which are close to the MAE values provided by MJD based CF. It can be noted that MAE of BCS based CF is computed over more number of valid predictions than the number of valid predictions used to compute MAE for MJD based CF[3].

Proposed BCS based CF make more reliable recommendations ($F1$ metric) to new users compared to MJD and PIP based measures. Under first situation (new user has less than four ratings), BCS based CF has $F1 = 0.512$, whereas MJD has $F1 = 0.485$ and PIP has $F1 = 0.481$ with $K = 200$ (Table 4). Similar trend is found under another cold-start situation (number of ratings of new user is less than 6).

We executed all three CFs on ML_2 and Net_2 subsets and results are shown in Fig. 3. Analysis of ML_2 subset shows that each user has on an average less than 20 neighbors who rate at most one co-rated item (Fig. 1). This might be a typical situation in a dataset with very large item space. As expected, PIP and MJD based CFs could not improve F1 metric with the increasing of K values (Fig. 3(a)). However, BCS measure is independent of number of co-rated items. As a results, BCS based CF keeps on improving $F1$ measures ($F1 = 0.151$ to $F1 = 0.471$) with the increasing of K values from 40 to 280 (Fig. 3(a)). The MJD and PIP based CFs provide F1 metric of less than 0.05 over wide range of K values. Proposed BCS based CF makes valid predictions (successful) for more than 62% of requested predictions ($1000 * 5 = 5000$) at $K = 280$ whereas MJD based CF could make predictions for only 3% of requested predictions. The BCS based CF also outperforms PIP and MJD based CFs in producing number of perfect predictions. Number of perfect predictions by BCS based CF is more than number of valid predictions (3%) by MJD based CF.

Experiments with Net_2 subset shows that our BCS based CF can provide reliable recommendations on highly sparse data with $\kappa = 0.10$. The $F1$ value of BCS based

---

[3] Experimentally, it is found that BCS based CF outperforms MJD based CF in MAE measure if MAE is computed over top T ($\leq$ # predictions by MJD) predictions.

CF can reach close to 0.45, whereas closest competitor MJD based CF can produce $F1 < 0.28$ (Fig. 3(b)). The BCS based CF and MJD based CFs make 49% and 24% valid predictions at $K = 600$, respectively. Similar trend is found in perfect prediction metric (4% for BCS based CF and 2% for MJD based CF at $K = 600$). Experiments with these subsets show that PIP and MJD measures are not suitable for providing suggestions to new users.

5 Conclusion

In this paper, we introduced a formulation for similarity measure, which is suitable in new user cold-start scenarios in sparse data. Proposed measure based CF can provide reliable recommendations to new users after receiving few ratings from them. Main advantage of the proposed measure is that it can find effective neighbors of a new user in the absence of co-rated items. Experiments with highly sparse data show that proposed measure based CF can produce significantly better accuracy in recommending items compared to the existing solutions.

Acknowledgment. This work is carried out during the tenure of an ERCIM "Alain Bensoussan" Fellowship Programme. First author has received funding from the European Union Seventh Framework Programme ($FP7/2007 - 2013$) under Grant agreement 246016.

References

1. Adomavicius, G., Tuzhilin, A.: Toward the next generation of recommender systems: a survey of the state-of-the-art and possible extensions. IEEE Transactions on Knowledge and Data Engineering 17(6), 734–749 (2005)
2. Billsus, D., Brunk, C.A., Evans, C., Gladish, B., Pazzani, M.: Adaptive interfaces for ubiquitous web access. Communication of the ACM 45(5), 34–38 (2002)
3. Linden, G., Smith, B., York, J.: Amazon.com Recommendations: Item-to-Item Collaborative Filtering. IEEE Internet Computing 7(1), 76–80 (2003)
4. Su, X., Khoshgoftaar, T.M.: A survey of collaborative filtering techniques. In: Advances in Artificial Intelligence, p. 4:2 (2009)
5. Hofmann, T.: Latent semantic models for collaborative filtering. ACM Transactions on Information Systems 22(1), 89–115 (2004)
6. Paterek, A.: Improving regularized singular value decomposition for collaborative filtering. In: Proceeding of KDD Cup Workshop at 13th ACM Int. Conf. on Knowledge Discovery and Data Mining, pp. 39–42 (2007)
7. Koren, Y.: Factorization Meets the Neighborhood: A Multifaceted Collaborative Filtering Model. In: Proceedings of the 14th ACM SIGKDD International Conference on Knowledge Discovery and Data Mining, pp. 426–434 (2008)
8. Koren, Y.: Factor in the Neighbors: Scalable and Accurate Collaborative Filtering. ACM Trans. Knowl. Discov. Data, 4(1), 1:1–1:24 (2010)
9. Desrosiers, C., Karypis, G.: A Comprehensive Survey of Neighborhood-based Recommendation Methods. In: Recommender Systems Handbook, pp. 107–144 (2011)

10. Kim, H.N., Ji, A.T., Ha, I., Jo, G.S.: Collaborative filtering based on collaborative tagging for enhancing the quality of recommendation. Electronic Commerce Research and Applications 9(1), 73–83 (2010)
11. Loh, S., Lorenzi, F., Granada, R., Lichtnow, D., Wives, L.K., de Oliveira, J.P.M.: Identifying similar users by their scientific publications to reduce cold start in recommender systems. In: Proceedings of the 5th International Conference on Web Information Systems and Technologies (WEBIST 2009), pp. 593–600 (2009)
12. Ahn, H.J.: A new similarity measure for collaborative filtering to alleviate the new user cold-starting problem. Inf. Sci. 178(1), 37–51 (2008)
13. Kim, H.N., El-Saddik, A., Jo, G.: Collaborative error-reflected models for cold-start recommender systems. Decision Support Systems 51(3), 519–531 (2011)
14. Bobadilla, J., Ortega, F., Hernando, A., Bernal, J.: A collaborative filtering approach to mitigate the new user cold start problem. Know.-Based Syst. 26, 225–238 (2012)
15. Bhattacharyya, A.: On A Measure of Divergence Between Two Statistical Populations Defined by their Probability Distributions. Bulletin of Calcutta Mathematical Society 35(1), 99–109 (1943)
16. Herlocker, J.L., Konstan, J.A., Borchers, A., Riedl, J.: An algorithmic framework for performing collaborative filtering. In: Proceedings of the 22nd Annual International ACM SIGIR Conference on Research and Development in Information Retrieval, SIGIR 1999, pp. 230–237 (1999)
17. Park, S.T., Pennock, D., Madani, O., Good, N., DeCoste, D.: Naive filterbots for robust cold-start recommendations. In: Proceedings of the 12th ACM SIGKDD International Conference on Knowledge Discovery and Data Mining, KDD 2006, pp. 699–705 (2006)
18. Kailath, T.: The Divergence and Bhattacharyya Distance Measures in Signal Selection. IEEE Transactions on Communication Technology 15(1), 52–60 (1967)
19. Comaniciu, D., Ramesh, V., Meer, P.: Real-Time Tracking of Non-Rigid Objects Using Mean Shift. In: Proceedings of the Conference on Computer Vision and Pattern Recognition (CVPR 2000), pp. 2142–2149 (2000)
20. Nielsen, F., Boltz, S.: The burbea-rao and bhattacharyya centroids. IEEE Transactions on Information Theory 57(8), 5455–5466 (2011)
21. Yu, K., Xu, X., Ester, M., Kriegel, H.P.: Feature weighting and instance selection for collaborative filtering: An information-theoretic approach. Knowledge and Information Systems 5(2), 201–224 (2003)
22. Bobadilla, J., Serradilla, F., Bernal, J.: A new collaborative filtering metric that improves the behavior of recommender systems. Knowl. -Based Syst. 23(6), 520–528 (2010)

Failure Prediction – An Application in the Railway Industry

Pedro Pereira[1,2], Rita P. Ribeiro[1,3], and João Gama[1,2]

[1] LIAAD-INESC TEC, University of Porto, Portugal
[2] Faculty of Economics, University Porto, Portugal
[3] Faculty of Sciences, University Porto, Portugal
pm.pereira.mail@gmail.com, rpribeiro@dcc.fc.up.pt, jgama@fep.up.pt

Abstract. Machine or system failures have high impact both at technical and economic levels. Most modern equipment has logging systems that allow us to collect a diversity of data regarding their operation and health. Using data mining models for novelty detection enables us to explore those datasets, building classification systems that can detect and issue an alert when a failure starts evolving, avoiding the unknown development up to breakdown. In the present case we use a failure detection system to predict train doors breakdowns before they happen using data from their logging system. We study three methods for failure detection: outlier detection, novelty detection and a supervised SVM. Given the problem's features, namely the possibility of a passenger interrupting the movement of a door, the three predictors are prone to false alarms. The main contribution of this work is the use of a low-pass filter to process the output of the predictors leading to a strong reduction in the false alarm rate.

1   Introduction

Predicting the future is an activity that has always captured the interest of humanity. As the Greek poet C. P. Cavalfy said: *'Ordinary mortals know what is happening now, the gods know what the future holds because They alone are totally enlightened. Wise men are aware of the future things just about to happen.'*

The ability to predict what is about to happen can make signicant changes in how to run a business. It is hoped that the practical demonstration of the improvements achievable through the application of a data mining system to a specific day-to-day problem can be a further contribution to this area of knowledge, pointing out the advantages at hand to a wide range of corporations once they embrace this kind of approach. This paper presents our study on a train door failure prediction problem, using outlier detection sequence analysis.

Doors are one of the most heavily used parts of a train. On a metro, a single door has to open and close more than 600 times in one day. Door failures cause delays, trip cancellation and other types of operational inefficiencies. In railways' early days, doors were locally and manually operated, but the challenges posed by the need to reduce on-board human resources, higher safety requirements and

S. Džeroski et al. (Eds.): DS 2014, LNAI 8777, pp. 264–275, 2014.
© Springer International Publishing Switzerland 2014

faster operation led to the sophistication of this equipment. Indeed, nowadays doors are a highly complex system, comprising electronic circuit control and pneumatic or electric drive systems, allowing opening cycles faster than two seconds and safety devices such as anti-pinch. The growing complexity of these functionalities has increased reliability and maintenance issues. In fact, modern doors comprise several pieces of equipment such as pneumatic valves, sensors, call buttons, microprocessors and others, which greatly contribute to the likelihood of a failure. The growing number of components and the increasing complexity of their control poses additional problems in terms of reliability. In the case of train doors, its failure often causes relevant damages to the operation, not only at service level, but also on the costs of operating the system, such as: delays, trip cancellation and operational inefficiencies. In this scenario, a great deal of attention has been paid to door maintenance. As a result of this effort, new methodologies have been tested, such as Reliability Centered Maintenance (RCM) [8], but its application is associated with some difficulties, like attribute selection or setting the right threshold. Data Mining tools in the field of Novelty Detection and more specifically Failure Prediction systems seem very promising opportunities to address some of the challenges that the railway industry must face to remain economically competitive.

The goal of this paper is to develop a system that signals an alarm when a sequence of door operations indicates a deterioration of the system. We must point out that we are not interested in signaling alarms when a single operation is abnormal. This is not an indication of a problem in the train opening system but, most probably, the interference of a passenger. Most of the predictive machine learning approaches for anomaly or failure prediction assume i.i.d. observations. They do not deal with sequential nor temporal information. In this study, we propose the use of a low-pass filter over the output of the predictive model to identify sequences of abnormal predictions that represent a deterioration of the train opening system.

The paper is organized as follows. In this Section we have explained the problem and the motivation. In Section 2 we discuss the anomaly detection techniques related to our target problem. In Section 3 we present our case study and, in Section 4, we show the results obtained. We conclude in Section 5.

2   Related Work

Automatic methods for fault detection [7,15] have been studied for a long time. In [7] techniques, such as expert systems, fuzzy logic and data mining, are used to address a diversity of application areas including aerospace, process controls, automotive, manufacturing, nuclear plants, etc. The methods we review here are the most common approaches using data mining: novelty, outlier detection and supervised learning [16].

Novelty Detection is often defined as the ability of a machine learning system to identify new or unknown concepts that were not present during the learning phase [9,12]. This feature is essential in a good classifier because in practical applications, especially when data streams are involved, the test examples contain information on concepts that were not known during the training of the decision model. The ability to identify what are the new concepts is vital if a classifier can learn continuously, which will require that: 1) the classifier represents the current state, i.e. normal behaviour and 2) systematically checks the compatibility between the current model and recent data.

Automatic anomaly detection to forecast potential failures on railway maintenance tasks appears in [10,11]. Authors start by characterizing normal behaviour taking into account the contextual criteria associated to railway data (itinerary, weather conditions, etc.). After that, they measure the compliance of new data, according to extracted knowledge, and provide information about the seriousness and possible causes of a detected anomaly.

One of the most globally accepted outlier definition states that an outlier is a data object that deviates significantly from the rest of the objects, as if it were generated by a different mechanism. According to [4], an outlier can be further divided into three different types: global, contextual and collective. In this paper our focus is on global outliers, also called point anomalies, data objects that are unlikely to follow the same distribution as the other objects in the data set.

Similarly to other learning tasks, depending on the existence of labeled instances, outlier detection techniques can be divided into three main groups: 1) unsupervised; 2) semi-supervised; 3) supervised. In the following subsections, we briefly describe these three different approaches and their application to our target problem.

Unsupervised Methods. According to [3], unsupervised outlier detection methods can be grouped into statistical methods, clustering methods, distance-based and density-based methods. The choice of the appropriate method relies on several factors, such as the number of dimensions of the data, data type, sample size, algorithms efficiency and, ultimately, on the user understanding of the problem.

Whenever the goal is to identify univariate outliers, such as in the context of our problem, the statistical methods are among the most simple methods. Assuming a Gaussian distribution and learning the parameters from the data, parametric methods identify the points with low probability as outliers. One of the methods used to spot such outliers is the boxplot method. Based on the first quartile (Q_1), the third quartile (Q_3) and the interquantile range $(IQR = Q_3 - Q_1)$ of data, it determines that the interval $[Q_1 - 1.5 * IQR, Q_3 + 1.5 * IQR]$ contains 99.3% of data. Therefore, points outside that interval are considered as mild outliers, and points outside the interval $[Q_1 - 3 * IQR, Q_3 + 3 * IQR]$ are considered extreme outliers.

We have decided to apply the boxplot method to identify extreme outliers in our target problem.

Semi-supervised Methods. Frequently, in day-to-day problems in the maintenance field, there are numerous examples belonging to the Normal class and very few from the Outlier class, maybe not even representing all the failure modes. As stated by [6], in engineering anomaly detection problems, often only examples from a single class, the normal, are available, whereas examples from the counter class might be very rare or expensive to obtain. This kind of problems is usually dubbed as one-class classification (OCC) or learning from positive-only examples [14]. There are various ways to address OCC, such as one-class SVMs [4], auto-associative neural networks, also known as *autoencoders* [6].

In this paper, we have chosen to use the OCC algorithm available in Weka by Hempstalk et al. [5], which combines density and class probability estimation. In this algorithm only the Normal class examples are used for training, as the learning phase is done without using any information from other classes. Firstly, a density approach is applied to training data so as to generate artificial data used to form an artificial outlier class. Then a classifier is built with examples from both Normal and Outlier classes.

Supervised Methods. Supervised outlier detection techniques assume the existence of historical information on all the normal and outlier instances from where predictive models for outliers can be built. Most of the work regarding this area focus on classification tasks and, in particular, on binary classification as it considers only two classes: Normal and Outlier. By the implicit definition of outlier, these classification tasks have an imbalanced class distribution, a well known problem and subject of research [1].

This paper covers the spectrum of supervised fault detection techniques by using a Support Vector Machine (SVM) [4]. In the scope of our problem, the SVM will search for an optimal hyperplane that can be used as decision boundary separating the examples from the Normal and Outlier classes.

3   The Case Study

The purpose of this paper is to develop a data mining system that issues an alarm whenever an automatic door is predicted to suffer a failure. In this case study we focus our attention on the behaviour of the pneumatic doors from one specific train. Each door is activated by a linear pneumatic actuator, equipped with one pressure transductor on both the inlet and outlet chamber, providing a pressure reading every 1/10 second whenever the door is commanded to move. The available data, representing operations from September to December 2012, consists of almost 500 thousand readings, corresponding to 4500 opening and closing cycles. We must note that current opening systems are equipped with sensors that react (inverse the operation) when a passenger interferes. This fact, a feature of the system, triggers false alarms in fault detection systems that we need to avoid.

To accomplish this task we have come up with a two-stage classification process. First, each cycle is classified as Normal or Abnormal, afterwards we use a low-pass filter on the output to decide if there is evidence that a door breakdown is about to happen. For the cycle classification problem we have experimented three different methods: 1) unsupervised learning based on boxplot; 2) semi-supervised learning with OneClassClassification; 3) supervised learning with Support Vector Machine.

3.1 Data Transformation

Considering that our plan involved working with classification algorithms, we defined an attribute value matrix, with each tuple representing a cycle, as our input data. For that purpose, we have created a new set of five variables, as described in detail below. In order to transform a time series dataset into an attribute value matrix, we started by calculating the difference between the inlet and outlet pressure at each moment. Then, for each cycle, we considered five bins of equal time length and calculated the average pressure for each one. Bearing in mind that the duration of each bin, and therefore the total cycle length, was vital information, we generated five new variables, multiplying the bin average pressure by its duration. Finally, we could rearrange our dataset, transforming 500 thousand pressure readings into 4950 door cycles, described by five variables. The new set of attributes was named B1 to B5, with B1 being the first bin, when the door has just started to move, and B5 representing the last bin, when the cycle has finished. Figure 1 shows the evolution of pressure and duration of the two types of cycles, opening and closing door movements, by the mean and standard deviation of each of these five bins.

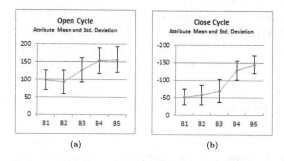

Fig. 1. Opening and closing door movements evolution

Bearing in mind that the temporal information was an important aspect of the dataset, daily averages were also calculated for each attribute. From the analysis of Figure 2 one can observe that, especially in closing movement, attributes average suffered an important shift in September, in what could be concept evolution or the development of a failure.

3.2 Labeling

Regarding labeling the dataset, two new attributes were introduced, one about the normality of the cycle itself, the cycle class, and another, the sequence class, with information on whether a particular sequence of cycles should be considered as Abnormal or Normal. As usual for these tasks, a domain expert was called in to classify each tuple in the data matrix. It is important to notice that a cycle can be labeled as Abnormal even though there is no failure associated. Such case may

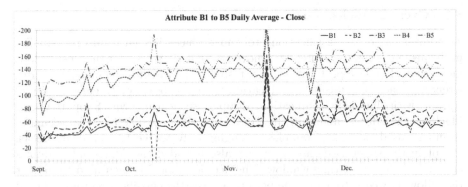

Fig. 2. Daily average of the five bins for closing door movements

arise from a door being blocked by a passenger, which must not be considered as a door failure. As for the sequence class label, door failure moments were identified from the Maintenance Reports, and failure windows were determined by encompassing the cycles that occurred before and that should be related to one specific failure.

In the end, after expert classification, there were 194 door cycles labeled as Abnormal, less than 3% of the total and three failure events, one of them occuring in both the opening and closing door movements (cf. Tables 1 and 2).

Table 1. Identification of Abnormal cycles by the domain expert

Month	Total Cycles		Abnormal Cycles		% Abnormal Cycles	
	Open	Close	Open	Close	Open	Close
Sept.	578	569	20	45	3%	8%
Oct.	628	612	20	9	3%	1%
Nov.	630	628	49	25	8%	4%
Dec.	480	465	19	7	4%	2%
Total	2316	2274	108	86	5%	4%

Table 2. Identification of Failures by the domain expert

Door Failure	Week	Date	Nr Abnormal Cycles	
			Open	Close
Failure 1	36	09/07/2012	52	0
Failure 2	48	11/28/2012	18	0
		11/29/2012	29	31
		11/30/2012	22	24
Failure 3	49	12/06/2012	33	0

3.3 Cycle Classification

To address the cycle classification problem we have divided it into two smaller problems: one for the opening door movements; and another for the closing door movements. For each problem, we have tested cycle classification under three different approaches: 1) unsupervised learning based on boxplot; 2) semi-supervised learning with OneClassClassification; 3) supervised learning with Support Vector Machine. In order to maintain the work as close as possible to a real scenario, training was done using examples belonging only to the two previous weeks, except in the supervised case where we also included the Abnormal examples occurring before that time window. The decision model is then used to predict the following week. To assess the performance of the classification task we used the approach suggested by Hempstalk [5]. For each classification we calculated two ratios: false alarm rate (FAR) and the impostor pass rate (IPR). The false alarm rate is the ratio of normal instances incorrectly identified as outliers. The impostor pass rate is the ratio of outlier instances that are wrongly classified as normal. These metrics are often used in outlier detection domains. A higher FAR results in a lower IPR and vice versa.

3.4 Sequence Classification

Once cycle classification was done, further treatment was applied to the dataset. In fact, the purpose of this work consists on the ability to issue an alarm whenever a door is about to have a breakdown, not to distinguish between normal and abnormal cycles. To achieve this part of the process, we use a low-pass filter, as described before. In each problem, we have tuned the low-pass filter with a specific parameterization, setting the threshold level and smoothing factor, in order to obtain the best possible result under that specific scenario.

Low-Pass Filter. A filter is a device that removes from a signal some unwanted component or feature [13]. The defining feature of filters is the complete or partial suppression of some aspect of the signal. Often, this means removing some frequencies and not others in order to suppress interfering signals and reduce background noise. There are several filters that can be designed to achieve specific goals taking application into account. A low-pass filter is a filter that smoothes abrupt changes in the signal attenuating (reducing the amplitude of) signals with frequencies higher than the cutoff frequency. The low-pass algorithm is detailed by the equation $y_i = y_{i-1} + \alpha * (x_i - y_{i-1})$, where y_i is the filter output for the original signal x_i for instant i and α is the smoothing parameter. The change from one filter output to the next is proportional to the difference between the previous output and the next input. This exponential smoothing property matches the exponential decay seen in the continuous-time system. As expected, as α decreases, the output samples respond more slowly to a change in the input samples: the system will have more inertia.

4   Experimental Evaluation

The main goal of this experimental study is to test the impact of the low-pass filter in the reduction of false alarms. We have evaluated three decision models in which results are post-processed by the low-pass filter. As already stated, we have trained a decision model using a sliding window of two weeks of data and evaluated using the following week. In a set of experiments, not reported here due to space limitations, training a static model with data from September resulted in a degradation of the performance in all models as the time horizon increased.

Results using Outlier detection. Applying the boxplot method for the two previous weeks for all five variables, outliers were detected if at least one of the variables value was an extreme outlier. Even though this approach could be considered too simplistic, assuming independent and Gaussian distributions, it turned out to work very well, especially when its output was post-processed with a low-pass filter. This method granted accurate results, with low IPR and manageable FAR from the cycle classification level. This performance (see Table 3 and Figure 3) was then enhanced with the low-pass filter setting the threshold at 0.5 and smoothing factor at 0.15. In the end, this system was able to correctly detect the three failures present in the dataset with small and acceptable lag, with just one incipient False Alarm at week 45. Overall, both failures on the end of week 48 and 49 could have been signaled with at least 24 hours in advance.

Table 3. Results using boxplot-based outlier detection

	Open			Close		
	Before Filter	After Filter		Before Filter	After Filter	
False Alarms	68	0		76	1	
Cycle Label	Abnormal	Normal	Total	Abnormal	Normal	Total
Abnormal	103	5	108	66	20	86
Normal	11	2197	2208	24	2164	2188

Cycle	Open			Close		
Classification	W49	W39	Other	W48	W38	Other
False Alarm Rate	44%	43%	4%	67%	0%	19%
Impostor Pass Rate	0%	0%	5%	0%	45%	13%

Results using Novelty Detection. As stated before, in a failure detection problem we must train the classifier with examples only belonging to Normal class or at least with unbalanced datasets. To deal with this challenge we have used the OCC algorithm available in Weka [5], which combines density and class probability estimation. The method we have chosen for the class probability estimation was a decision tree with pruning. To reduce variance, bagging was applied with 10 time iterations and bag size set to 100%. Training was done with normal cycles from the preceding two weeks, previously labeled by an expert.

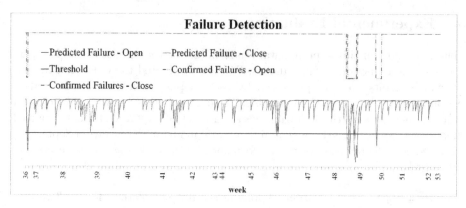

Fig. 3. The impact of the low-pass filter using boxplot-based outlier detection

Aggregated cycle classifications looked acceptable, comparing to unsupervised classification, but a closer look at a week level showed an enormous concentration on False Alarms (see Table 4 and Figure 4). In fact, the false alarms recorded from week 38 to 41 could not be attenuated enough by the low-pass filter.

Table 4. Results using OCC for novelty detection

	Open			Close		
	Before Filter	After Filter		Before Filter	After Filter	
False Alarms	171	5		319	18	
Cycle Label	Abnormal	Normal	Total	Abnormal	Normal	Total
Abnormal	105	3	108	81	5	86
Normal	133	2075	2208	280	1908	2188

Cycle	Open			Close		
Classification	W41	W49	Other	W38	W39	Other
False Alarm Rate	69%	82%	49%	56%	98%	70%
Impostor Pass Rate	0%	0%	3%	0%	0%	9%

As mentioned before, the false alarm concentration in week 39 could not be sufficiently attenuated by the low-pass filter, causing several incorrect door failure alarms. In this scenario, we can distinguish between the performance achieved before week 43 and after. Until week 43 the system was unable to correctly identify abnormal cycles, raising incorrect false alarms, whereas afterwards the level of accuracy increased significantly, even though clearly worse than that obtained with the unsupervised method. One justification for this behaviour might come from concept evolution. When looking at the evolution of the attributes along the time window, there is a strong change in the daily average from week 36 to 40. Nevertheless, there were no records of door failures. As one might expect, the OCC classifier was not able to spot the new outliers when the data model was evolving at that pace.

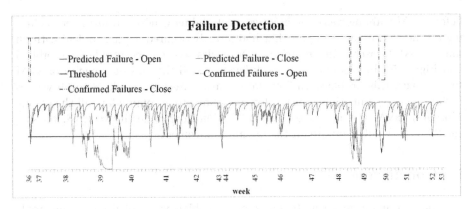

Fig. 4. The impact of the low-pass filter using OCC for novelty detection

Table 5. Results using SVM supervised learning

	Open			Close		
	Before Filter	After Filter		Before Filter	After Filter	
False Alarms	77	3		170	11	
Cycle Label	Abnormal	Normal	Total	Abnormal	Normal	Total
Abnormal	66	42	108	47	39	86
Normal	45	2163	2208	137	2051	2188

Cycle	Open			Close		
Classification	W48	W49	Other	W38	W39	Other
False Alarm Rate	0%	86%	19%	11%	99%	33%
Impostor Pass Rate	88%	0%	0%	74%	50%	28%

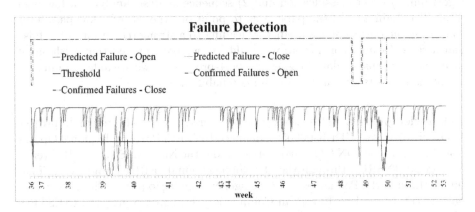

Fig. 5. The impact of the low-pass filter using SVM supervised learning

Supervised Classification - SVM. The last experiment conducted in this study consisted on testing the application of a supervised classifier. After initial trials, we chose a Support Vector Machine, as implemented in Knime [2]. As mentioned before SVM training was done in a two-week sliding window training set, including all the Abnormal examples that had already occured, trying to deal with an unbalanced dataset. Once more, overall results seemed acceptable (see Table 5 and Figure 5), but looking at a week level showed the lack of capacity of SVM to deal with this problem. In fact, false alarm rate from week 38 to 39 was so high that it was impossible for the low-pass filter to accommodate all the errors. Moreover, SVM was not able to spot 30 Abnormal cycles in week 48, missing the 29-30 November failure on the opening movement.

5   Conclusions

The paper's main goal was to point out the high value of data mining tools for maintenance management, with a specific study case in the railway field. We started by applying an unsupervised technique, spotting outliers as examples with at least one attribute with an outlier value determined by the boxplot method. Classification results were processed with a low-pass filter, enabling us to anticipate door failures by 24 hours. One Class Classification algorithm, an approach widely used for novelty detection, was also tested. Unfortunately, we were not able to guarantee an adequate level of reliability, presumably due to concept evolution. Finally, a more standard, supervised, method was experimented, trying a two class classification problem using a Support Vector Machine. SVM granted good aggregated results, but on the opening cycles it totally missed the most important door failure at the end of week 48 and on the closing movements. The False Alarms issued at week 38 and 39 must be considered an important weakness. To conclude, we must stress out that we have demonstrated that, at least in specific problems, failure prediction can be achieved with a two stage algorithm: 1) event classification and 2) sequence analysis applying a low-pass filter. The main goal of the paper was to point out the use of a low-pass filter to process the output of the predictors leading to a strong reduction in the false alarm rate. As final conclusion, we could say we have proved that with small investment in sensors, data logging and post-processing, we are able to minimize maintenance costs and increase systems reliability.

Acknowledgments. This work was supported by Sibila research project (NORTE-07-0124-FEDER-000059), financed by North Portugal Regional Operational Programme (ON.2 O Novo Norte), under the National Strategic Reference Framework (NSRF), through the Development Fund (ERDF), and by national funds, through the Portuguese funding agency, Fundação para a Ciência e a Tecnologia (FCT), and by European Commission through the project MAESTRA (Grant number ICT-2013-612944).

References

1. Aggarwal, C.C.: Outlier Analysis. Springer (2013)
2. Berthold, M.R., Cebron, N., Dill, F., Gabriel, T.R., Kötter, T., Meinl, T., Ohl, P., Sieb, C., Thiel, K., Wiswedel, B.: KNIME: The Konstanz Information Miner. In: Studies in Classification, Data Analysis, and Knowledge Organization (GfKL 2007). Springer (2007)
3. Chandola, V., Banerjee, A., Kumar, V.: Anomaly detection: A survey. ACM Comput. Surv. 41(3) (2009)
4. Han, J., Kamber, M., Pei, J.: Data Mining: Concepts and Techniques, 3rd edn. Morgan Kaufmann Publishers Inc., San Francisco (2011)
5. Hempstalk, K., Frank, E., Witten, I.H.: One-class classification by combining density and class probability estimation. In: Daelemans, W., Goethals, B., Morik, K. (eds.) ECML PKDD 2008, Part I. LNCS (LNAI), vol. 5211, pp. 505–519. Springer, Heidelberg (2008)
6. Japkowicz, N., Myers, C., Gluck, M.A.: A novelty detection approach to classification. In: IJCAI, pp. 518–523. Morgan Kaufmann (1995)
7. Katipamula, S., Michael, P., Brambley, R.: Methods for fault detection, diagnostics, and prognostics for building systems–a review, part i (2004)
8. Nowlan, F.S., Heap, H.F.: Reliability-centered Maintenance. Dolby Access Press (1978)
9. Petsche, T., Marcantonio, A., Darken, C., Hanson, S.J., Kuhn, G.M., Santoso, I.: A neural network autoassociator for induction motor failure prediction, pp. 924–930. MIT Press (1996)
10. Rabatel, J., Bringay, S., Poncelet, P.: SO_MAD: SensOr mining for anomaly detection in railway data. In: Perner, P. (ed.) ICDM 2009. LNCS, vol. 5633, pp. 191–205. Springer, Heidelberg (2009)
11. Rabatel, J., Bringay, S., Poncelet, P.: Anomaly detection in monitoring sensor data for preventive maintenance. Expert Syst. Appl. 38(6), 7003–7015 (2011)
12. Saxena, A., Saad, A.: Evolving an artificial neural network classifier for condition monitoring of rotating mechanical systems. Appl. Soft Comput. 7(1), 441–454 (2007)
13. Shenoi, B.A.: Introduction to Digital Signal Processing and Filter Design. John Wiley & Sons (2005)
14. Tax, D.: One-class classification: Concept learning in the absence of counter-examples. PhD thesis, Technische Universiteit Delft (2001)
15. Yilboga, H., Eker, O.F., Guculu, A., Camci, F.: Failure prediction on railway turnouts using time delay neural networks. In: IEEE International Conference on Computational Intelligence for Measurement Systems and Applications, pp. 134–137 (2010)
16. Zhang, J., Yan, Q., Zhang, Y., Huang, Z.: Novel fault class detection based on novelty detection methods. In: Huang, D., Li, K., Irwin, G. (eds.) Intelligent Computing in Signal Processing and Pattern Recognition. LNCIS, vol. 345, pp. 982–987. Springer, Heidelberg (2006)

Wind Power Forecasting
Using Time Series Cluster Analysis

Sonja Pravilovic[1,2], Annalisa Appice[1],
Antonietta Lanza[1], and Donato Malerba[1]

[1] Dipartimento di Informatica, Università degli Studi di Bari Aldo Moro
via Orabona, 4 - 70126 Bari, Italy
[2] Faculty of Information Technology, Mediterranean University
Vaka Djurovica b.b. 81000 Podgorica, Montenegro
{sonja.pravilovic,annalisa.appice,
antonietta.lanza,donato.malerba}@uniba.it

Abstract. The growing integration of wind turbines into the power grid
can only be balanced with precise forecasts of upcoming energy produc-
tions. This information plays as basis for operation and management
strategies for a reliable and economical integration into the power grid.
A precise forecast needs to overcome problems of variable energy production caused by fluctuating weather conditions. In this paper, we define a
data mining approach, in order to process a past set of the wind power
measurements of a wind turbine and extract a robust prediction model.
We resort to a time series clustering algorithm, in order to extract a
compact, informative representation of the time series of wind power
measurements in the past set. We use cluster prototypes for predicting
upcoming wind powers of the turbine. We illustrate a case study with
real data collected from a wind turbine installed in the Apulia region.

1 Introduction

The capacity of renewable energy sources constantly increases world-wide and
challenges the maintenance of the electric balance between power demand and
supply. In Italy, with an installed capacity of more than 8.700 MWp at the end of
2012, wind power prediction services are becoming an essential part of the grid
control. On the local scale, storage management and smart grid applications
define a sector with increasing power for renewable energy power forecasting.
As the benefit of using a forecast is directly related to the forecast accuracy,
continuous research is performed to enhance wind turbine power predictions.

Several wind power forecasting models have been reported in the literature
over the past few years (see [16] and [1] for recent surveys of these models). They
vary widely in their time horizons, factors determining actual outcomes, types
of data patterns and many other aspects. In this study, we decide to work with
a cluster data pattern that is exclusively based on the time series of the past
wind power measurements of a turbine. The cluster model is learned, in order
to summarize day-ahead wind power values of a turbine. It is noteworthy that

S. Džeroski et al. (Eds.): DS 2014, LNAI 8777, pp. 276–287, 2014.

this summarization goal is out of the scope of the plethora of predictive methods traditionally developed for the time series forecasting. Daily powers, collected over several days, are grouped in clusters. A cluster will group day-ahead wind power time series that have a similar efficiency over the day. Distinct clusters will group day-ahead wind power time series that have different daily efficiency. The efficiency is measured by the area under the time series clustered. The existence of different daily models can be caused by several external factors (e.g. weather conditions), which may keep happening repeatedly over the time. The main contribution of this paper is that of leveraging the power of the cluster analysis in the time series analysis, in order to gain insights into clusters of day-ahead wind power time series measured by a turbine. We use the cluster model associated with a turbine, in order to yield day-ahead predictions of the wind power measures outcoming at the same turbine. We select the cluster prototype that matches the wind power measurements collected up to the present. This selection is done, in order to derive a robust prediction of the upcoming wind powers until dying day. The advantage of integrating the cluster analysis in a forecasting process is twofold. First, the discovery of a cluster pattern allows us to produce a compact, but knowledgeable model of the observed wind power of a turbine. This descriptive model, which is unavailable with traditional forecasting models (e.g. exponential smoothing models or autoregressive models), can be used as a descriptive summary of the past performances of a turbine. Second, the cluster model can be computed offline once, and used online, repeatedly, in order to produce forecasts in (near) real time.

The paper is organized as follows. In the next Section, we review the state of art of related literature. In Section 3, we illustrate the characteristics of the power wind time series. In Section 4, we describe a data mining system that allows us to compute a cluster model of the wind power time series data. In Section 5, we illustrate how a forecasting phase can be built on a clustering wind power modeling phase. Finally, in Section 6, we illustrate a case study with data collected by a wind turbine installed in the Apulia region.

2   Related Work

Several wind power forecasting methods are reported in the literatures. They can be classified based on the model into physical models, statistical models and hybrid models. Physical models (e.g. [2,11]) describe a physical relationship between wind speed, atmospheric conditions, local topography and the output from the wind power turbine. These models consist of several sub models that, together, deliver the translation from both the wind forecast at a specific site and the model level to the power forecast at the considered site and at the turbine hub height. This consists of two main steps: downscaling and conversion to power. The downscaling step scales the wind speed and direction to the turbine hubs height. It consists of finding the best performing numeric weather prediction model. The conversion-to-power step consists of converting the wind speed to power by using a power curve. Statistical models (e.g. [15,6,14,10,13]) estimate

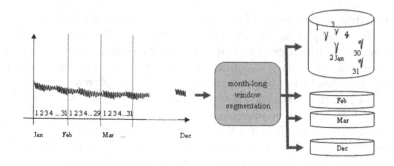

Fig. 1. The month-long window decomposition of day-ahead time series

a statistical relationship between relevant input data and the wind power generation. They involve a direct transformation of the input variables into the wind generation using a statistical model. With these models, a direct estimation of the wind power from the input parameters is possible in a single step. The output models can include most of the data mining based models (e.g. ANN, SVM, fuzzy model, model trees), as well as time series analysis models (e.g. ARIMA, fractional ARIMA). Finally, hybrid models (e.g. [12,7,9]) are based on the combination of the physical and statistical models, the combination of models with several time horizons and the combination of alternative statistical models.

3   Basics and Data Setting

The data mining problem we address here is described by the following premises. First, the statistical model of the wind power may depend on the characteristics and location of a turbine. Second, the statistical model of a turbine can be mined by processing the past measurements of the wind power collected by the turbine under analysis. Third, the statistical model may change with time. In the following, we build on such premises to define the data setting we consider, in order to predict the power wind of a turbine.

A wind power time series is a chronological sequence of observations of the wind power variable routinely measured by a wind turbine. As documented in [1], this time series is, in general, characterized by seasonal as well as diurnal effects. The diurnal effect corresponds to possible daily periodic behavior of wind. The seasonal effect corresponds to possible seasonally periodic behavior of wind. The existence of a seasonal effect inspires the idea of a wind model that changes with seasons. In this study, we account for the expected seasonal effect of the wind power by dividing the yearlong time series of the wind power measured by a turbine into twelve month-long data windows. We decide to split data into month-long learning periods as they are a reasonable compromise between fine-grained week periods and coarse-grained season periods. Then, for each calendar month, we account for the expected daily effect by dividing month-ahead data

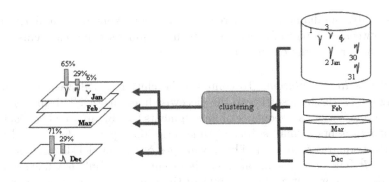

Fig. 2. A cluster based model of wind power time series

into day-ahead time series (see Figure 1). In this way, we can compute a distinct (cluster) model for each calendar month. This model describes the expected behavior of a day-ahead time series observable in the specific month. The turbine may use this model to forecast day-ahead wind powers expected for a calendar day falling in the same month of the model (e.g. the July's day-ahead data model is considered, in order to predict upcoming wind powers of a July's day).

4   Cluster Data Model

The modeling phase uses a clustering algorithm, in order to compute the data model of a turbine. We organize day-ahead time series observed in a calendar month into homogeneous groups where the within-group-object similarity is minimized and the between-group-object dissimilarity is maximized. For each computed cluster C, we determine the cluster prototype $\overline{z}$ and the cluster strength s (see Figure 2). The cluster prototype is a time series summarizing training times series grouped in the cluster. The cluster strength is the percentage of the number of training time series that are grouped in the cluster. The strength represents the frequency of a model over the month.

Clustering algorithm. To determine the cluster model, we use the Quality Threshold (QT) algorithm [8]. This is a partitioning clustering algorithm, originally defined to cluster gene, that is here used, in order to group time series in high quality clusters. Quality is ensured by finding a dense cluster whose diameter does not exceed a given user-defined diameter threshold. We opt for this clustering algorithm since it prevents dissimilar time series from being forced into the same cluster and ensures that only good quality clusters will be formed. The basic idea of QT is as follows. We select a random, unclustered time series as cluster center and add this center to the candidate cluster. We iteratively add other unclustered time series, with each iteration adding the time series that minimizes the increase in cluster diameter. The diameter is the maximum dissimilarity between the cluster center and each time series grouped in the cluster.

In this study, the diameter is a percentage of the maximum dissimilarity computed between each pair of times series of the training set. For each cluster, the center time series is the cluster prototype.

Time Series Representation. To compute the cluster model, we represent the training time series by resorting to the piecewise linear model. This model represents the time series as a sequence of line segments (see Figure 3) where each line segment represents a small subset of time series data points, determined using linear least-squares regression. The advantages of this model data are summarized in [17]. The piecewise linear model is succinct, since only few line segments are needed to represent a large amount of time series data. It is representative as essential information (e.g. significant patterns) in the data is captured. It is robust to changes in the time series model parameters as well as to faults and noise in time series measurements. Formally, let Z be a time series, the piecewise linear model $plm(Z)$ is defined as follows:

$$plm(Z) = \begin{matrix} (t_{start}^1, \hat{z}_{start}^1) \ (t_{end}^1, \hat{z}_{end}^1) \\ \cdots \qquad \cdots \\ (t_{start}^k, \hat{z}_{start}^k) \ (t_{end}^k, \hat{z}_{end}^k) \end{matrix}, \tag{1}$$

where t is the time at which a data point was collected, while $\hat{z}$ represents the piecewise linear estimate of the actual time series data value collected at the time t. In particular $(t_{start}^i, \hat{z}_{start}^i)$ is the starting point of a line segment, while $(t_{end}^i, \hat{z}_{end}^i)$ is the ending point of the line segment i. We note that t_{start}^1 is the starting time point of the time series, t_{end}^k is the ending time point of the time series and t_{start}^i is the time point consecutive to t_{end}^{i-1} in the time series Z. We compute the piecewise linear model of a time series by determining the linearization error between a time series data point and the line segment covering it. We define this error as the perpendicular distance between the point and the line segment. The linearization error for a piecewise linear model representing a time series is the maximum linearization error across all the data points in the time series. For a fixed choice of the maximum linearization error, we start with the first two data points of the time series and fit a line segment to them by using the linear least-square regression theory [4]. Then we consider the data points one at a time and recompute the line segment, in order to fit the new data point. We compute the perpendicular distance of the new data point from the new line segment. If this distance is greater than ϵ, we start a new line segment. We keep repeating this process until we exhaust all data points in the time series.

Dissimilarity Measure. We compute a (dis)similarity measure, based on the area under curve, in order to measure the diameter of a QT cluster. This measure is used, in order to group days, which display similar daily efficiency of the turbine, in the same cluster. Let T_1 and T_2 be two day-ahead time series of n equally spaced wind power measures, the dissimilarity $d(\cdot, \cdot)$ is computed as follows:

$$d(Z_1, Z_2) = \|(area(plm_{Z_1}) - area(plm_{Z_2})\|, \tag{2}$$

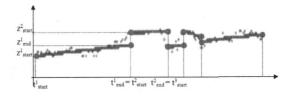

Fig. 3. The piecewise linear model of a time series

where $plm_T(t)$ is the piecewise linear model of the time series Z so that
$area(plm_T) = \sum_{t=1}^{k} \frac{(\hat{z}_{start}^k + \hat{z}_{end}^k) * (t_{end}^k - t_{start}^k)}{2}$ is the area under the curve
$plm(Z)$.

5 Data Forecasting

The prediction phase uses the cluster model $\mathcal{P}(\mathcal{C})$ associated with the same month of the current calendar day, as well as the wind powers $z[1], z[2], \dots z[t-1], z[t]$ measured on the current day, up to the present time t. Both information are used to produce forecasts $\hat{z}[t+1], \hat{z}[t+2], \dots \hat{z}[n]$ till dying day. Operatively, forecasts are produced by predicting $\hat{z}[i] = \bar{z}[i]$ with $i > t$, where $\bar{z}$ is the cluster prototype of $\mathcal{P}(\mathcal{C})$ that best fits observed data up to t. Formally,

$$\bar{z} = \underset{(\bar{z},s) \in \mathcal{P}(\mathcal{C})}{\operatorname{argmin}} \ d(z|_{1\dots t}, \bar{z}|_{1\dots t}), \tag{3}$$

where $d(\cdot, \cdot)$ is the dissimilarity measure according to $\mathcal{P}(\mathcal{C})$ was computed (see Formula 2). In Formula 3, we compute the dissimilarity between the actual wind powers measured by the turbine until the time t strikes (denoted as $z|_{1\dots t}$) and the corresponding data points of a selected cluster prototype $\bar{z}$ (denoted as $\bar{z}|_{1\dots t}$). The selected cluster prototype is that minimizing the dissimilarity with actual measures collected up to t (see Figures 4(a)-4(c)). According to this formulation, when a turbine records a new actual measure of the wind power, remaining forecasts are updated accordingly.

6 Case Study

We evaluate the accuracy of the model in a real case study. Experiments are run on an Intel(R) Core(TM) i7-2670QM CPU@2.20GHz running Windows 7 Professional.

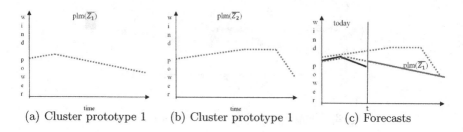

(a) Cluster prototype 1   (b) Cluster prototype 1   (c) Forecasts

Fig. 4. Forecasting phase: 4(a)-4(b) cluster prototypes and 4(c) forecasts. Forecasts (red colored time series in Figure 4(c)) are produced by selecting the cluster prototype (dotted red line in Figure 4(a)) that best fits data up to t (black line in Figure 4(c)) according to Formula 3.

Data, Metrics and Competitors. We consider the wind power measured by an anemometer installed in Foggia province (Apulia, Italy). Training data are collected every ten minutes on 2008 and 2009. Testing data are available for three calendar days on 2007 (04-10-2007, 15-11-2007 and 18-12-2007) and three calendar days on 2010 (01-01-2010, 16-03-2010 and 23-04-2010). We learn the cluster model from the training data, and use this model to produce predictions for testing data. The goals of this empirical study is to evaluate the efficacy of the clustering phase, as well as the accuracy of the forecasting phase.

We analyze the quality of the detected clusters discovered by varying the QT diameter threshold. We compute the Davis Bouldein index, in order to measure the efficacy of clustering. It is a function of the ratio of the within cluster scatter, to the between cluster separation. Thus, the best cluster model minimizes the Davies Bouldin value. We analyze the accuracy of the series of day-ahead forecasts produced until dying day as they are updated at the consecutive time points of a testing day. We compute the root mean square error. We start forecasting from the time point 2:00 on each testing day. For each testing time point t, we compute the error of forecasting the day-ahead wind powers after t. Therefore, we measure $rmse(t)$, that is, the root mean square error over the forecasts $\hat{z}[t+1], \ldots z[t_n]$ of the day.

The forecasting ability of our cluster-based model is compared to the forecasting ability of a regression model. For this comparative study, we consider the data mining system CLUS [3] as a competitor. CLUS (`http://dtai.cs.kuleuven.be/clus/`) is a decision tree induction system that implements the predictive clustering framework. This framework unifies unsupervised clustering and predictive modeling and allows for a natural extension to more complex prediction settings such as multi-target learning. For each time point t of a day, we formulate a multi target regression problem where each wind power measured before t in the day is dealt as a descriptive (independent) variable, while each wind power measured after t in the day is dealt as target variable. By using the monthly-defined training set we learn a predictive clustering tree for each time point t. We use the predictive clustering tree learned with descriptive variables associated to the daily measures before t, in order to forecast the wind power values after t.

Results and Discussion We describe results of our evaluation of both clustering and forecasting phases.

Clustering Evaluation. The clustering phase is repeated by varying the radius threshold of the QT clustering. The radius is computed as a percentage of the maximum dissimilarity between pairs of time series in the training set. Formally,

$$r(M) = r\% * \max_{z_1, z_2 \in D(M)} d(Z_1, Z_2), \qquad (4)$$

where M is a calendar month, $D(M)$ is a training data set collected for M, $r(M)$ is the QT radius chosen for discovering a cluster model associated with the month M and $d(\cdot, \cdot)$ is the dissimilarity measure computed as reported in Formula 2. Figures 5(a)-5(f) 6(a)-6(f) show the number of clusters detected, monthly, by varying the radius percentage of Formula 4 between 5% and 100%. They also show the Davies Bouldein value for each computed cluster model. We can observe that, in general, a local minimum of the Davis Boldein index is achieved when the QT radius is between 20% and 30% of maximum dissimilarity between the training time series. The number of clusters in the clustering model achieved with these thresholds is always between 2 and 3. This means that few day-ahead time series prototypes can allow us to summarize efficaciously observed data. Based on this analysis, we select the cluster model discovered with $r\% = 25\%$ as radius percentage, in order to forecast wind powers in new days. This cluster model is used for the forecasting evaluation. As example, the January-stamped model learned with this parameter set-up, is reported in Figure 7. We can observe that this model provides also a descriptive pattern of the wind power produced in the training period.

Table 1. Cluster-based forecasting model vs CLUS model: The result of the pairwise Wilcoxon signed rank test comparing errors of the two models is reported in the second column. + (-) means that the cluster model is better (worse) than the predictive model learned by CLUS. ++ (−) is reported in the case H0 (hypothesis of equal performance) is rejected at the 0.05 significance level.

testing day	Wilcoxon test (rmse)	p value	testing day	Wilcoxon test (rmse)	p value 1
04-10-2007	-	.20	01-01-2010	−	< .05
15-11-2007	++	< .05	16-03-2010	+	.71
18-12-2007	++	< .05	23-04-2010	+	.98

Forecasting Evaluation. Figures 8(a)-8(f) report the root means squared errors of the forecasts computed until dying day. In order to compare the predictive capabilities of the cluster-based forecasting model and CLUS model, we use the non-parametric Wilcoxon two-sample paired signed rank test. Results of this statistical test are collected in Table 1. Finally, the computation times spent

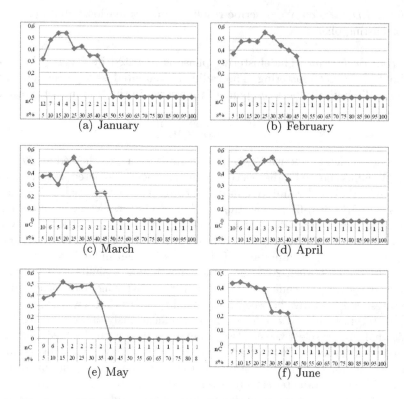

Fig. 5. Cluster models January-June: Number of cluster (nC on X axis) and Davies Bouldein value (Y axis) by varying the QT radius percentage (r% on X axis)

to compute the yearlong model used for the forecasting phase are reported in Table 2. We observe that, for this comparison, we computed one cluster model monthly with out algorithm, while we computed 132 predictive clustering trees monthly with CLUS. The analysis of these results reveals that our algorithm gives statistically better forecasts than CLUS in two out of six testing days (++ in Table 1), whereas CLUS performs statistically better than our algorithm in only one testing day (– in Table 1). For the other testing days, results obtained by our algorithm are generally better than results obtained by CLUS, but not significantly (+ in Table 1). On the other hand, the cluster model computed by our algorithm provides an interpretable description of the expected day-ahead trend of the wind power, which is missed by CLUS (see Figure 7). Finally, computing a single cluster model monthly is less expensive than computing 132 predictive clustering trees monthly.

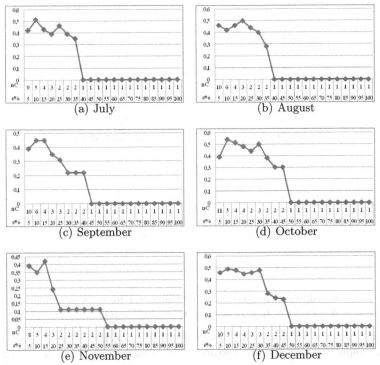

Fig. 6. Cluster models July-December: Number of cluster (nC on X axis) and Davies Bouldein value (Y axis) by varying the QT radius percentage (r% on X axis)

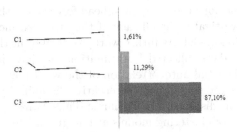

Fig. 7. The January-stamped cluster model mined from the wind power data collected on January 2008 and January 2009

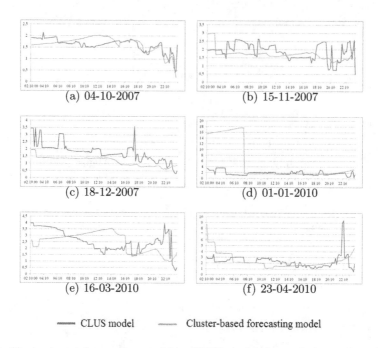

(a) 04-10-2007   (b) 15-11-2007

(c) 18-12-2007   (d) 01-01-2010

(e) 16-03-2010   (f) 23-04-2010

—— CLUS model   —— Cluster-based forecasting model

Fig. 8. Cluster-based forecasting model vs CLUS model: for each time point t (X axis) of the calendar day ($t \geq 02:00$), $rmse(t)$ is plotted (Y axis)

Table 2. Cluster-based forecasting model vs CLUS model: The total time (in ms) spent to compute the yearlong forecasting model. Times are averaged on five trials

Cluster-based forecasting model	CLUS model
580	551.49e3

7   Conclusion

This paper presents a data mining system that leverages the power of the time series cluster analysis for producing day-ahead forecasts of the wind power. The presented study investigates the efficacy of the proposed model compared to a state of art regression model. As future work, we plan to fit the cluster modeling phase to (multiple) data collected from multiple turbines, in order to account for the property of spatial autocorrelation of the wind power. Furthermore, we intend to investigate the performance alternative dissimilarity measures (e.g. the traditional Euclidean distance, the Dynamic Time Warping) in the clustering phase, the discovery of clustering models at fine-grained (e.g. week) and coarse-grained (e.g. quarter or season) level, as well as the use of predictive clustering framework defined for time series in the forecasting phase [5].

Acknowledgments. Authors wish to thank Matteo Gagliardi for his support in developing of the algorithm presented. This work is carried out in fulfillment of the research objectives of both the Startup project "VIPOC: Virtual Power Operation Center" funded by the Italian Ministry of University and Research (MIUR) and the European project "MAESTRA - Learning from Massive, Incompletely annotated, and Structured Data (Grant number ICT-2013-612944)" funded by the European Commission.

References

1. Aggarwal, S., Gupta, M.: Wind power forecasting: A review of statistical models-wind power forecasting. International Journal of Energy Science 3(1), 1–10 (2013)
2. Albert, W., Chi, S., Chen, J.H.: An improved grey-based approach for electricity demand forecasting. Electric Power Systems Research 67, 217–224 (2003)
3. Blockeel, H., De Raedt, L., Ramon, J.: Top-down induction of clustering trees. In: Proceedings of ICML, pp. 55–63. Morgan Kaufmann (1998)
4. Draper, N.R., Smith, H.: Applied regression analysis. Wiley (1982)
5. Džeroski, S., Gjorgjioski, V., Slavkov, I., Struyf, J.: Analysis of time series data with predictive clustering trees. In: Džeroski, S., Struyf, J. (eds.) KDID 2006. LNCS, vol. 4747, pp. 63–80. Springer, Heidelberg (2007)
6. Fan, S., Liao, J., Yokoyama, R., Chen, L., Lee, W.-J.: Forecasting the wind generation using a two-stage network based on meteorological information. IEEE Transactions on Energy Conversion 24(2), 474–482 (2009)
7. Giebel, G., Badger, J., Perez, I.M., Louka, P., Kallos, G.: Short-term forecasting using advanced physical modelling-the results of the anemos project. Results from mesoscale, microscale and cfd modelling. In: Proc. of the European Wind Energy Conference 2006 (2006)
8. Heyer, L.J., Kruglyak, S., Yooseph, S.: Exploring expression data: Identification and analysis of coexpressed genes. Genome Res. 9(11), 1106–1115 (1999)
9. Hui, Z., Bin, L., Zhuo-qun, Z.: Short-term wind speed forecasting simulation research based on arima-lssvm combination method. In: ICMREE 2011, vol. 1, pp. 583–586 (2011)
10. Kavasseri, R.G., Seetharaman, K.: Day-ahead wind speed forecasting using f-arima models. Renewable Energy 34(5), 1388–1393 (2009)
11. Lange, M., Focken, U.: New developments in wind energy forecasting. In: 2008 IEEE Power and Energy Society General Meeting - Conversion and Delivery of Electrical Energy in the 21st Century, pp. 1–8 (2008)
12. Negnevitsky, M., Johnson, P., Santoso, S.: Short term wind power forecasting using hybrid intelligent systems. In: IEEE Power Engineering Society General Meeting, pp. 1–4 (2007)
13. Ohashi, O., Torgo, L.: Wind speed forecasting using spatio-temporal indicators. In: De Raedt, L., Bessière, C., Dubois, D., Doherty, P., Frasconi, P., Heintz, F., Lucas, P.J.F. (eds.) ECAI 2012 - PAIS-2012) System Demonstrations Track. Frontiers in Artificial Intelligence and Applications, vol. 242, pp. 975–980. IOS Press (2012)
14. Potter, C., Negnevitsky, M.: Very short-term wind forecasting for tasmanian power generation. IEEE Transactions on Power Systems 21(2), 965–972 (2006)
15. Sideratos, G., Hatziargyriou, N.: An advanced statistical method for wind power forecasting. IEEE Transactions on Power Systems 22(1), 258–265 (2007)
16. Wang, X., Guo, P., Huang, X.: A review of wind power forecasting models. Energy Procedia 12, 770–778 (2011)
17. Yao, Y., Sharma, A., Golubchik, L., Govindan, R.: Online anomaly detection for sensor systems: A simple and efficient approach. Performance Evaluation 67(11), 1059–1075 (2010)

Feature Selection in Hierarchical Feature Spaces

Petar Ristoski and Heiko Paulheim

University of Mannheim, Germany
Research Group Data and Web Science
{petar.ristoski,heiko}@informatik.uni-mannheim.de

Abstract. Feature selection is an important preprocessing step in data mining, which has an impact on both the runtime and the result quality of the subsequent processing steps. While there are many cases where hierarchic relations between features exist, most existing feature selection approaches are not capable of exploiting those relations. In this paper, we introduce a method for feature selection in hierarchical feature spaces. The method first eliminates redundant features along paths in the hierarchy, and further prunes the resulting feature set based on the features' relevance. We show that our method yields a good trade-off between feature space compression and classification accuracy, and outperforms both standard approaches as well as other approaches which also exploit hierarchies.

Keywords: Feature Subset Selection, Hierarchical Feature Spaces, Feature Space Compression.

1 Introduction

In machine learning and data mining, data is usually described as a vector of *features* or *attributes*, such as the age, income, and gender of a person. Based on this representation, predictive or descriptive models are built.

For many practical applications, the set of features can be very large, which leads to problems both with respect to the performance as well as the accuracy of learning algorithms. Thus, it may be useful to reduce the set of features in a preprocessing step, i.e., perform a *feature selection* [2,8]. Usually, the goal is to compress the feature space as good as possible without a loss (or even with a gain) in the accuracy of the model learned on the data.

In some cases, external knowledge about attributes exist, in particular about their hierarchies. For example, a product may belong to different categories, which form a hierarchy (such as *Headphones < Accessories < Consumer Electronics*). Likewise, hyponym and hyperonym relations can be exploited when using bag-of-words features for text classification [3], or hierarchies defined by ontologies when generating features from Linked Open Data [10].

In this paper, we introduce an approach that exploits hierarchies for feature selection in combination with standard metrics, such as *information gain* or *correlation*. With an evaluation on a number of synthetic and real world datasets,

S. Džeroski et al. (Eds.): DS 2014, LNAI 8777, pp. 288–300, 2014.
© Springer International Publishing Switzerland 2014

we show that using a combined approach works better than approaches not using the hierarchy, and also outperforms existing approaches for feature selection that exploit the hierarchy.

The rest of this paper is structured as follows. In section 2, we formally define the problem of feature selection in hierarchical feature spaces. In section 3, we give an overview of related work. Section 4, we introduce our approach, followed by an evaluation in section 5. We conclude with a summary and an outlook on future work.

2   Problem Statement

We describe each instance as an n-dimensional binary feature vector $\langle v_1, v_2, ..., v_n \rangle$, with $v_i \in \{0, 1\}$ for all $1 \leq i \leq n$. We call $V = \{v_1, v_2, ..., v_n\}$ the *feature space*.

Furthermore, we denote a hierarchic relation between two features v_i and v_j as $v_i < v_j$, i.e., v_i is more specific than v_j. For hierarchic features, the following implication holds:

$$v_i < v_j \rightarrow (v_i = 1 \rightarrow v_j = 1), \tag{1}$$

i.e., if a feature v_i is set, then v_j is also set. Using the example of product categories, this means that a product belonging to a category also belongs to that product's super categories. Note that the implication is not symmetric, i.e., even if $v_i = 1 \rightarrow v_j = 1$ holds for two features v_i and v_j, they do not necessarily have to be in a hierarchic relation. We furthermore assume transitivity of the hierarchy, i.e.,

$$v_i < v_j \wedge v_j < v_k \rightarrow v_i < v_k \tag{2}$$

The problem of *feature selection* can be defined as finding a projection of V to V', where $V' \subseteq V$. Ideally, V' is much smaller than V.

Feature selection is usually regarded with respect to a certain problem, where a solution S using a subset V' of the features yields a certain performance $p(V')$, i.e., p is a function

$$p : \mathcal{P}(V) \rightarrow [0, 1], \tag{3}$$

which is normalized to $[0, 1]$ without loss of generality. For example, for a classification problem, the accuracy achieved by a certain classifier on a feature subset can be used as the performance function p. Besides the quality, another interesting measure is the *feature space compression*, which we define as

$$c(V') := 1 - \frac{|V'|}{|V|} \tag{4}$$

Since there is a trade-off between the feature set and the performance, an overall target function is, e.g., the harmonic mean of p and c.

For most problems, we expect the optimal features to be somewhere in the *middle* of the hierarchy, while the most general features are often too general for predictive models, and the most specific ones are too specific. The hierarchy level of the most valuable features depends on the task at hand. Fig. 1 shows a small part of the hierarchical feature space extracted for dataset Sports Tweets T (see section 5.1). If the task is to classify tweets into sports and non sports

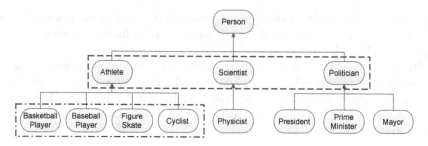

Fig. 1. An example hierarchy of binary features

related, the optimal features are those in the upper rectangle, if the task is to classify them by different kinds of sports, then the features in the lower rectangle are more valuable.

3   Related Work

Feature selection is a very important and well studied problem in the literature. The objective is to identify features that are correlated with or predictive of the class label. Generally, all feature selection methods can be divided into two broader categories: wrapper methods and filter methods (John et al. [4] and Blum et al. [1]). The wrapper methods use the predictive accuracy of a predetermined learning method to evaluate the relevance of the feature sub set. Because of their large computational complexity, the wrapper methods are not suitable to be used for large feature spaces. Filter methods are trying to select the most representative sub-set of features based on a criterion used to score the relevance of the features. In the literature several techniques for scoring the relevance of features exist, e.g., Information Gain, χ^2 measure, Gini Index, and Odds Ratio. However, standard feature selection methods tend to select the features that have the highest relevance score without exploiting the hierarchical structure of the feature space. Therefore, using such methods on hierarchical feature spaces, may lead to the selection of redundant features, i.e., nodes that are closely connected in the hierarchy and carry similar semantic information.

While there are a lot of state-of-the-art approaches for feature selection in standard feature space [8], only few approaches for feature selection in hierarchical feature space are proposed in the literature. Jeong et al. [3] propose the *TSEL* method using a semantic hierarchy of features based on WordNet relations. The presented algorithm tries to find the most representative and most effective features from the complete feature space. To do so, they select one representative feature from each path in the tree, where path is the set of nodes between each leaf node and the root, based on the *lift* measure, and use χ^2 to select the most effective features from the reduced feature space.

Wang et al. [13] propose a *bottom-up hill climbing* search algorithm to find an optimal subset of concepts for document representation. For each feature in the initial feature space, they use a kNN classifier to detect the k nearest neighbors of each instance in the training dataset, and then use the purity of those

---

Algorithm 1. Algorithm for initial hierarchy selection strategy.

Data: H: Feature hierarchy, F: Feature set, t: Importance similarity threshold,
      s:= Importance similarity measurement {"Information Gain",
      "Correlation"}
Result: F: Feature set

```
 1  L := leaf nodes from hierarchy H
 2  foreach leaf l ∈ L do
 3  │   D := direct ascendants of node l
 4  │   foreach node d ∈ D do
 5  │   │   similarity = 0
 6  │   │   if s == "Information Gain" then
 7  │   │   │   similarity = 1-ABS(IGweight(d)-IGweight(l))
 8  │   │   else
 9  │   │   │   similarity =Correlation(d,l)
10  │   │   end
11  │   │   if similarity ≥ threshold then
12  │   │   │   remove l from F
13  │   │   │   remove l from H
14  │   │   │   break
15  │   │   end
16  │   end
17  │   add direct ascendants of l to L
18  end
```

---

instances to assign scores to features. As shown in section 5.3, the approach is computationally expensive, and not applicable for datasets with a large number of instances. Furthermore, the approach uses a strict policy for selecting features that are as high as possible in the feature hierarchy, which may lead to selecting low-value features from the top levels of the hierarchy.

Lu et al. [6] describe a *greedy top-down* search strategy for feature selection in a hierarchical feature space. The algorithm starts with defining all possible paths from each leaf node to the root node of the hierarchy. The nodes of each path are sorted in descending order based on the nodes' information gain ratio. Then, a greedy-based strategy is used to prune the sorted lists. Specifically, it iteratively removes the first element in the list and adds it to the list of selected features. Then, removes all ascendants and descendants of this element in the sorted list. Therefore, the selected features list can be interpreted as a mixture of concepts from different levels of the hierarchy.

4   Approach

Following the implication shown in Eq. 1, we can assume that if two features subsume each other, they are usually highly correlated to each other and have similar relevance for building the model. Following the definition for "relevance" by Blum et al. [1], two features v_i and v_j have similar relevance if $1 - |R(v_i) - R(v_j)| \geq t$, $t \to [0, 1]$, where t is a user specified threshold.

The core idea of our SHSEL approach is to identify features with similar relevance, and select the most valuable abstract features, i.e. features from as high as possible levels of the hierarchy, without losing predictive power. In our approach, to measure the similarity of relevance between two nodes, we use the standard correlation and information gain measure. The approach is implemented in two steps, i.e, initial selection and pruning. In the first step, we try to identify, and filter out the ranges of nodes with similar relevance in each branch of the hierarchy. In the second step we try to select only the most valuable features from the previously reduced set.

The initial selection algorithm is shown in Algorithm 1. The algorithm takes as input the feature hierarchy H, the initial feature set F, a relevance similarity threshold t, and the relevance similarity measure s to be used by the algorithm. The relevance similarity threshold is used to decide whether two features would be similar enough, thus it controls how many nodes from different levels in the hierarchy will be merged. The algorithm starts with identifying the leaf nodes of the feature hierarchy. Then, starting from each leaf node l, it calculates the relevance similarity value between the current node and its direct ascendants d. The relevance similarity value is calculated using the selected relevance measure s. If the relevance similarity value is greater or equal to the similarity threshold t, then the node from the lower level of the hierarchy is removed from the feature space F. Also, the node is removed from the feature hierarchy H, and the paths in the hierarchy are updated accordingly. For the next iteration, the direct ascendants of the current node are added in the list L.

The algorithm for pruning is shown in Algorithm 2. The algorithm takes as input the feature hierarchy H and the previously reduced feature set F. The algorithm starts with identifying all paths P from all leaf nodes to the root node of the hierarchy. Then, for each path p it calculates the average information gain of all features on the path p. All features that have lower information gain than the average information gain on the path, are removed from the feature space F, and from the feature hierarchy H. In cases where a feature is located on more than one path, it is sufficient that the feature has greater information gain than the average information gain on at least one of the paths. This way, we prevent removing relevant features. Practically, the paths from the leafs to the root node, as well as the average information gain per path, can already be precomputed in the initial selection algorithm. The loop in the lines $3 - 6$ is only added for illustrating the algorithm.

Fig. 2a shows an example hierarchical feature set, with the information gain value of each feature. Applying the initial selection algorithm on that input hierarchical feature set, using information gain as a relevance similarity measurement, would reduce the feature set as shown in Fig. 2b. We can see that all feature pairs that have high relevance similarity value, are replaced with only one feature. However, the feature set still contains features that have a rather

Algorithm 2. Algorithm for pruning strategy.

Data: H: Feature hierarchy, F: Feature set
Result: F: Feature set

1 $L :=$ leaf nodes from hierarchy H
2 $P := \emptyset$
3 **foreach** *leaf* $l \in L$ **do**
4 | $p =$ paths from l to root of H
5 | add p to P
6 **end**
7 **foreach** *path* $p \in P$ **do**
8 | $avg =$ Information gain average of path p
9 | **foreach** *node* $n \in$ *path* p **do**
10 | | **if** *IGweight(n)* $< avg$ **then**
11 | | | remove n from F
12 | | | remove n from H
13 | | **end**
14 | **end**
15 **end**

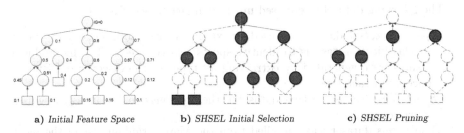

a) *Initial Feature Space*   b) *SHSEL Initial Selection*   c) *SHSEL Pruning*

Fig. 2. Illustration of the two steps of the proposed hierarchical selection strategy

small relevance value. In Fig. 2c we can see that running the pruning algorithm, removes the unnecessary features.

For n features and m instances, iterating over the features, and computing the correlation or information gain with each feature's ancestor takes $O(am)$, given that a feature has an average of a ancestors.[1] Thus, the overall computational complexity is $O(amn)$. It is, however, noteworthy that the selection of the features in both algorithms can be executed in parallel.

5   Evaluation

We perform an evaluation, both on real and on synthetic datasets, and compare different configurations of our approach to standard approaches for feature selection, as well as the approaches described in Section 3.

---

[1] a is 1 in the absence of multiple inheritance, and close to 1 in most practical cases.

5.1 Datasets

In our evaluation, we used five real-world datasets and six synthetically generated datasets. The real-world datasets cover different domains, and are used for
different classification tasks. Initially, the datasets contained only the instances
with a given class label, which afterwards were extended with hierarchical features.

For generating the hierarchical features, we used the RapidMiner Linked Open
Data extension [11], which is able to identify Linked Open Data resources inside
the given datasets, and extract different types of features from any Linked Open
Data source. In particular, we used DBpedia Spotlight [7], which annotates a
text with concepts in DBpedia, a structured data version of Wikipedia [5]. From
those, we can extract further features, such as the types of the concepts found
in a text. For example, when the concept *Kobe Bryant* is found in a text, we can
extract a hierarchy of types (such as *Basketball Player* < *Athlete* < *Person*),
as well as a hierarchy of categories (such as *Shooting Guards* < *Basketball* <
Sports). The generation of the features is independent from the class labels of
the instances (i.e., the classification task), and it is completely unbiased towards
any of the feature selection approaches.

The following datasets were used in the evaluation (see Table 1):

- *Sports Tweets T* dataset was used for existing Twitter topic classifier[2], where
  the classification task is to identify sports related tweets. The hierarchical
  features were generated by extracting all types of the discovered DBpedia
  concepts in each tweet.
- *Sports Tweets C* is the same dataset as the previous one, but using categories
  instead of types.
- The *Cities* dataset was compiled from the Mercer ranking list of the most
  and the least livable cities, as described in [9]. The classification task is
  to classify each city into high, medium, and low livability. The hierarchical
  features were generated by extracting the types for each city.
- The *NY Daily* dataset is a set of crawled news texts, which are augmented
  with sentiment scores[3]. Again, the hierarchical features were generated by
  extracting types.
- The *StumbleUpon* dataset is the training dataset used for the StumbleUpon
  Evergreen Classification Challenge[4]. To generate the hierarchical features,
  we used the Open Directory Project[5] to extract categories for each URL in
  the dataset.

To generate the synthetic datasets, we start with generating features in a
flat hierarchy, i.e. all features are on the same level. The initial features were

---

[2] https://github.com/vinaykola/twitter-topic-classifier/blob/master/
training.txt
[3] http://dws.informatik.uni-mannheim.de/en/research/identifying-disputed-
topics-in-the-news
[4] https://www.kaggle.com/c/stumbleupon
[5] http://www.dmoz.org/

Table 1. Evaluation Datasets

Name	Features	# Instances	Class Labels	# Features
Sports Tweets T	DBpedia Direct Types	1,179	positive(523); negative(656)	4,082
Sports Tweets C	DBpedia Categories	1,179	positive(523); negative(656)	10,883
Cities	DBpedia Direct Types	212	high(67); medium(106); low(39)	727
NY Daily Headings	DBpedia Direct Types	1,016	positive(580); negative(436)	5,145
StumbleUpon	DMOZ Categories	3,020	positive(1,370); negative(1,650)	3,976

generated using a polynomial function, and then discretizing each attribute into a binary one. These features represent the *middle layer* of the hierarchy, which are then used to build the hierarchy upwards and downwards. The hierarchical feature implication (1) and the transitivity rule (2) hold for all generated features in the hierarchy. By merging the predecessors of two or more neighboring nodes from the middle layer, we are able to create more complex branches inside the hierarchy. We control the depth and the branching factor of the hierarchy with two parameters D and B, respectively. Each of the datasets that we use for the evaluation contains 1000 instances, and contains 300 features in the middle layer. The datasets are shown in Table 2.

5.2   Experiment Setup

In order to demonstrate the effectiveness of our proposed feature selection in hierarchical feature space, we compare the proposed approach with the following methods:

- *CompleteFS*: the complete feature set, without any filtering.
- *SIG*: standard feature selection based on information gain value.
- *SC*: Standard feature selection based on feature correlation.
- *TSEL Lift*: tree selection approach proposed in [3], which selects the most representative features from each hierarchical branch based on the *lift* value.
- *TSEL IG*: this approach follows the same algorithm as *TSEL Lift*, but uses information gain instead of lift.
- *HillClimbing*: bottom-up hill-climbing approach proposed in [13].We use $k = 10$ for the kNN classifier used for scoring.
- *GreedyTopDown*: greedy based top-down approach described in [6], which tries to select the most valuable features from different levels of the hierarchy.
- *initialSHSEL IG* and *initialSHSEL C*: our proposed initial selection approach shown with Algorithm 1, using information gain and correlation as relevance similarity measurement, respectively.
- *pruneSHSEL IG* and *pruneSHSEL C*: our proposed pruning selection approach shown with Algorithm 2, applied on previously reduced feature set, using *initialSHSEL IG* and *initialSHSEL C*, respectively.

For all algorithms involving a threshold (i.e., SIG, SC, and the variants of SHSEL), we use thresholds between 0 and 1 with a step width of 0.01.

For conducting the experiments, we used the RapidMiner machine learning platform and the RapidMiner development library. For SIG and SC, we used the built-in RapidMiner operators. The proposed approach for feature selection, as

Table 2. Synthetic Evaluation Datasets

Name	Feature Generation Strategy	# Instances	Classes	# Features
S-D2-B2	D=2; B=2	1,000	positive(500); negative(500)	1,201
S-D2-B5	D=2; B=5	1,000	positive(500); negative(500)	1,021
S-D2-B10	D=2; B=10	1,000	positive(500); negative(500)	961
S-D4-B2	D=4; B=2	1,000	positive(500); negative(500)	2,101
S-D4-B5	D=4; B=5	1,000	positive(500); negative(500)	1,741
S-D4-B10	D=4; B=10	1,000	positive(500); negative(500)	1,621

well as all other related approaches, were implemented in a separate operator as part of the RapidMiner Linked Open Data extension. All experiments were run using standard laptop computer with 8GB of RAM and Intel Core i7-3540M 3.0GHz CPU. The RapidMiner processes and datasets used for the evaluation can be found online[6].

5.3 Results

To evaluate how well the feature selection approaches perform, we use three classifiers for each approach on all datasets, i.e., Naïve Bayes, k-Nearest Neighbors (with $k = 3$), and Support Vector Machine. For the latter, we use Platt's sequential minimal optimization algorithm and a polynomial kernel function [12]. For each of the classifiers we were using the default parameters values in RapidMiner, and no further parameter tuning was undertaken. The classification results are calculated using stratified 10-fold cross validation, where the feature selection is performed separately for each cross-validation fold. For each approach, we report accuracy, feature space compression (4), and their harmonic mean.

Results on Real World Datasets. Table 3 shows the results of all approaches. Because of the space constrains, for the *SIG* and *SC* approaches, as well as for our proposed approaches, we show only the best achieved results. The best results for each classification model are marked in bold. As we can observe from the table, our proposed approach outperforms all other approaches in all five datasets for both classifiers in terms of accuracy. Furthermore, we can conclude that our proposed approach delivers the best feature space compression for four out of five datasets. When looking at the harmonic mean, our approach also outperforms all other approaches, most often with a large gap. From the results for the harmonic mean we can conclude that the *pruneSHSEL IG* approach, in most of the cases, delivers the best results

Additionally, we report the runtime of all approaches on different datasets in Fig. 3. The runtime of our approaches is comparable to the standard feature selection approach, *SIG*, runtime. The *HillClimbing* approach has the longest runtime due to the repetitive calculation of the kNN for each instance. Also, the standard feature selection approach *SC* shows a long runtime, which is due to the computation of correlation between all pairs of features in the feature set.

---

[6] http://dws.informatik.uni-mannheim.de/en/research/feature-selection-in-hierarchical-feature-spaces

Table 3. Results on real world datasets

	Sports Tweets T			Sports Tweets C			StumbleUpon			Cities			NY Daily Headings		
	NB	KNN	SVM	NB	KNN	SVM	NB	KNN	SVM	NB	KNN	SVM	NB	KNN	SVM
Classification Accuracy															
CompleteFS	.655	.759	.797	.943	.920	.946	.582	.699	.730	.625	.562	.684	.534	.586	.577
initialSHSEL IG	**.836**	.768	**.824**	**.974**	.768	**.953**	.661	.709	.733	.671	.609	.674	**.688**	.629	.635
initialSHSEL C	.819	.765	.811	.946	**.937**	.953	.689	**.723**	.732	.640	.671	.683	.547	.580	.596
pruneSHSEL IG	.791	**.793**	.773	.909	.909	.946	**.717**	.695	**.737**	**.687**	.669	**.689**	**.688**	**.659**	**.671**
pruneSHSEL C	.786	.791	.772	.946	.918	.935	.711	.707	.732	.656	**.687**	.646	.665	**.659**	.661
SIG	.819	.788	.814	.966	.936	.940	.681	.707	.729	.656	.640	.671	.675	.652	.668
SC	.816	.765	.813	.937	.918	.932	.587	.711	.726	.625	.656	.677	.534	.583	.606
TSEL Lift	.641	.740	.787	.836	.855	.893	.570	.613	.690	0	0	0	.498	.544	.565
TSEL IG	.632	.734	.782	.923	.909	.935	.579	.661	.724	.640	.580	.580	.521	.560	.610
HillClimbing	.528	.647	.742	.823	.836	.876	.548	.653	.683	.622	.562	.551	.573	.583	.530
GreedyTopDown	.658	.788	.800	.943	.929	.944	.582	.698	.727	.625	.562	.679	.534	.570	.595
Feature Space Compression															
initialSHSEL IG	.456	.207	.222	.318	.708	.288	.672	.843	.642	.781	**.902**	.779	.858	.322	.631
initialSHSEL C	.231	.173	.290	.321	.264	.228	**.993**	.445	.644	.184	.121	.116	.285	.572	.790
pruneSHSEL IG	**.985**	**.986**	**.969**	.895	**.907**	**.916**	.976	.957	**.975**	**.823**	.466	.452	.912	.817	.817
pruneSHSEL C	.971	.965	.965	**.897**	.857	.861	.966	**.968**	.959	.305	.265	.308	.519	.586	.566
SIG	.360	.741	.038	.380	.847	.574	.940	.615	.604	.774	.775	04	.240	.289	.565
SC	.667	.712	.635	.887	.710	.792	.585	.821	.712	.631	.704	.598	.632	.927	.620
TSEL Lift	.247	.247	.247	.511	.511	.511	.412	.412	.412	0	0	0	**.956**	**.956**	**.956**
TSEL IG	.920	.920	.920	.522	.522	.522	.471	.471	.471	.126	.126	.126	.926	.926	.926
HillClimbing	.770	.770	.770	.748	.748	.748	.756	.756	.756	.817	.817	**.817**	.713	.713	.713
GreedyTopDown	.136	.136	.136	.030	0.030	.030	.285	.285	.285	.048	.048	.048	.135	.135	.135
Harmonic Mean of Classification Accuracy and Feature Space Compression															
initialSHSEL IG	.590	.326	.350	.480	.737	.442	.666	.770	.684	.722	**.727**	**.723**	**.764**	.426	.633
initialSHSEL C	.360	.282	.427	.479	.412	.368	.814	.551	.686	.286	.205	.199	.375	.576	.679
pruneSHSEL IG	**.877**	**.879**	**.860**	.902	**.908**	**.931**	**.827**	.805	**.840**	**.749**	.383	.546	.784	**.729**	**.737**
pruneSHSEL C	.869	.869	.858	**.921**	.886	.896	.820	**.817**	.830	.416	.383	.417	.583	.620	.610
SIG	.500	.764	.073	.545	.889	.713	.789	.658	.660	.710	.701	08	.354	.401	.612
SC	.734	.738	.713	.911	.801	.856	.586	.762	.719	.628	.679	.635	.579	.716	.613
TSEL Lift	.356	.370	.376	.634	.640	.650	.479	.493	.516	0	0	0	.655	.693	.711
TSEL IG	.749	.817	.846	.667	.663	.670	.520	.550	.571	.211	.207	.207	.667	.698	.735
HillClimbing	.626	.703	.756	.784	.790	.807	.636	.701	.718	.706	.666	.658	.635	.641	.608
GreedyTopDown	.225	.232	.232	0.058	.058	.058	.383	.405	.409	.089	.088	.089	.216	.219	.221

Results on Synthetic Datasets. Table 4 shows the results for the different synthetic datasets. Our approaches achieve the best results, or same results as the standard feature selection approach *SIG*. The results for the feature space compression are rather mixed, while again, our approach outperforms all other approaches in terms of the harmonic mean of accuracy and feature space compression. The runtimes for the synthetic datasets, which we omit here, show the same characteristics as for the real-world datasets.

Overall, *pruneSHSEL IG* delivers the best results on average, with an *importance similarity threshold* t in the interval $[0.99; 0.9999]$. When using correlation, the results show that t should be chosen greater than 0.6. However, the selection of the approach and the parameters' values highly depends on the given dataset, the given data mining task, and the data mining algorithm to be used.

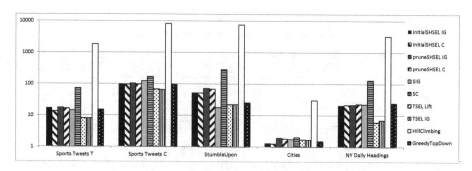

Fig. 3. Runtime (seconds) - Real World Datasets

Table 4. Results on synthetic datasets

	S_D2_B2			S_D2_B5			S_D2_B10			S_D4_B2			S_D4_B5			S_D4_B10		
	NB	KNN	SVM	NB	KNN	SVM	NB	KNN	SVM	NB	KNN	SVM	NB	KNN	SVM	NB	KNN	SVM
Classification Accuracy																		
CompleteFS	.565	.500	.500	.700	.433	.530	.600	.466	.610	.666	.600	.560	.566	.566	.600	.533	.466	.630
initialSHSEL IG	1.0	.833	.880	1.0	.766	.850	1.0	**.866**	.890	1.0	**.866**	.880	1.0	**.936**	.870	.956	.733	.910
initialSHSEL C	.666	.633	.833	.700	.633	.740	.666	.633	.780	.766	.666	.740	.600	.633	.730	.633	.533	.860
pruneSHSEL IG	1.0	**.933**	**.920**	1.0	**.800**	**.910**	1.0	.833	**.960**	1.0	**.866**	**.960**	1.0	.866	**.980**	1.0	**.800**	**.986**
pruneSHSEL C	.866	.666	.910	.866	.700	.900	.800	.766	.900	.800	.766	.910	.933	.833	.880	.933	.666	.930
SIG	.960	.900	.830	1.0	**.800**	.900	.930	.766	.933	1.0	.833	.933	1.0	.900	.966	1.0	.733	.966
SC	.700	.700	.733	.700	.666	.733	.730	.600	.700	.733	.666	.700	.700	.666	.766	.700	.700	.733
TSEL Lift	.553	.500	.540	.633	.666	.630	.400	.500	.540	.566	.533	.540	.500	.566	.510	.466	.533	.480
TSEL IG	.866	.533	.810	.666	.566	.700	.733	.500	.770	.766	.666	.720	.533	.600	.700	.500	.566	.710
HillClimbing	.652	.633	.630	.633	.636	.580	.633	.566	.640	.676	.566	.620	.676	.534	.586	.689	.523	.590
GreedyTopDown	.666	.600	.800	.703	.633	.780	.633	.466	.830	.703	.566	.820	.752	.700	.850	.833	.500	.830
Feature Space Compression																		
initialSHSEL IG	.846	.572	.864	**.948**	.907	.880	.861	.810	**.886**	.914	.789	.740	.929	.868	.746	.912	.750	.918
initialSHSEL C	.267	.557	.875	.104	.888	**.938**	.279	**.956**	.656	.441	.893	.890	.831	.742	.786	.627	.805	.805
pruneSHSEL IG	**.930**	.911	.796	.925	**.933**	.824	**.899**	.944	.850	**.956**	.877	.877	.955	**.969**	.863	.956	.873	.791
pruneSHSEL C	.697	.896	.800	.639	.636	.667	.781	.696	.823	.795	.776	.849	.692	.726	.742	.731	.826	.750
SIG	.922	**.922**	.861	.842	.842	.753	.865	.930	.865	.886	.595	.708	.891	.719	.525	.900	.704	.704
SC	.717	.909	**.880**	.693	.900	.159	.750	.692	.869	.379	.493	.769	.628	.736	.742	.667	.727	.289
TSEL Lift	.750	.750	.750	.706	.706	.706	.687	.687	.687	.857	.857	.857	.827	.827	.827	.814	.814	.814
TSEL IG	.836	.836	.836	.866	.866	.866	.856	.856	.856	.926	**.926**	**.926**	.965	.965	**.965**	**.970**	**.970**	**.970**
HillClimbing	.770	.770	.770	.751	.751	.751	.805	.805	.805	.792	.792	.792	.776	.776	.776	.795	.795	.795
GreedyTopDown	.399	.399	.399	.370	.370	.370	.356	.356	.356	.470	.470	.470	.404	.404	.404	.438	.438	.438
Harmonic Mean of Classification Accuracy and Feature Space Compression																		
initialSHSEL IG	.917	.679	.872	.973	.831	.865	.925	.837	.888	.955	.826	.804	.963	.901	.803	.933	.741	**.914**
initialSHSEL C	.381	.592	.853	.182	.739	.827	.394	.762	.713	.560	.763	.808	.697	.683	.757	.630	.641	.832
pruneSHSEL IG	**.964**	**.922**	**.854**	**.961**	**.861**	**.865**	**.946**	**.885**	**.901**	**.977**	**.871**	**.916**	**.977**	**.915**	**.918**	**.977**	**.835**	.878
pruneSHSEL C	.773	.764	.851	.736	.666	.766	.790	.729	.859	.797	.771	.878	.795	.776	.805	.819	.737	.830
SIG	.940	.911	.845	.914	.820	.820	.896	.840	.898	.940	.694	.805	.942	.799	.680	.947	.718	.815
SC	.708	.791	.800	.696	.766	.262	.740	.642	.775	.500	.567	.733	.662	.700	.754	.683	.713	.415
TSEL Lift	.636	.600	.628	.667	.685	.665	.505	.579	.605	.682	.657	.662	.623	.672	.631	.593	.644	.604
TSEL IG	.851	.651	.822	.753	.685	.774	.790	.631	.810	.839	.775	.810	.687	.740	.811	.660	.715	.820
HillClimbing	.706	.695	.693	.687	.689	.654	.709	.665	.713	.730	.660	.695	.723	.633	.668	.738	.631	.677
GreedyTopDown	.499	.479	.533	.485	.467	.502	.456	.404	.499	.564	.514	.598	.526	.513	.548	.574	.467	.573

6 Conclusion and Outlook

In this paper, we have proposed a feature selection method exploiting hierarchic relations between features. It runs in two steps: it first removes redundant features along the hierarchy's paths, and then prunes the remaining set based on the features' predictive power. Our evaluation has shown that the approach outperforms standard feature selection techniques as well as with recent approaches which use hierarchies.

So far, we have only considered classification problems. A generalizing of the pruning step to tasks other than classification would be an interesting extension. While a variant for regression tasks seems to be rather straight forward, other problems, like association rule mining, clustering, or outlier detection, would probably require entirely different pruning strategies.

Furthermore, we have only regarded simple hierarchies so far. When features are organized in a complex ontology, there are other relations as well, which may be exploited for feature selection. Generalizing the approach to *arbitrary* relations between features is also a relevant direction of future work.

Acknowledgements. The work presented in this paper has been partly funded by the German Research Foundation (DFG) under grant number PA 2373/1-1 (Mine@LOD).

References

1. Blum, A.L., Langley, P.: Selection of relevant features and examples in machine learning. Artificial Intelligence 97, 245–271 (1997)
2. Dash, M., Liu, H.: Feature selection for classification. Intelligent Data Analysis 1(3), 131–156 (1997)
3. Jeong, Y., Myaeng, S.-H.: Feature selection using a semantic hierarchy for event recognition and type classification. In: International Joint Conference on Natural Language Processing (2013)
4. John, G.H., Kohavi, R., Pfleger, K.: Irrelevant features and the subset selection problem. In: ICML 1994, pp. 121–129 (1994)
5. Lehmann, J., Isele, R., Jakob, M., Jentzsch, A., Kontokostas, D., Mendes, P.N., Hellmann, S., Morsey, M., van Kleef, P., Auer, S., Bizer, C.: DBpedia – A Large-scale, Multilingual Knowledge Base Extracted from Wikipedia. Semantic Web Journal (2013)
6. Lu, S., Ye, Y., Tsui, R., Su, H., Rexit, R., Wesaratchakit, S., Liu, X., Hwa, R.: Domain ontology-based feature reduction for high dimensional drug data and its application to 30-day heart failure readmission prediction. In: International Conference on Collaborative Computing (Collaboratecom), pp. 478–484 (2013)
7. Mendes, P.N., Jakob, M., García-Silva, A., Bizer, C.: Dbpedia spotlight: Shedding light on the web of documents. In: Proceedings of the 7th International Conference on Semantic Systems, I-Semantics (2011)
8. Molina, L.C., Belanche, L., Nebot, À.: Feature selection algorithms: A survey and experimental evaluation. In: International Conference on Data Mining (ICDM), pp. 306–313. IEEE (2002)
9. Paulheim, H.: Generating possible interpretations for statistics from linked open data. In: Simperl, E., Cimiano, P., Polleres, A., Corcho, O., Presutti, V. (eds.) ESWC 2012. LNCS, vol. 7295, pp. 560–574. Springer, Heidelberg (2012)

10. Paulheim, H., Fürnkranz, J.: Unsupervised Generation of Data Mining Features from Linked Open Data. In: International Conference on Web Intelligence, Mining, and Semantics, WIMS 2012 (2012)
11. Paulheim, H., Ristoski, P., Mitichkin, E., Bizer, C.: Data mining with background knowledge from the web. In: RapidMiner World (to appear, 2014)
12. Platt, J.C.: Sequential minimal optimization: A fast algorithm for training support vector machines. Technical report, Advances in Kernel Methods - Support Vector Learning (1998)
13. Wang, B.B., Bob Mckay, R.I., Abbass, H.A., Barlow, M.: A comparative study for domain ontology guided feature extraction. In: Australasian Computer Science Conference (2003)

Incorporating Regime Metrics into Latent Variable Dynamic Models to Detect Early-Warning Signals of Functional Changes in Fisheries Ecology

Neda Trifonova[1], Daniel Duplisea[2], Andrew Kenny[3], David Maxwell[3], and Allan Tucker[1]

[1] Department of Computer Science, Brunel University, London, UK
[2] Fisheries and Oceans, Canada
[3] Centre for Environment, Fisheries and Aquaculture Science, Lowestoft, UK

Abstract. In this study, dynamic Bayesian networks have been applied to predict future biomass of geographically different but functionally equivalent fish species. A latent variable is incorporated to model functional collapse, where the underlying food web structure dramatically changes irrevocably (known as a *regime shift*). We examined if the use of a hidden variable can reflect changes in the trophic dynamics of the system and also whether the inclusion of recognised statistical metrics would improve predictive accuracy of the dynamic models. The hidden variable appears to reflect some of the metrics' characteristics in terms of identifying regime shifts that are known to have occurred. It also appears to capture changes in the variance of different species biomass. Including metrics in the models had an impact on predictive accuracy but only in some cases. Finally, we explore whether exploiting expert knowledge in the form of diet matrices based upon stomach surveys is a better approach to learning model structure than using biomass data alone when predicting food web dynamics. A non-parametric bootstrap in combination with a greedy search algorithm was applied to estimate the confidence of features of networks learned from the data, allowing us to identify pairwise relations of high confidence between species.

1 Introduction

Some spectacular collapses in fish stocks have occurred in the past 20 years but the most notable is the once largest cod (*Gadus morhua*) stock in the world, the Northern cod stock off eastern Newfoundland, which experienced a 99% decline in biomass (the total quantity or weight of organisms in a given area or volume). Such regions have experienced a "regime shift" or moved to an "alternative stable state" and are unlikely to return to a cod dominated community without some influence beyond human control [9]. The main question for environmental management is whether such changes could have been detected by early-warning signals. There is a growing literature that addresses indicators that can be used as early-warning signals of an approaching critical transition (or regime shift) [3].

S. Džeroski et al. (Eds.): DS 2014, LNAI 8777, pp. 301–312, 2014.

Regime (functional) changes can affect the abundance and distribution of fish populations, either directly or by affecting prey or predator populations [9]. Different species may have similar functional roles (the functional status of an organism) within a system depending on the region. For example, one species may act as a predator of another which regulates a population in one location, but another species may perform an almost identical role in another location. If we can model the function of the interaction rather than the species itself, data from different regions can be used to confirm key functional relationships, to generalise over systems and to predict impacts of forces such as fishing and climate change.

We explored functional relationships (such as predator, prey) that are generalizable between different oceanic regions allowing predictions to be made about future biomass. In particular, we exploited multiple fisheries datasets in order to identify species with similar functional roles in different fish communities. The species were then used to predict functional collapse in their respective regions through the use of Dynamic Bayesian Networks (DBNs) with latent variables. Formally, a Bayesian Network (BN) exploits the conditional independence relationships over a set of variables, represented by directed acyclic graphs (DAG) [6]. Modelling time series is achieved by the DBN, where nodes represent variables at particular time slices [6]. Closely related to the DBN is the Hidden Markov Model (HMM) which models the dynamics of a dataset through the use of a latent or hidden variable. This latent variable is used to infer some underlying state of the series and can be applied through an autoregressive link which can capture relationships of a higher order. Hidden variables can also be incorporated to model unobserved variables and missing data by using the EM algorithm [2]. This represents the most challenging inference problem here as we make computationally complex predictions involving dynamic processes. However, the hidden variable is chosen to most easily reflect such complex interdependencies between the acting variables. See Fig. 1 for an illustration of the architecture of the DBN used in this paper including a hidden variable.

In this paper, we investigate the reliability of our modelling approach in detecting *early-warning* signals of functional change across *different geographic regions*. We explore how the latent variable reflects the *regime metrics* (the applied statistical indicators of functional changes in the study) and the variability of exploited fisheries and to what extent including them in our models impacts the expected values of the latent variable. We also explore how these models can be used to identify species that are key to regime shifts in different regions. An earlier work by [12] explores functionally equivalent species but here we further adopt the approach to predict functional collapse by investigating *early-warning signals* and comparing learned BN topology prior to and after suspected regime changes. At larger spatial scales, although fishing can still be the dominant driver of regime changes, the consequences of fishing are not predictable without understanding the trophic (relating to the feeding habits of different organisms in a food chain) dynamics [9]. A clear example is the Scotian Shelf, where fishing has led to a restructuring of the ecosystem [9]. We investigate whether exploiting expert knowledge (in the form of diet matrix, that represents the prey-predator functional relationships between species) of this region or learning model structure from the data alone is a better approach when predicting food web dynamics.

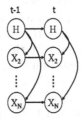

Fig. 1. The Dynamic Bayesian Model with N variables used in this study where H denotes the unmeasured hidden variable

2 Methods

We apply our modelling approach to predict species biomass and functional change in three different geographical regions: North Sea (NS), Georges Bank (GB) and East Scotian Shelf (ESS) (Fig. 2). For all of the datasets, the biomass was determined from research vessel fish trawling surveys assuring consistent sampling from year to year, resulting in 44 species for NS (1967-2009), 44 species for GB (1963-2008), and 42 for ESS (1970-2006). Large groundfish declines occurred on GB and ESS which resulted in the year 1988 being designated as a collapse year for GB and 1992 for ESS. Despite the extremely high fishing pressure in NS and complex climate-ocean interactions, it is difficult to distinguish a radical switch in the system that might be termed a regime shift. However, experts refer to some ecosystem changes in the period of late 1980s to mid-1990s. In addition to survey data on fish abundance, grey seal abundance and plankton time series were also included in the analysis.

The experiments involve the prediction of a pre-selected variable (here functional collapse, represented by the latent variable) based on the values of other

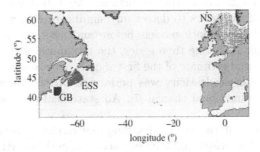

Fig. 2. Regions of the three surveys (shaded area) corresponding to the three datasets: Georges Bank (GB), East Scotian Shelf (ESS) and the North Sea (NS)

variables (here species biomass). We select a number of species that are associated with cod collapse by using wrapper feature selection with a Bayesian Network Classifier on GB data where the class node is a binary variable that represents functional collapse in GB. The greedy K2 search algorithm [4] is used to build the BN classifiers. A bootstrap approach is employed to repeat the following 1000 times: learn BN structure with the K2 algorithm and score the proportion of times that links are associated with the class node during the bootstrap. This is a form of wrapper feature selection [8] and scores each variable by taking into account their interaction with other variables through the use of a classifier model. Next, we identify the equivalent species in the other two datasets using the features discovered using Algorithm 1. The *functional equivalence* search algorithm [12] works by using a BN model, where the given function is in the form of a predefined structure, BN_1, and a set of variables, $vars_1$, parameterised on $data_1$, (here a BN model parameterised on the GB data). Simulated annealing [10] is applied to identify variables in another dataset, $data_2$, (here species in the ESS and NS datasets) that best fit this model. We set *iterations* = 1000 and t_{start} = 1000 as these were found through experimentation to allow convergence to a good solution. The fit is scored using the log likelihood score [4]. In Algorithm 1 *UnifRand* represents a random value generated from uniform distribution with limits between (0,1).

After choosing the species, we want to predict their biomass and the functional collapse in the relevant geographic region. For example to predict functional collapse we compute $P(H^t|X^t, X^{t-1})$, where H^t represents the hidden variable (functional collapse) and X^t represents all observed variables at times t. First, we infer the biomass at time t (Fig.1) by using the observed evidence and then use the predicted variable states to infer the hidden state at time t. The hidden variable was parameterised using the EM algorithm.

The metrics: variance and autocorrelation were calculated on a window of data, set to size 10, so that each metric captures the value of interest over the previous 10 years. Two sets of experiments were then conducted: one that excludes the regime metrics to examine the expected state of the fitted hidden variable (HDBN) and in the other, metrics were included in the model (HDBN + metrics) to see if they improve prediction of species biomass. Non-parametric bootstrap analysis [6] was applied 250 times for each variant of the model to obtain statistical validation in the predictions. An F-test was performed over a sliding window of five years to detect any significant changes in the slope of the hidden variable from both models before and after the expected collapse [7]. Given a breakpoint in the time series, the minimum of this sequence of p-values gives a potential estimate of the first signals of ecosystem change in time. Levene's test on homoscedasticity was performed on the variance before and after the predicted functional change [7]. All statistical tests were reported at 5% significance level.

For the next part of the study- learning the model structure, the species biomass data was discretised and a greedy search algorithm: REVEAL [11] was applied to learn the structure of the DBN model for each region. The nonparametric bootstrap was also applied 250 times to identify statistical confidence in the discovered network links with threshold ≥ 0.5. Features with statistical confidence above the threshold are labelled "positive" or "negative" if the confidence is below the threshold. We measure the number of "true positives", correct features of the generating network (based upon a pre-defined diet matrix, established by stomach content surveys for the relevant region) or "false negatives", correct features labelled as negatives [6].

Algorithm 1. The *functional equivalence* search algorithm.

1: Input: $t_{start}, iterations, data_1, data_2, vars_1, BN_1$
2: Parameterize Bayesian Network, BN_1, from $data_1$
3: Generate randomly selected variables in $data_2 : vars_2$
4: Use $vars_2$ to score the fit with selected model $BN_1 : score$
5: Set $bestscore = score$
6: Set initial temperature: $t = t_{start}$
7: **for** $i = 1$ to *iterations* **do**
8:     Randomly replace one selected variable in $data_2$ and rescore: *rescore*
9:     $dscore = rescore - bestscore$
10:    **if** $dscore \geq 0$ OR *UnifRand and* $(0,1) < \exp^{(dscore/t)}$ **then**
11:        $bestscore = rescore$
12:    **else**
13:        Undo variable switch in $vars_2$
14:    **end if**
15:    Update the temperature: $t = t$ x 0.9
16: **end for**
17: Output: $vars_2$

---

3   Results

The wrapper feature selection approach managed to identify the species likely to be associated with cod collapse on GB, Table 1 illustrates the resulting ordered list of most relevant variables (BN wrapper confidence reported in brackets). For example, herring (*Clupea harengus*) was identified as a key species and it is known that there were large abundance changes in the late 1980s [1].

The species from ESS and NS that were identified by the *functional equivalence* search algorithm are ranked based upon the confidence associated with their equivalent species in GB (Table 2, confidence reported in brackets). A striking feature of the identified ESS species is the presence of many deepwater species like argentine (*Argenti silus*) and grenadier (*Nezumia bairdi*). That could be an indication of the water cooling that occurred in the late 1980s and early 1990s. In the NS, most of the selected species are commercially desirable and some experienced large declines in biomass in this period, though the nature of the species is not dissimilar to GB when compared with ESS, which showed the appearance of some qualitatively different species. For example, megrim (*Lepidorhombus whiffiagonis*) and solenette (*Buglossidium luteum*), not fished commercially, are also selected as being implicated by other groundfish decline. Such species would probably be less likely to

Table 1. Wrapper feature selection results for GB region

GB Wrapper Feature Selection	
1. Thorny skate (1.0)	14. Lady crab (0.24)
2. Blackbelly rosefish (0.98)	15. Spotted flounder (0.23)
3. Herring (0.97)	16. *Calanus* spp. (0.20)
4. Fourbeard rockling (0.82)	17. American lobster (0.13)
5. Cusk (0.75)	18. American plaice (0.13)
6. *Pseudocalanus* spp. (0.65)	19. Ocean pout (0.09)
7. Gulf stream flounder (0.47)	20. Little skate (0.07)
8. *Centropages typicus* (0.44)	21. Sea scallop (0.07)
9. Atlantic rock crab (0.41)	22. Sand lance (0.05)
10. Witch flounder (0.29)	23. Winter flounder (0.03)
11. American angler (0.28)	24. Moustache sculpin (0.02)
12. White hake (0.26)	25. Silver hake (0.02)
13. Krill (0.25)	26. Longfin hake (0.02)

Table 2. The functionally equivalent species to GB dataset for ESS and NS. These are each ordered based upon their relevance to species in Table 1.

Functionally Equivalent Species	
ESS	**NS**
1. Cod (1.0)	1. European plaice (0.98)
2. Pollock (0.58)	2. Atlantic halibut (0.93)
3. Grenadier (0.51)	3. Cod (0.87)
4. White hake (0.50)	4. Lumpfish (0.78)
5. Mackerel (0.23)	5. Thorny skate (0.53)
6. Rockfish (0.22)	6. Whiting (0.50)
7. Grey seals (0.20)	7. Argentine (0.42)
8. Argentine (0.15)	8. Megrim (0.35)
9. Atlantic halibut (0.12)	9. Haddock (0.29)
10. Spiny dogfish (0.11)	10. Atlantic wolfish (0.24)
11. Little skate (0.09)	11. American plaice (0.21)
12. Atlantic wolfish (0.07)	12. Common dragonet (0.20)
13. American plaice (0.04)	13. Solenette (0.16)
14. Red hake (0.04)	14. Poor cod (0.13)
15. Silver hake (0.03)	15. Sprat (0.05)
16. Little hake (0.01)	16. Pollock (0.02)

be considered as indicator species of regime shifts elsewhere. However, here the *functional equivalence* search algorithm performed well in terms of identifying key species, associated with functional changes in the relevant regions which would be potentially beneficial when investigating the reliability of our modelling approach in terms of detecting signals of functional change.

We now explore the latent variable models for ESS and NS learnt from the selected functionally equivalent species. We also focus on the relationship of the latent variable to the two regime metrics. The expected value of the hidden variable for ESS managed to capture some of the key predictive qualities of the metrics in terms of identifying a regime shift that is known to have occurred. The ESS latent variable model (HDBN) in Fig. 3a demonstrates a large fluctuation between 1980 and 1990 with a steep increase in 1984 and 1989 prior to the time of the expected regime shift and it was then followed by a consistent decline following the collapse in 1992. The hidden variable increase coincides with a steep increase in variance (Fig.3c) in 1985, all above the 95% confidence upper interval. However, lowest p-value ($F_{21,13} = 11.59, p < .0001$) was reported at the time of the known collapse in 1992 (balanced design of the sliding window). The assumption of homoscedasticity was not met ($F_{1,25} = 4.05, p < .05$), indicating variance inequality before and after the collapse. The autocorrelation showed little variation and it remained close to 1, as already illustrated for ecosystems undergoing a transition [5]. According to theoretical expectations of critical slowing down, both latent variable and variance appear to increase prior to the expected regime shift and follow a consistent decline throughout time following the collapse correctly resulting in a clear early-warning signal to forewarn a major ecosystem change. After the addition of the metrics in the model, the latent variable was more stable and still reflective of capturing the correct dynamics and characterised by rising trends in time prior to the expected transition (HDBN + metrics in Fig. 3b - Note this starts from 1980 due to the windowing required for calculating the metrics). In 1990, the lowest p-value was recorded ($F_{9,15} = 23.90, p < .0001$) which was actually lower than the p-value reported by the HDBN, suggesting that the latent variable in combination with the metrics might be performing better in earlier detection of change in the time series, though having a negative impact on the predictive performance of biomass (SSE HDBN: 4.83 and SSE HDBN+ metrics: 13.65) (Fig.4).

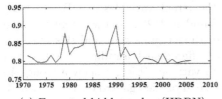

(a) Expected hidden value (HDBN)

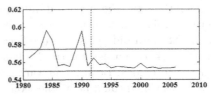

(b) Expected hidden value (HDBN+ metrics)

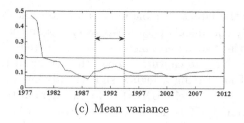

(c) Mean variance

Fig. 3. The expected values of the discovered hidden variable from HDBN (a), HDBN+ metrics (b) and mean variance (c) for ESS. The dashed line indicates the time of the regime shift in 1992. The solid line indicates upper and lower 95% confidence intervals, obtained from bootstrap predictions' mean and standard deviation.

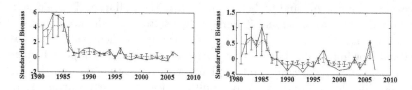

Fig. 4. Biomass predictions generated by HDBN+ metrics of cod (left), and silver hake (right) for ESS region. 95% confidence intervals report bootstrap predictions' mean and standard deviation. Dashed line indicates predictions by the model, whilst solid indicates standardised observed biomass for the time period 1980-2006.

The expected value of the hidden variable for NS (Fig.5a) was characterised by some fluctuation up to early 1980s followed by a small decrease below the lower confidence level coinciding with the time around the functional changes in late 1980s to mid-1990s. Nevertheless, the F-test did not detect any significantly different changes in the slope of the hidden variable. These values are much smaller than for the expected values of the latent variables in GB and ESS. Perhaps this is not surprising as it was found in [12] that the latent variable in the NS data did not seem to reflect a distinct regime shift and this fits with the general consensus that the NS has not suffered such a radical switch as the other two regions. Both latent variable and variance (Fig.5c) show a trough in late 1980s which could be a reflection of the end of the "gadoid outburst" where groundfish were very abundant for about the previous 25 years [1]. Here, the condition for equality of variance before and after the predicted functional change was fulfilled ($F_{1,31} = 1.40, p = 0.08$). The latent variable from the HDBN+ metrics (Fig.5b) was more explicit and clear, finding the lowest p-value ($F_{10,12} = 0.27, p < .05$) in 1988 when first functional changes are believed to have occurred in the system according to experts. NS is a diverse system, subject to external anthropogenic forcing and internal environmental variation and as such, it is suggested that it seems to exhibit a range of discontinuous disturbances which would be more difficult to interpret by the hidden variable alone [3]. However, the effect of the metrics on the latent variable assisted in the correct identification of the time period where we would expect some functional change or disturbance in NS. Results for NS showed reliable prediction of species biomass, with improved ability of the dynamic models when used in combination with the published metrics (SSE model: 5.50, SSE model+ metrics: 2.82).

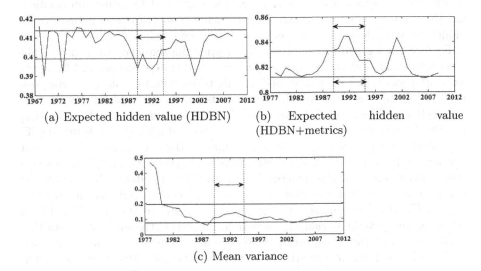

(a) Expected hidden value (HDBN)   (b)   Expected   hidden   value
                                                 (HDBN+metrics)

(c) Mean variance

Fig. 5. The expected value of the discovered hidden variable from HDBN (a), HDBN+ metrics (b) and mean variance (c) for NS. The dashed lines indicate the time period of the regime shift. The solid line indicates upper and lower 95% confidence intervals, obtained from bootstrap predictions' mean and standard deviation.

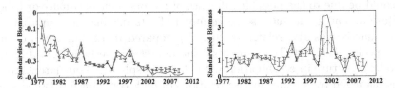

Fig. 6. Biomass predictions generated by HDBN+ metrics of cod (left), and haddock (right) for NS region. 95% confidence intervals report bootstrap predictions' mean and standard deviation. Dashed line indicates predictions by the model, whilst solid indicates standardised observed biomass for the time period 1977-2009.

To summarise, the models that included the regime metrics performed better in terms of capturing the correct dynamics earlier in the time series. The latent variable alone managed to reflect the ecosystem dynamics but that was more evident in the ESS region with a larger regime shift.

We now turn to the analysis of the learned networks by separating the data before and after the regime shift according to experts and comparing them to the networks generated by data split from our latent variable models (timing identified from F-test significant results in the first part of the study). Some high confidence relationships were identified which represent likely models of the functional interactions between species. The direction of the discovered significant links did not mean causation and it was not considered in the comparison with the generating diet matrix as we were interested in finding correctly identified species *associations*. Note that some of the discovered links, not directly relating to the diet matrix, could have been explained by either intermediate variables not included in the model or common observed effects acting on the model variables, however this was not the purpose here.

For ESS the learned network before the regime shift based on the experts' split was complex, identifying 7 significant features (four true positives) whilst the network after was rather simplified, finding only two significant links (one true positive), suggesting the influence of a radical switch in the system following the fisheries collapse. The network before 1990 (Fig.7b) (as found by HDBN+metrics) identified 8 significant links (four true positives) and after (Fig.7c)- five significant links (four true positives). When comparing the networks of experts' split and data split, three of the significant links were preserved, one of them was a true positive. Learning the structure before and after the data split for ESS was a much better case in terms of detecting more correct associations with the diet matrix (Fig.7a). To recap, species selected in ESS were based on a regime shift in GB using the *functional equivalence* search, suggesting the successful algorithm performance in terms of capturing the correct structure and food web dynamics.

For NS, the learned network before the experts' split identified five significant features (one true positive) and the network after- 7 significant features but none of them were true positives. The network before 1988 (as found by HDBN+metrics) identified four significant links and after: one significant link, no true positives. The relative simplicity of the NS networks and much lower number of correctly identified associations with the diet matrix compared to ESS, could be due to the possible influence of factors such as climate or fisheries exploitation that might have some common effects on different variables. The NS diet matrix was also relatively "poor" compared to ESS in terms of quantity of species recorded. To summarise, the bootstrap methodology of learning the model structure in combination with the data split from our latent variable models managed to detect pairwise relations of high confidence between species providing us with assumptions about the relevant food web structure and dynamics. Also, in both regions, significant links found before the data split, were generally reduced after, implying a signal of functional changes in the ecosystems.

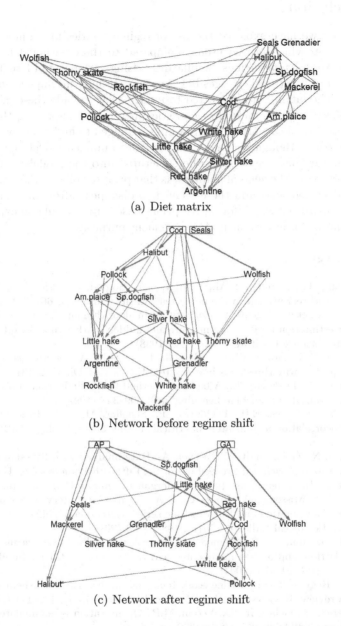

(a) Diet matrix

(b) Network before regime shift

(c) Network after regime shift

Fig. 7. Diet matrix (a) with the network before (b) and after (c) the regime shift for ESS, generated by the data split using REVEAL. The width of edges corresponds to the computed confidence level (bold line: 0.5 and light line: 0.1). The squared nodes are significant themselves. For the diet matrix direction of links represents predator-prey interactions. In bottom network (c): AP- American plaice and GA- Greater argentine.

4   Conclusion

In this paper we have explored the use of regime metrics in conjunction with latent variables which proved useful (compared to these without) in terms of detecting *early-warning* signals (significantly rising variance and latent variable fluctuations) of functional changes but it seemed to have an impact on biomass prediction. The latent variables fitted to models that exclude these metrics appear to reflect some of their characteristics in terms of capturing the correct trophic dynamics. The learned network links managed to find some overlap with the diet matrices, though not many, maybe due to implicit correlations (and so more latent variables may need to be structured into the models to deal with this). Nevertheless, the general finding was that prior to collapse there were more correctly identified links and these seemed to disappear after the regime shift. Further work will involve informative priors based upon available expertise to create scenarios for environmental management purposes.

References

1. Beaugrand, G., Brander, K.M., Lindley, J.A., Souissi, S., Reid, P.C.: Plankton effect on cod recruitment in the north sea. Nature 426(6967), 661–664 (2003)
2. Bilmes, J.A., et al.: A gentle tutorial of the em algorithm and its application to parameter estimation for gaussian mixture and hidden markov models. International Computer Science Institute 4(510), 126 (1998)
3. Carpenter, S.R., Brock, W.A., Cole, J.J., Pace, M.L.: A new approach for rapid detection of nearby thresholds in ecosystem time series. Oikos (2013)
4. Cooper, G.F., Herskovits, E.: A bayesian method for the induction of probabilistic networks from data. Machine Learning 9(4), 309–347 (1992)
5. Dakos, V., Van Nes, E.H., D'Odorico, P., Scheffer, M.: Robustness of variance and autocorrelation as indicators of critical slowing down. Ecology 93(2), 264–271 (2012)
6. Friedman, N., Goldszmidt, M., Wyner, A.: Data analysis with bayesian networks: A bootstrap approach. In: Proceedings of the Fifteenth Conference on Uncertainty in Artificial Intelligence, pp. 196–205. Morgan Kaufmann Publishers Inc. (1999)
7. Gröger, J.P., Missong, M., Rountree, R.A.: Analyses of interventions and structural breaks in marine and fisheries time series: Detection of shifts using iterative methods. Ecological Indicators 11(5), 1084–1092 (2011)
8. Inza, I., Larrañaga, P., Blanco, R., Cerrolaza, A.J.: Filter versus wrapper gene selection approaches in dna microarray domains. Artificial Intelligence in Medicine 31(2), 91–103 (2004)
9. Jiao, Y.: Regime shift in marine ecosystems and implications for fisheries management, a review. Reviews in Fish Biology and Fisheries 19(2), 177–191 (2009)
10. Kirkpatrick, S., Gelatt Jr., D., Vecchi, M.P.: Optimization by simmulated annealing. Science 220(4598), 671–680 (1983)
11. Liang, S., Fuhrman, S., Somogyi, R., et al.: Reveal, a general reverse engineering algorithm for inference of genetic network architectures. In: Pacific Symposium on Biocomputing, vol. 3, p. 2 (1998)
12. Tucker, A., Duplisea, D.: Bioinformatics tools in predictive ecology: applications to fisheries. Philosophical Transactions of the Royal Society B: Biological Sciences 367(1586), 279–290 (2012)

An Efficient Algorithm for Enumerating Chordless Cycles and Chordless Paths

Takeaki Uno and Hiroko Satoh

National Institute of Informatics, 2-1-2 Hitotsubashi,
Chiyoda-ku, Tokyo 101-8430, Japan
{uno,hsatoh}@nii.ac.jp

Abstract. A chordless cycle (induced cycle) C of a graph is a cycle without any chord, meaning that there is no edge outside the cycle connecting two vertices of the cycle. A chordless path is defined similarly. In this paper, we consider the problems of enumerating chordless cycles/paths of a given graph $G = (V, E)$, and propose algorithms taking $O(|E|)$ time for each chordless cycle/path. In the existing studies, the problems had not been deeply studied in the theoretical computer science area, and no output polynomial time algorithm has been proposed. Our experiments showed that the computation time of our algorithms is constant per chordless cycle/path for non-dense random graphs and real-world graphs. They also show that the number of chordless cycles is much smaller than the number of cycles. We applied the algorithm to prediction of NMR (Nuclear Magnetic Resonance) spectra, and increased the accuracy of the prediction.

1 Introduction

Enumeration is a fundamental problem in computer science, and many algorithms have been proposed for many problems, such as cycles, paths, trees and cliques[5, 8, 10, 12, 13, 17]. However, their applications to real world problems have not been studied intensively, due to the handling needed for the huge amount of and the high computational cost. However, this situation is now changing, thanks to the rapid increase in computational power, and the emergence of data centric science. For example, the enumeration of all substructures frequently appearing in a database, i.e., frequent pattern mining, has been intensively studied. This method is adopted for capturing the properties of databases, or for discovering new interesting knowledge in databases. Enumeration is necessary for such tasks because the objectives cannot be expressed well in mathematical terms. The use of good models helps to reduce the amount of output, and the use of efficient algorithms enables huge databases to be more easily handled[9, 1, 11]. More specifically, introducing a threshold value for the frequency, which enables controlling the number of solutions. In such areas, minimal/maximal solutions are also enumerated to reduce the number of solutions. For example, enumerating all cliques is usually not practical while enumerating all maximal cliques, i.e. cliques included in no other cliques, is often practical[13, 10]. In real-world

S. Džeroski et al. (Eds.): DS 2014, LNAI 8777, pp. 313–324, 2014.

sparse graphs, the number of maximal cliques is not exponential, so, even in large-scale graphs, the maximal cliques can often be enumerated in a practically short time by a stand alone PC even for graphs with millions of vertices. However, the enumeration of maximum cliques, that have the maximum number of vertices among all cliques, is often not acceptable in practice, since the purpose of enumeration is to find all locally dense structures, and finding only maximum cliques will lose relatively small dense structures, thus it does not cover whole the data.

Paths and cycles are two of the most fundamental graph structures. They appear in many problems in computer science and are used for solving problems, such as optimizations (e.g. flow problems) and information retrieval (e.g. connectivity and movement of objects). Paths and cycles themselves are also used to model other objects. For example, in chemistry, the size and the fusing pattern of cycles in chemical graphs, representing chemical compounds, are considered to be essential structural attributes affecting on several important properties of chemical compounds, such as spectroscopic output, physical property, chemical reactivity, and biological activity.

For a path/cycle P, an edge connecting two vertices of P but not included in P is called a chord. A path/cycle without a chord is called a chordless path/cycle. Since a chordless cycle includes no other cycle as a vertex set, it is considered minimal. Thus, chordless cycles can be used to represent cyclic structures. For instance, the size and fusing pattern of chordless cycles in chemical graphs as well as other properties of chemical structures are taken into account when selecting data for prediction of nuclear magnetic resonance (NMR) chemical shift values[14]. Most chemical compounds contain cycles. In chemistry, the term 'ring' is used instead of 'cycle', for example a cycle consisting of 5 vertices is called 5-membered ring. Since the character of ring structures of chemical compounds is assumed to be important to study the nature of the structure-property relationships, the ring perception is one of classical questions [2–4, 7] in the context of chemical informatics, so called chemoinformatics. Several kinds of ring structures, such as all rings and the smallest set of smallest ring (SSSR) are usually included in a basic dataset of chemical information. NMR chemical shift prediction is a successful case where the information about chordless cycles is employed to improve the accuracy of the prediction. The path/cycle enumeration is supposed to be useful also for analysis of network systems such as Web and social networks.

In this paper, we consider the problem of enumerating all chordless paths (resp., cycles) of the given graph. While optimization problems for paths and cycles have been studied well, their enumeration problems have not. This is because there are huge numbers of paths and cycles even in small graphs. However, we can reduce the numbers so that the problem becomes tractable by introducing the concept of chordless. The first path/cycle enumeration algorithm was proposed by Read and Tarjan in 1975[12]. Their algorithm takes as input a graph $G = (V, E)$ and enumerates all cycles, or all paths connecting given vertices s and t, in $O(|V| + |E|)$ time for each. The total computation time is $O((|V| + |E|)N)$

Fig. 1. Left bold cycle is a chordless cycle, right bold cycle has two chords

where N is the number of output cycles/paths. Ferreira et al. [6] recently proposed a faster algorithm, that takes time linear in the output size, that is the sum of the lengths of the paths.

The chordless version was considered by Wild[19]. An algorithm based on the principle of exclusion is proposed, but the computational efficiency was not considered deeply. In this paper, we propose algorithms for enumerating chordless cycles and chordless paths connecting two vertices s and t (reported in 2003[18]). Note that chordless cycles can be enumerated by chordless path enumeration. The running time of the algorithm is $O(|V|+|E|)$ for each, the same as the Read and Tarjan algorithm.

We experimentally evaluate the practical performance of the algorithms for random graphs and real-world graphs. The results show that its practical computation time is much smaller than $O(|V|+|E|)$, meaning that the algorithms can be used for large-scale graphs with non-huge amount of solutions. The results also show that the number of chordless cycles is drastically small compared to the number of usual cycles.

2   Preliminaries

A *graph* is a combination of a vertex set and an edge set such that each edge is a pair of vertices. A graph G with vertex set V and edge set E is denoted by $G = (V, E)$. An edge e of pair v and u is denoted by $\{u, v\}$. We say that the edge *connects* u and v, e is *incident* to u and v, and v and u are *adjacent* to each other, and call u and v *end vertices* of e. An edge with end vertices that are the same vertex is called a *self-loop*. Two edges having the same end vertices u and v are called *multi-edges*. We deal only with graphs with neither a self-loop nor a multi-edge. This restriction does not lose the generality of the problem formulation.

A path is a graph of vertices and edges composing a sequence $v_1, \{v_1, v_2\}, v_2, \{v_2, v_3\}, \ldots, \{v_{k-1}, v_k\}, v_k$ satisfying $v_i \neq v_j$ and $i \neq j$. The v_1 and v_k are called the *end vertices* of the path. If the end vertices of P are s and t, the path is called an *s-t* path. When $v_1 = v_k$ holds, a path is called a cycle. Here we represent paths and cycles by vertex sequences, such as $(v_1, \ldots, v_k)$. An edge connecting two vertices of a path/cycle P and not included in P is called a *chord* of P. A path/cycle P such that the graph includes no chord of P is called *chordless*.

Figure 1 shows examples. In a set system composed of the vertex sets of cycles (resp., s-t paths), the vertex set of a chordless cycle (resp., s-t path) is a minimal element.

For a graph G and a vertex subset S of G, $G \setminus S$ denotes the graph obtained from G by removing all vertices of S and all edges incident to some vertices in S. For a vertex v, $N(v)$ denotes the *neighbor* of v, that is, the set of vertices adjacent to v. For a vertex set S and a vertex v, $S \setminus v$ and $S \cup v$ denote $S \setminus \{v\}$ and $S \cup \{v\}$, respectively. For a path P and its end vertex v, $P \setminus v$ denotes the path obtained by removing v from P.

Property 1. There is a chordless s-t path if and only if there is an s-t path.

Proof. A chordless s-t path is an s-t path, thus only if part is true. If an s-t path exists, a shortest path from s to t is a chordless s-t path, and thus it always exists.   □

Property 2. A vertex v is included in a cycle if and only if v is included in a chordless cycle.

Proof. If v is not included in any cycle, it obviously is not included in any chordless cycle. Hence, we investigate the case in which v is included in a cycle C. If C is chordless, we are done. If C has a chord, the addition of the chord splits C into two smaller cycles, and v is always included in one of them. We then consider the cycle as C. The cycle with three vertices can not have a chord, thus we always meet a chordless cycle including v.   □

For a recursive algorithm, an iteration means the computation from the beginning of a recursive call to its end, excluding any computation done in recursive calls generated in the iteration. If an iteration I recursively calls an iteration I', I' is called a *child* of I, and I is called the *parent* of I'.

3   Algorithm for Chordless s-t Path Enumeration

Our enumeration problem is formulated as follows.

Chordless s-t Path Enumeration Problem
For a given graph $G = (V, E)$ and two vertices s and t, enumerate all chordless s-t paths included in G.

We first observe that chordless cycle enumeration is done with chordless s-t path enumeration by repeating steps; (1) for a vertex s, enumerate chordless s-t paths in $G \setminus \{s, t\}$ for each vertex t adjacent to s, and (2) remove s from the graph. Here $G \setminus \{s, t\}$ is the graph obtained from G by removing the edge $\{s, t\}$. This implies that we only have to consider chordless s-t path enumeration.

Lemma 1. *For a vertex $v \in N(s)$, P is a chordless s-t path including v if and only if $P \setminus s$ is a chordless v-t path of the graph $G \setminus (N(s) \setminus v)$.*

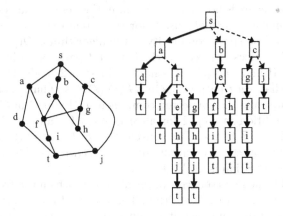

Fig. 2. Tree on the right represents recursive structure of s-t path enumeration in the graph on left; bold lines correspond to recursive calls in step 2, and dotted lines correspond to those in step 6

Proof. If $P \setminus s$ is a chordless v-t path in $G \setminus (N(s) \setminus v)$, P is an s-t path all whose chords are incident to s. Since P has no vertex in $N(s) \setminus v$, no vertex of P other than v is adjacent to s. Thus, P has no chord incident to s, and is chordless.

If P is a chordless s-t path including v, no vertex $u \in N(s) \setminus v$ is included in P, since the edge $\{s, u\}$ would be a chord if was included. Thus, $P \setminus s$ is a chordless v-t path in $G \setminus (N(s) \setminus v)$. ☐

Lemma 2. *The set of chordless s-t paths of G is partitioned into disjoint sets of chordless s-t paths in the graphs $G \setminus (N(s) \setminus v)$ for each v.*

Proof. Suppose that P is a chordless s-t path in G. Then, from lemma 1, P includes exactly one vertex among $N(s)$. If P includes $v \in N(s)$, $P \setminus s$ is a chordless v-t path in $G \setminus (N(s) \setminus v)$, thus P is a chordless s-t path in $G \setminus (N(s) \setminus v)$. Since P is not an s-t path in $G \setminus (N(s) \setminus u)$ for any $u \in N(s) \setminus v$, the statement holds. ☐

From the lemma, we obtain the following algorithm. The Q is the sequence of vertices attached to the paths in the ancestor iterations, and set to be empty at the start of the algorithm.

Algorithm EumChordlessPath $(G = (V, E), s, t, Q)$
1. **if** edge $\{s, t\}$ exists in E **then output** $Q \cup t$; **return**
2. **for** each $v \in N(s)$ s.t. a v-t path exists in $G \setminus (N(s) \setminus v)$ **do**
3.   call **call** EnumChordlessPath $((G \setminus (N(s) \setminus v)) \setminus s, v, t, Q \cup v)$
4. **end for**

When a recursive call is generated in an iteration of the algorithm, $G \setminus (N(s) \setminus v)$ is generated from G by removing vertices and edges. The removed vertices and edges are kept in memory so that G can be reconstructed from the graph. A removed edge on vertex is not removed again in the descendants of the iteration. Thus, the accumulated memory usage for these removed vertices and edges is $O(|V| + |E|)$, and the space complexity of the algorithm is $O(|V| + |E|)$.

In step 2, all vertices $v \in N(s)$ such that a v-t path exists in $G \setminus (N(s) \setminus v)$ must be listed. If and only if the condition in step 2 holds, there is a vertex $u \in N(v)$ such that a u-t path exists in $(G \setminus N(s)) \setminus s$. Thus, those vertices can be listed by computing the connected component including t in $G \setminus N(s) \setminus s$ and checking the condition in step 2 for all $u \in N(v)$ for all $v \in N(s)$. This can be done in $O(|V| + |E|)$ time. The construction of $(G \setminus (N(s) \setminus v)) \setminus s$ is done in $O(|N(v)|)$ time by constructing it from $(G \setminus N(s)) \setminus s$. Therefore, the time complexity of an iteration is $O(|V| + |E|)$.

Let us consider the recursion tree of the algorithm which is a tree representing the recursive structure of the algorithm. The vertex of the recursion tree corresponds to an iteration, and each iteration and its parent are connected by an edge. The leaves correspond to the iterations generating no recursive calls, and the algorithm outputs a solution on each leaf. Because of the condition given placed on vertices in step 2, there is always at least one s-t path in the given graph. This implies that at least one recursive call occurs when step 2 is executed. Hence, the algorithm outputs a solution at every leaf of the recursion tree. The depth of the recursion tree is $O(|V|)$ since at least one vertex is removed from the graph to generate a recursive call. We can conclude from these observations that the time complexity of the algorithm is $O(N|V|(|V| + |E|))$ where N is the number of chordless s-t paths in G. Next, we discuss the reduction of the time complexity to $O(N(|V| + |E|))$.

We first rewrite the above algorithm as follows. We denote the vertex next to v in path P by $nxt(v)$. Note that although we introduce several variables, the algorithms are equivalent.

Algorithm EnumChordlessPath2 $(G = (V, E), s, t, Q)$
1. **if** s is adjacent to t **then output** $Q \cup t$; **return**
2. $P :=$ a chordless s-t path in G
3. **call** EnumChordlessPath2 $(G \setminus (N(s) \setminus nxt(s)), nxt(s), t, Q \cup nxt(s))$
4. **for** each $v \in N(s), v \neq nxt(s)$ **do**
5.   **if** there is a v-t path in $G \setminus (N(s) \setminus v)$ **then**
6.   **call** EnumChordlessPath2 $(G \setminus (N(s) \setminus v), v, t, Q \cup v)$
7. **end for**

We further rewrite the algorithm as follows. We compute the chordless s-t path P computed in step 2 of the above algorithm, before the start of the iteration, i.e., in its parent, and give it as a parameter to the recursive call.

Algorithm EnumChordlessPath3 $(G = (V, E), s, t, Q, P)$
1. **if** s is adjacent to t **then output** $Q \cup t$; **return**
2. **call** EnumChordlessPath3 $(G \setminus (N(s) \setminus nxt(s)), nxt(s), t, Q \cup nxt(s), P \setminus s)$
3. **for** each $v \in N(s), v \neq nxt(s)$ **do**
4.   **if** there is an v-t path in $G \setminus (N(s) \setminus v)$ **then**
5. $P :=$ a chordless v-t path in $G \setminus (N(s) \setminus v)$ (found by a breadth first search)

6.   **call** EnumChordlessPath3 $(G \setminus (N(s) \setminus v), v, t, Q \cup v, P)$
7.   **end if**
8. **end for**

Figure 2 illustrates an example of the recursive structure of this algorithm. The tail of an arrow is a parent and the head is its child. We call the child generated in step 2 *first child*, and the arrow pointing at the first child is drawn with a bold line. We can make a path by following the bold-arrows, and we call a maximal such path a *straight path*. Since the bottom of a straight path is a leaf, the number of straight paths is bounded by the number of chordless paths. Since the head of a non-bold arrow always points an end of a straight path, the number of non-bold arrows, that correspond to the recursive calls done in step 6, is bounded by the number of straight paths.

From these observations, we infer the following points regarding time complexity.

- An iteration takes $O(|V| + |E|)$ time when a chordless path is output. This computation time is $O(|V| + |E|)$ per chordless path.
- Steps 1 and 2 take $O(N^*(s))$ time where $N^*(s)$ is the number of edges adjacent to vertices in $N(s)$. This time is spent checking the adjacency of s and t and constructing $G \setminus (N(s) \setminus v)$ for all $v \in N(s)$. This comes from that $G \setminus (N(s) \setminus v)$ can be constructed from $G \setminus N(s)$ by adding edges adjacent to v in $O(|N(v)|)$ time.
- The number of executions of the for loop in step 3 is bounded by $|N(s)|$. Their sum over all iterations in a straight path does not exceed the number of edges.
- Steps 5 and 6 take $O(|V| + |E|)$ time to find a chordless v-t path, and to construct $G \setminus (N(s) \setminus v)$. Since the recursive call in step 6 corresponds to a straight path, this computation time is $O(|V| + |E|)$ per chordless path.
- The execution time for step 4 is $O(|V| + |E|)$.

We see from the above that the bottle neck in terms of time complexity is step 4. The other parts of the algorithm takes $O(|V| + |E|)$ time per chordless s-t path. We speed up step 4 by using the following property.

Property 3. $G \setminus \{v\}$ includes a v-t path for $v \in N(s)$ if and only if there is a vertex $u \in N(v) \setminus N(s)$ such that $G \setminus N(s)$ includes a u-t path.   □

In each iteration we put mark on vertices u such that there is a u-t path in $G \setminus N(s)$. Step 4 is then done in $O(|N(v)|)$ time by looking at the marks on the vertices in $N(v)$. The marks can be put in short time, by updating the marks put in the first child. The condition of step 4 is checked by finding all vertices in $G \setminus N(s)$ from which going to t is possible. This also takes $O(|V| + |E|)$ time, but the time is reduced by re-using the results of the computation done for the first child. In the first child, marks are put according to the reachability to t in $G \setminus (N(s) \cup N(nxt(s)))$. To put marks for $G \setminus N(s)$, we find all vertices u such that any u-t path in $G \setminus N(s)$ includes a vertex of $N(nxt(s)) \setminus N(s)$. This is done by using a graph search starting from the vertices of $N(nxt(s)) \setminus N(s)$ that are adjacent to a marked vertex, and visiting only unmarked vertices. The time taken is linear in the number of edges adjacent to newly marked vertices.

Consider the computation time with respect to step 4, for the iterations in a straight path. In these operations, a vertex (resp., an edge) gets a mark at most once, i.e., it never gets a mark twice. Thus, the total computation time for this computation is linear in the sum of the degrees of marked vertices and vertices in $N(nxt(s))$, and is bounded by $O(|V| + |E|)$. The computation time for step 4 is thus reduced to $O(|V| + |E|)$ per chordless s-t path. When a recursive call for a non-first child is made, all marks are deleted. We then perform a graph search starting from t to put the marks. Both steps take $O(|V| + |E|)$ time. Since this computation is done only when generating non-first child, the total number of occurrences of this computation is bounded by the number of maximal paths, i.e., the number of chordless paths. Thus, this computation takes $O(|V| + |E|)$ time for each chordless path. The algorithm is written as follows.

Algorithm EnumChordlessPath4 $(G = (V, E), s, t, Q, P)$
1. **if** s is adjacent to t **then output** $Q \cup t$; **go to** 11
2. **call** EnumChordlessPath4 $(G \setminus (N(s) \setminus nxt(s)), nxt(s), t, Q \cup nxt(s), P \setminus s)$
3. put mark by graph search on $G \setminus N(s)$ from vertices in $N(nxt(s))$
4. **for each** $v \in N(s), v \neq nxt(s)$ **do**
4.   **if** a vertex adjacent to v is marked **then**
5.   delete marks from all vertices in G
6. $P :=$ a chordless v-t path in $G \setminus (N(s) \setminus v)$
7.   **call** EnumChordlessPath4 $(G \setminus (N(s) \setminus v), v, t, Q \cup v, P)$
8.   recover the marks deleted in step 5, by graph search starting from t on $G \setminus N(s)$
9.   **end if**
10. **end for**

Theorem 1. *The chordless s-t paths in a given graph $G = (V, E)$ can be enumerated in $O(|V| + |E|)$ time per chordless path, in particular, polynomial time delay.*

Proof. We can see the correctness in the above. The time complexity of an iteration is $O(|V| + |E|)$, and each iteration outputs an s-t-path. Moreover, the height of the recursion tree is at most $|V|$, thus the time between two consecutive output paths is bounded by $O(|V|+|E|)+O(|V|) = O(|V|+|E|)$. This concludes the theorem. □

Theorem 2. *The chordless cycles in a given graph $G = (V, E)$ can be enumerated in $O(|V| + |E|)$ time per chordless cycle, in particular, polynomial time delay.* □

4   Computational Experiments

The practical efficiency of the proposed algorithms is evaluated by computational experiments. The results were compared with those of the cycle enumeration

Table 1. Computational time (in seconds) for randomly generated graphs

edge density	10%	20%	30%	40%	50%	60%	70%	80%	90%
no. of vertices 50	0.18	0.12	0.098	0.089	0.082	0.08	0.085	0.1	0.11
75	0.17	0.12	0.099	0.088	0.079	0.074	0.077	0.088	0.1
100	0.17	0.12	0.099	0.09	0.083	0.081	0.089	0.095	0.12
150	0.2	0.12	0.099	0.098	0.077	0.075	0.083	0.103	0.14
200	0.18	0.12	0.1	0.088	0.081	0.078	0.085	0.11	0.17
300	0.19	0.12	0.1	0.087	0.082	0.083	0.091	0.12	0.21
400	0.17	0.12	0.1	0.089	0.08	0.086	0.1	0.15	0.26
600	0.18	0.11	0.12	0.12	0.11	0.1	0.13	0.23	0.42
800	0.2	0.12	0.14	0.13	0.11	0.11	0.13	0.26	0.54
1200	0.23	0.17	0.17	0.13	0.12	0.12	0.15	0.28	1
1600	0.24	0.19	0.14	0.13	0.13	0.14	0.21	0.29	1.3
2400	0.25	0.19	0.17	0.15	0.16	0.16	0.19	0.44	1.4
3200	0.29	0.23	0.2	0.19	0.18	0.2	0.25	0.61	1.79
4800	0.28	0.28	0.27						

algorithm proposed in [12]. The difference between the number of cycles and of chordless cycles was also compared. The program was coded in C, and compiled using gcc. The experiments were done on a PC with a Core i7 3GHz CPU. The code is available at the author's web site (http://research.nii.ac.jp/~uno/codes. html). We did not use multiple cores, and the memory usage by the algorithm was less than 4MB. The instance graphs were random graphs and the real-world graphs taken from the UCI machine learning repository[16]. All the test instances shown here are downloadable from the author's web site, except for those from UCI repository. Tables 1 to 4 summarize the computation time, number of cycles, and number of chordless cycles for each instance, and clarify the effectiveness of the chordless cycle model and our algorithm.

The computation time results for randomly generated graphs are shown in Table 1. The edge density means the probability of being connected by an edge for any two vertices. Execution of an enumeration algorithm involves many iterations with different input graphs, thus we thought that there are sufficiently many samples even in one execution of the algorithm. Therefore, we generated just one instance for each parameter. Each cell represents the computation time needed for 10,000 cycles or chordless cycles. When the computation time was too long so that the number of output cycles exceeded one million, we stopped the computation.

When the edge density was close to 100%, almost all the chordless cycles were triangles. In this case, intuitively, the algorithm spent $O(|V||E|) = O(|V|^3)$ time to find $O(|V|^2)$ chordless cycles. In contrast, it took almost constant time for each chordless cycle in sparse graphs. This is because the graph was reduced by repeated recursive calls, and at the bottom levels, the graph sizes were usually constant.

Table 2 shows that the number of chordless cycles exponentially increased against with the edge density, but not as much as usual cycles. Table 3 shows

Table 2. Number of chordless cycles (upper) and of cycles (lower)

edge density	10%	20%	30%	40%	50%	60%	70%	80%	90%	100%
no. of vertices 10	1	3	14	20	41	45	63	81	120	
	0	1	4	116	352	2302	3697	24k	108k	
15	0	10	34	116	165	193	247	297	350	455
	0	36	1470	613k	6620k	55525k	-	-	-	-
20	1	78	298	523	637	752	771	846	908	1140
	1	56k	9114k	-	-	-	-	-	-	-
25	8	1218	2049	2387	2099	1891	1775	1854	1928	2300
50	64k	395k	267k	146k	82k	49k	34k	23k	18k	19k
75	119379k	69357k	14504k	3679k	1158k	465k	221k	119k	73810	67525
100	-	-	-	40436k	8269k	2395k	877k	393k	199k	161k
150	-	-	-	-	149022k	27483k	6641k	2167k	854k	551k
200	-	-	-	-	-	167408k	30334k	7755k	2466k	1313k
300	-	-	-	-	-	-	-	51043k	11457k	4455k
400	-	-	-	-	-	-	-	-	35154k	10586k

Table 3. Computation time and number of chordless cycles for sparse graphs

graph size (no. of vertices)	10	20	30	40	50	60	70	80	90
no. of chordless cycles	12	90	743	5371	89164	853704	4194491	45634757	-
time for 10,000 chordless cycles	16.6	3.33	0.53	0.22	0.18	0.2	0.23	0.24	-

the experimental results for sparse graphs. The graphs were generated by adding chords randomly to a cycle of n vertices so that the average degree was four. These sparse graphs included so many chordless cycles. The graphs with at most 100 vertices were solved in a practically short time, and the computation time for each chordless cycle were almost the same.

Table 4 shows the number of chordless cycles with limited lengths including a vertex (the first vertex) for the real-world data, taken from the UCI repository. The number of all chordless cycles is shown at the bottom for reference. The graphs were basically sparse, and globally well connected, and thus included a large number of cycles. Even in such cases, by giving an upper bound of the length, Some graphs can be made tractable in such cases by placing an upper bound on the length. These results show the possibility of using chordless cycles with limited lengths for practical data analysis of real-world graphs such as those for social networks.

4.1   Application to NMR Prediction

Chordless cycle enumeration has already been implemented as a part of a database system of chemoinformatics[14], composed of structural data of chemical compounds. In this system, the number of chordless cycles in the chemical graph of a chemical compound is considered to be an attribute of the compound. In response to a query about the chemical structure of a compound, the system searches in the database for structures partially similar to the structure of query compound, and predict some functions of the query compound. A chemical graph

Table 4. Number of chordless cycles including a vertex (of ID 0), for real-world graphs

	adjnoun	astro-ph	breast	celegen	cond-mat-2005	cond-mat_large	dolphins
(no. of vertices)	114	16708	7539	298	40423	30561	64
(no. of edges)	425	121251	5848	2359	175693	24334	159
length < 5	8	327	-	3342	393	6	26
length < 8	251		-	1738k	-	6	320
length < 16	65350	-	-	-	-	6	1780
#chord. cyc.	66235k	-	-	-	-	-	6966

	football	human_ppi	karate	lesmis	netscience	polblogs	polboopks	power
(no. of vertices)	117	10347	36	79	1591	1492	107	4943
(no. of edges)	616	5418	78	254	2742	19090	441	6594
length < 5	81	1838	37	3	1	35881	21	0
length < 8	11869	-	38	3	1	-	187	4
length < 16	256664k	-	38	3	1	-	34742	60
#chord. cyc.	-	-	103	594	5760	-	2273k	-

is usually sparse and is globally a tree or a combination of several large cycles. Small components can be attached to the large cycles. Thus, the number of chordless cycles is not so huge and is tractable in most cases.

The program code was implemented in the CAST/CNMR system for predicting the ^{13}C-NMR chemical shift[14, 15]. The codes and a more precise description of this system are available at http://research.nii.ac.jp/~hsatoh/subjects/NMR-e.html. The information obtained about chordless cycles is used to improve the accuracy for the predicted values when the ring attributes affects the NMR spectrum. The CAST/CNMR system predicts chemical shifts by using a chemical structure-spectrum database, containing mainly natural organic products and their related synthetic compounds. Since most of the compounds include chains of fused rings, enumerating all rings for these compounds would greatly increase the output size, with lots of data useless for NMR prediction. Therefore, the chordless cycle was adopted as a relevant ring attribute for the CAST/CNMR system. For accurate NMR prediction for carbon atoms, an error within 1.0 ppm (parts per million) is generally required. Use of chordless cycle information reduced error values of -4.1 to 1.6 ppm for some problematic carbon atoms to less than 1.0 ppm[14].

5   Conclusion

We proposed an algorithm for enumerating all chordless s-t paths, that is applicable to chordless cycle enumeration without increasing the time complexity. By reusing the results of the subroutines, the computation time is reduced to $O(|V| + |E|)$ for each chordless path. The results of the computational experiments showed that the algorithm works well for both random graphs and real-world graphs; the computation time was $O(|V|)$ in dense graphs, and almost constant for sparse graphs. The results also showed that the number of chordless cycles is small compared to the number of usual cycles. This algorithm thus paves the way to efficient use of cycle enumeration in data mining.

References

1. Asai, T., Arimura, H., Uno, T., Nakano, S.-I.: Discovering Frequent Substructures in Large Unordered Trees. In: Grieser, G., Tanaka, Y., Yamamoto, A. (eds.) DS 2003. LNCS (LNAI), vol. 2843, pp. 47–61. Springer, Heidelberg (2003)
2. Balaban, A.T., Filip, P., Balaban, T.S.: Computer Program for Finding All Possible Cycles in Graphs. J. Comput. Chem. 6, 316–329 (1985)
3. Downs, G.M., Gillet, V.J., Holiday, J.D., Lynch, M.F.: Review of Ring Perception Algorithms for Chemical Graphs. J. Chem. Inf. Comp. Sci. 29, 172–187 (1989)
4. Downs, G.M.: Ring perception. In: The Encyclopedia of Computational Chemistry, vol. 4, John Wiley & Sons, Chichester (1998)
5. Eppstein, D.: Finding the k Smallest Spanning Trees. In: Gilbert, J.R., Karlsson, R. (eds.) SWAT 1990. LNCS, vol. 447, pp. 38–47. Springer, Heidelberg (1990)
6. Ferreira, R., Grossi, R., Marino, A., Pisanti, N., Rizzi, R., Sacomoto, G.: Optimal Listing of Cycles and st-Paths in Undirected Graphs. In: SODA 2013 (2013)
7. Hanser, T., Jauffret, P., Gaufmann, G.: A New Algorithm for Exhaustive Ring Perception in a Molecular Graph. J. Chem. Inf. Comp. Sci. 36, 1146–1152 (1996)
8. Kapoor, S., Ramesh, H.: An Algorithm for Enumerating All Spanning Trees of a Directed Graph. Algorithmica 27, 120–130 (2000)
9. Inokuchi, A., Washio, T., Motoda, H.: Complete Mining of Frequent Patterns from Graphs. Machine Learning 50, 321–354 (2003)
10. Makino, K., Uno, T.: New Algorithms for Enumerating All Maximal Cliques. In: Hagerup, T., Katajainen, J. (eds.) SWAT 2004. LNCS, vol. 3111, pp. 260–272. Springer, Heidelberg (2004)
11. Parthasarathy, S., Zaki, M.J., Ogihara, M., Dwarkadas, S.: Incremental and Interactive Sequence Mining. In: CIKM 1999, pp. 251–258 (1999)
12. Read, R.C., Tarjan, R.E.: Bounds on Backtrack Algorithms for Listing Cycles, Paths, and Spanning Trees. Networks 5, 237–252 (1975)
13. Tomita, E., Tanaka, A., Takahashi, H.: The Worst-case Time Complexity for Generating all Maximal Cliques and Computational Experiments. Theo. Comp. Sci. 363, 28–42 (2006)
14. Satoh, H., Koshino, H., Uno, T., Koichi, S., Iwata, S., Nakata, T.: Effective consideration of ring structures in CAST/CNMR for highly accurate ^{13}C NMR chemical shift prediction. Tetrahedron 61, 7431–7437 (2005)
15. Satoh, H., Koshino, H., Uzawa, J., Nakata, T.: CAST/CNMR: Highly Accurate ^{13}C NMR Chemical Shift Prediction System Considering Stereochemistry. Tetrahedron 59, 4539–4547 (2003)
16. UCI Machine Learning Repository, http://archive.ics.uci.edu/ml/
17. Uno, T.: A Fast Algorithm for Enumerating Bipartite Perfect Matchings. In: Eades, P., Takaoka, T. (eds.) ISAAC 2001. LNCS, vol. 2223, pp. 367–379. Springer, Heidelberg (2001)
18. Uno, T.: An Output Linear Time Algorithm for Enumerating Chordless Cycles, IPSJ, SIG-AL 92 (2003) (in Japanese, technical report, non-refereed)
19. Wild, M.: Generating all Cycles, Chordless Cycles, and Hamiltonian Cycles with the Principle of Exclusion. J. Discrete Alg. 6, 93–102 (2008)

Algorithm Selection on Data Streams

Jan N. van Rijn[1], Geoffrey Holmes[2],
Bernhard Pfahringer[2], and Joaquin Vanschoren[3]

[1] Leiden University, Leiden, Netherlands
`j.n.van.rijn@liacs.leidenuniv.nl`
[2] University of Waikato, Hamilton, New Zealand
`{geoff,bernhard}@cs.waikato.ac.nz`
[3] Eindhoven University of Technology, Eindhoven, Netherlands
`j.vanschoren@tue.nl`

Abstract. We explore the possibilities of meta-learning on data streams, in particular algorithm selection. In a first experiment we calculate the characteristics of a small sample of a data stream, and try to predict which classifier performs best on the entire stream. This yields promising results and interesting patterns. In a second experiment, we build a meta-classifier that predicts, based on measurable data characteristics in a window of the data stream, the best classifier for the next window. The results show that this meta-algorithm is very competitive with state of the art ensembles, such as OzaBag, OzaBoost and Leveraged Bagging. The results of all experiments are made publicly available in an online experiment database, for the purpose of verifiability, reproducibility and generalizability.

Keywords: Meta Learning, Data Stream Mining.

1 Introduction

Modern society produces vast amounts of data coming, for instance, from sensor networks and various text sources on the internet. Various machine learning algorithms are able to capture general trends and make predictions for future observations with a reasonable success rate. The number of algorithms is large, and most of these work well on a varying range of data streams. However, there is not much knowledge yet about on which kinds of data certain algorithms perform well and when a certain algorithm should be preferred over another.

In this work we investigate how to predict what algorithm will perform well on a given data stream. This problem is generally known as the algorithm selection problem [14]. For each data stream, we measure the performance of various data stream classifiers and we calculate measurable data characteristics, called *meta-features*. In addition to many existing meta-features, we introduce a new type of meta-feature, based on concept drift detection methods. Next, we build a model that predicts which algorithms will work well on a given data stream based on its characteristics. Indeed, having knowledge about which classifier to apply on what data could greatly increase the performance of predictive tools in

S. Džeroski et al. (Eds.): DS 2014, LNAI 8777, pp. 325–336, 2014.

real world applications. For example, it is common for the performance curves of data stream algorithms to cross as the stream evolves. This means that at certain points in the stream, the data is best modelled with algorithm A, while at a later point, the data is better modelled with algorithm B. As such, selecting the right algorithm at each point in the stream has the potential to increase performance. In this work we focus on data streams with a nominal target. Although much work has been done on both data streams and meta-learning, to the best of our knowledge, this is the first effort to do meta-learning on data streams.

The remainder of this paper is structured as follows. Section 2 contains a description of related work. Section 3 describes how the meta-dataset was created, the data streams that it contains, how they are characterized, and how we have measured the performance of data stream algorithms. Section 4 describes an experiment in which we calculate the characteristics of a *small sample* of a data stream, and predict which classifier performs best on the entire stream. Next, Section 5 describes a second experiment, in which we continuously measure the characteristics of a *sliding window* on the data stream, and use a meta-algorithm to predict the best classifier for the next window. In Section 6 we present and discuss some emerging patterns from the data. Section 7 concludes.

2   Related Work

It has been recognized that mining data streams differs from conventional batch data mining [4,13]. In the conventional batch setting, usually a limited amount of data is provided and the goal is to build a model that fits the data as well as possible, whereas in the data stream setting, there is a possibly infinite amount of data, with concepts possibly changing over time, and the goal is to build a model that captures general trends.

More specifically, the requirements for processing streams of data are: process one example at a time (and inspect it only once), use a limited amount of time and memory, and be ready to predict at any point in the stream [3,13]. These requirements inhibit the use of most batch data mining algorithms. However, some algorithms can trivially be used or adapted to be used in a data stream setting, for example, NaiveBayes, k Nearest Neighbour, and Stochastic Gradient Descent, as done in [13]. Also, many algorithms have been created specifically to operate on data streams. Most notably, the Hoeffding Tree [6] is a tree based algorithm that splits based on information gain, but using only a small sample of the data determined by the Hoeffding bound. The Hoeffding bound gives an upper bound on the difference between the mean of a variable estimated after a number of observations and the true mean, with a certain probability.

Conventional batch data mining methods can also be adapted for use in the data streams setting by training them on a set of instances sampled from recent data. Typically, a set of w (window size) training instances is formed. Every w instances form a batch and are provided to the learner, which builds a model based on these instances. The disadvantages of this approach are that the most recent examples are not used until a batch is complete, and old models need to be deleted to make room for new models. Read et al. [13] distinguish between instance incremental methods and batch incremental methods and compare the performance of both approaches. Their main conclusion is that the performance in terms of accuracy is equivalent. However, the instance incremental algorithms use fewer resources.

Ensembles of classifiers are among the best performing learning algorithms in the traditional batch setting. Multiple models are produced that all vote for the label of a certain instance. The final prediction is made according to a predefined voting schema, e.g., the class with the most votes wins. In [10] it is proven that the error rate of an ensemble in the limit goes to zero if two conditions are met: first, the individual models must do better than random guessing, and second, the individual models must be diverse, meaning that their errors should not be correlated. Popular ensemble methods in the traditional batch setting are Bagging [5], Boosting [17] and Stacking [7]. Bagging and Boosting have equivalents in the data stream setting, e.g., OzaBag [11], OzaBoost [11] and Leveraging Bagging [4]. The Average Weighted Ensemble [20] tracks which individual classifiers perform well on recent data, and uses this information to weight the votes.

The field of meta-learning addresses the question what machine learning algorithms work well on what data. The algorithm selection problem, formalised by Rice in [14], is a natural problem from the field of meta-learning. According to the definition of Rice, the problem space P consists of all machine learning tasks from a certain domain, the feature space F contains measurable characteristics calculated upon this data (called meta-features), the algorithm space A is the set of all considered algorithms that can execute these tasks and the performance space Y represents the mapping of these algorithms to a set of performance measures. The task is for any given $x \in P$, to select the algorithm $\alpha \in A$ that maximizes a predefined performance measure $y \in Y$, which is a classification problem. Similar ranking and regression problems are derived from this.

Much effort has been devoted to the development of meta-features that effectively describe the characteristics of the data (called meta-features). Commonly used meta-features are typically categorised as one of the following: simple (number of instances, number of attributes, number of classes), statistical (mean standard deviation of attributes, mean kurtosis of attributes, mean skewness of attributes), information theoretic (class entropy, mean entropy of attributes, noise-signal ratio) or landmarkers [12] (performance of a simple classifier on the data). The authors of [18] give an extensive description of many meta-features. Furthermore, they propose a new type of meta-feature, pair-wise meta-rules.

Another recent development is the concept of experiment databases [15,19], databases which contain detailed information about a large range of experiments. Experiment databases enable the reproduction of earlier results for verification and reusability purposes, and make much larger studies (covering more algorithms and parameter settings) feasible [19]. Above all, experiment databases allow a variety of studies to be executed by a database look-up, rather than setting up new experiments. An example of such an online experiment database is OpenML [15].

3   Meta-dataset

For the purpose of meta-learning, we need to obtain data on previous experiments, i.e., runs of various algorithms on various data streams (from here on referred to as *base datasets*, or *base data streams*). From this we construct a so-called *meta-dataset*, in which each instance is a data stream characterized by a number of measurable characteristics, the *meta-features*, as well as performance scores of various machine learning algorithms.

3.1   Bayesian Network Generator

Unfortunately, the data stream literature contains few publicly available data streams. In order to obtain a reasonable number of experiments on data streams, we propose a new type of data generator that generates data streams based on real world data [16]. It takes a dataset as input, preferably consisting of real world data and a reasonable number of features, and builds a Bayesian Network over it, which is then used to generate instances based on the probability tables. These streams can also be combined together to simulate concept drift, similar to what is commonly done with the Covertype, Pokerhand and Electricity dataset [4].

The generator takes a dataset as input, and outputs a data stream containing a similar concept, with a predefined number of instances. The input dataset is preprocessed with the following operations: all missing values are first replaced by the majority value of that attribute, and numeric attributes are discretized using Weka's binning algorithm [9]. Values for attributes that are numeric in the original dataset can be determined using two strategies. The *nominal strategy* assigns one of the bins as the attribute value, determining the bin based on the probability tables. The *numeric strategy* takes the bin with the highest probability value and draws a number from this bin based on its normal distribution. The generated data streams are denoted as BNG(*data, strategy, num_instances*), with *data* denoting the original dataset, *strategy* denoting the chosen strategy for numeric values, and *instances* denoting the number of generated instances. The algorithm is implemented in the OpenML MOA package[1].

Figure 1 shows the meta-features as calculated over the two Bayesian Network Generated data streams based on the *glass* dataset, compared to the values of the original dataset. As many of these qualities are quite similar, there is some indication that there is a similar concept underlying the data. The *dimensionality* indicates the ratio between the number of instances and the number of attributes. The decrease of this value can be explained by the fact that the number of attributes is the same, yet the number of instances has increased. Furthermore, data streams generated using the nominal strategy do not have any numeric attributes, hence meta-features like *Mean Skewness Of Numeric Attributes* and *Mean StDev Of Numeric Attributes* are zero. The meta-features indicating the J48 landmarkers (with varying confidence factors) have better values on the generated data streams, hinting at a slightly easier concept represented by the data. Similar patterns can be found for other generated data streams.

---

[1] Can be obtained from http://www.openml.org/

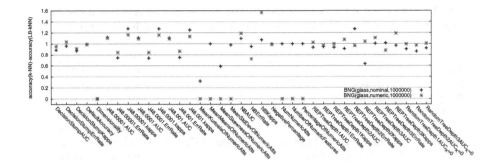

Fig. 1. Meta-features of two Bayesian Network Generated data streams, where each value is divided by the corresponding meta-feature value of the original dataset

3.2 Base Data Streams

The data streams generated by the Bayesian Network Generator form the basis of our meta-dataset. We took datasets from the UCI repository [1] as input for the Bayesian Network generator. We aimed to generate data streams of 1,000,000 instances, or less if the original dataset did not have enough attributes to obtain a million different instances. We used both the nominal and the numeric strategy in those cases where the original dataset had numeric attributes. In addition to these, we also generate data streams by commonly used data generators from the literature. We used the following data generators, as implemented in the MOA workbench [3]: the SEA Concepts Generator, Rotating Hyperplane Generator, Random RBF Generator and the LED Generator. For the generation of these data streams we used the same parameters as described in [13].

Additionally, we also included the commonly used datasets Covertype, Electricity and Pokerhand, and a combination of these in order to generate concept drift, as is done in [4]. We also included large text datasets, i.e., the IMDB dataset and the 20 Newsgroups dataset. We converted the IMDB dataset into a binary classification problem, having the *drama* genre as target. The 20 Newsgroups dataset is first converted into 20 binary classification problems (one for every Newsgroup) and then appended again into one big binary-class dataset, as is done in [13]. This simulates a data stream with 19 concept drifts.

3.3 Meta-Features

We have characterized the base data streams using a wide variety of meta-features. Most of the features are already described in the literature, e.g., in [18], and are either of the type simple, statistical, information theoretic or landmarker. We also introduce stream specific meta-features, based on a change detector. We have run both Hoeffding Trees and NaiveBayes with both the ADWIN [2] and DDM [8] change detectors over all data streams, and have recorded the number

Table 1. Algorithms used in the experiments. Batch incremental classifiers have a window of size 1,000; ensembles contain 10 base classifiers; k-NN is used with $k = 10$.

Key	Classifier	Type
NB	NaiveBayes	Instance incremental
SGD	Stochastic Gradient Descent	Instance incremental
SPeg	SPegasus	Instance incremental
k-NN	k Nearest Neighbour	Instance incremental
HT	Hoeffding Tree	Instance incremental
SMO	Support Vector Machine / Polynomial Kernel	Batch incremental
J48	C4.5 Decision Tree	Batch incremental
REP	Reduced-Error Pruning Decision Tree	Batch incremental
OneR	One Rule	Batch incremental
LB-kNN	Leveraging Bagging / k-NN	Ensemble
LB-HT	Leveraging Bagging / Hoeffding Tree	Ensemble
Bag-HT	OzaBag / Hoeffding Tree	Ensemble
Boost-HT	OzaBoost / Hoeffding Tree	Ensemble

of changes detected (both ADWIN and DDM) and the number of warnings (DDM only). All these meta-features are calculated for each window of 1,000 instances on each data stream.

3.4 Algorithms

The algorithms included in this study are shown in Table 1, and can be grouped into three types: instance incremental classifiers, batch incremental classifiers and ensembles. For all classifiers, we have recorded the predictive accuracy, the runtime, and RAM Hours on each data stream. The predictive accuracy was determined using the Interleaved Test Then Train procedure, where each instance is first used as a test instance, before it can be used to train the classifier.

4   Algorithm Selection in the Classical Setting

In the classical setting of the algorithm selection problem, the goal is to predict, for a certain dataset, what algorithm would perform best.

4.1   Experimental Setup

In prior studies, the algorithm selection problem was treated as either a classification problem, regression problem, or ranking problem [18]. To investigate what kind of information can be obtained from the data, we treat the algorithm prediction problem on data streams as a classification problem. We create a meta-dataset[2] using the experimental data described in Section 3. For each base

---

[2] All meta-datasets that were created for this study can be obtained from http://www.openml.org/d/

Table 2. Results of the algorithm selection problems in the classical setting

Task	A	Majority	%	Decision Stump	Random Forest
Instance incremental	5	HT	71.25	79.63	88.46
Batch incremental	4	SMO	67.86	65.07	68.73
Ensembles	4	LB-HT	62.25	63.98	67.11
All classifiers	13	LB-HT	61.43	60.14	69.57

data stream, the meta-features are recorded in the first window of 1,000 observations. By the definition of Rice [14], this means that we are working on the algorithm selection problem with $A = 13$ (algorithms), $P = 75$ (data streams), $Y = predictive\ accuracy$ (performance measure) and $F = 58$ (meta-features).

The goal is to predict which algorithm performs best, measured over the whole data stream. In order to obtain deeper insight into what kind of targets we can predict, we also defined three sub tasks, i.e., predicting the best instance incremental classifier ($A = 5$), predicting the best batch incremental classifier ($A = 4$) and predicting the best ensemble ($A = 4$). We have selected the "Decision Stump" and "Random Forest" classifiers (as implemented in Weka 3.7.11 [9]) as meta-algorithms. The Random Forest algorithm has proven to be a useful meta-algorithm in prior work [18], while models obtained from a single decision tree or stump are especially easy to interpret. Both classifiers are tree-based, which guards them against modeling irrelevant features. We ran the Random Forest algorithm with 100 trees and 10 attributes. We estimate the performance of the meta-algorithm by doing 10 times 10 fold cross-validation, and compare its performance against predicting the majority class. For each meta-dataset, we have filtered out the instances that contain a unique class value; since they will either only appear in the training or test set, these do not form a reliable source for estimating the accuracy.

4.2   Results

Table 2 shows the results obtained from the various tasks. It shows for every task which classifier performs best in most cases (and therefore is the majority class of the task) and in what percentage of the data streams that is the case. Column "A" denotes the number of classes that were distinguished in the tasks; column "Majority" and "%" denote the majority class and its size. The other columns denote the accuracy score of the respective meta-algorithms. The Random Forest algorithm performs better than the baseline on all defined tasks. The Decision Stump algorithm performs well only at predicting both the best instance incremental classifiers and ensembles. The results of both meta-algorithms on the task of predicting the best instance incremental were marked as significantly better than predicting majority class, tested using a Paired T-Test with a confidence of 95%. Since the data characteristics were obtained over only the first 1,000 instances of each stream, algorithm selection on data streams can improve results of classifiers at already very low computational cost.

5 Algorithm Selection in the Stream Setting

In data stream prediction, it is common for performance curves of different algorithms to cross each other multiple times. Whereas, at a certain interval of the stream, one of the algorithms performs best, this might be different for other intervals of the stream. By applying algorithm selection in a stream setting, our goal is to predict what algorithm will perform best for the next window of data.

5.1 Experimental Setup

In this experiment we want to determine whether meta-knowledge can improve the predictive performance of data stream algorithms in the following setting. Consider an ensemble of algorithms that are all trained on the same data stream. For each window of size w, an abstract meta-algorithm determines which algorithm will be used to predict the next window of instances, based on data characteristics measured in the previous window and the meta-knowledge. Note that the performance of the meta-algorithm depends on the size of this window. Meta-features calculated over a small window size are probably not able to adequately represent the characteristics of the data, whereas calculating meta-features over large windows is computationally expensive. Since our previous experiment obtained good results with a window size of 1,000, we perform our experiments with the same window size.

We have constructed a new meta-dataset, containing for each window the meta-features measured over the previous interval, and the performance of all base-classifiers trained on the entire data stream up to that window. We only include a subset of the generated data streams. Indeed, since the Bayesian Network Generator was used to generate multiple data streams based upon the same source data (using the nominal strategy and the numeric strategy), it is likely that these data streams contain a similar concept. We therefore remove all data streams generated using the nominal strategy, if a version generated by the numeric strategy also exists. After filtering out these data streams, there are still 49 data streams left. The meta-dataset consists of roughly 45,000 instances, each describing a window of 1,000 observations from one of these base data streams.

For each of the base data streams, a meta-algorithm is trained using only the intervals of the other data streams. We use a `Random Forest` (100 trees, 10 attributes) as meta-algorithm, since it proved to be a reasonable choice in the previous experiment. We measure how it performs on the meta-learning task of predicting the right algorithm for a given interval, as well as how it would actually perform on the base data streams. To the best of our knowledge, `Leveraged Bagging Hoeffding Trees` is the state of the art algorithm on these data streams, so we will compare it against our abstract algorithm. As in the previous experiment, we distinguish the tasks of selecting the best instance incremental classifier, the best batch incremental classifier and the best ensemble.

Table 3. Results of the algorithm selection problems in the stream setting

Task	A	Majority	%	RF_{meta}	$ZeroR_{base}$	RF_{base}	MAX_{base}
Instance incremental	5	HT	59.75	80.78	80.98	84.07	84.59
Batch incremental	4	SMO	65.56	68.17	74.38	75.33	76.02
Ensembles	4	LB-HT	57.78	56.20	84.27	85.15	86.12
All classifiers	13	LB-HT	50.97	50.92	84.27	85.31	86.30

5.2 Results

Table 3 shows the results obtained from this experiment. As with the results of the previous experiment, column A indicates the number of classes in the classification problem, "Majority" denotes which classifier is the majority class (i.e., the classifier that performs best in most observations), and "%" shows the size of the majority class. We measure two types of accuracy, *meta-level* accuracy and *base-level* accuracy. Meta-level accuracy records the performance using a zero-one loss function. Consequently, it indicates for how many windows the meta-algorithm predicted the best classifier. Column RF_{meta} shows the meta-level accuracy of a Random Forest. Base-level accuracy records the performance using a loss function equivalent to the performance of the predicted classifier. For example, when the meta-algorithm predicts k-NN to be the best classifier on a certain interval, the accuracy of k-NN on this interval will be used as loss. Accordingly, base-level accuracy indicates the performance of the meta-classifier on the base data streams when dynamically selecting the base-classifier. The base-level score of the Random Forest meta-algorithm is shown in column RF_{base}. Column $ZeroR_{base}$ shows the base-level accuracy when the majority class is always predicted. Note that this is the score obtained by the majority class base-classifier, measured over all base data streams. Column MAX_{base} shows what the base-level score would be, if for any interval the best classifier would have been predicted.

As in the classical setting, it appears that determining the best instance-incremental classifier yields good results. In more than 80% of the cases, the correct classifier is predicted. This also results in a notable increase in base level performance, in such a way that it is comparable with Leveraged Bagged Hoeffding Trees (84.27), and outperforms the scores obtained by OzaBag (82.58) and OzaBoost (80.55). The results also show a consistent increase in performance. For all defined tasks, the meta-algorithm outperforms the use of the single best classifier in its pool, even though, on the ensemble task, RF_{meta} performs no better than predicting the majority class. Apparently, the meta-algorithm was able to avoid the use of Leveraged Bagged Hoeffding Trees on windows where the performance is very low. Furthermore, the base level performances are in many cases close to the maximum possible value given the pool of classifiers. This indicates that the main challenge is to find ways to improve this limit. This could be done by using a larger set of algorithms, or by using other techniques such as parameter optimisation.

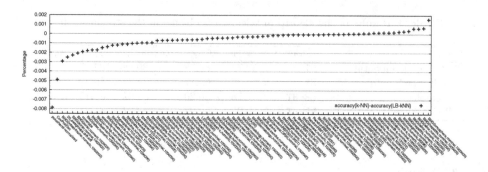

Fig. 2. The difference in predictive accuracy of k-NN and `Leveraged Bagging` k-NN. ($k = 10$, 10 base classifiers).

6   Discoveries

In this section we discuss interesting patterns that were obtained while analysing the meta-datasets, some of which corroborate earlier findings in the literature. Other findings will form the basis of our future work.

Discovery 1. An abstract meta-classifier consisting of 5 instance incremental classifiers (`NaiveBayes`, 10 `Nearest Neighbour`, `SPegasus`, `SGD` and a `Hoeffding Tree`) is competitive with state of the art ensembles, tested over a large range of data streams. It is likely that this result generalizes to an even larger number of data streams, because both conditions for an ensemble of classifiers to be successful are met [10]: the base classifiers are diverse and do better than random guessing. Moreover, the experiment in Section 4 shows that it is possible to predict which one will perform well based on prior data.

Discovery 2. In contrast to the increase in performance that `Leveraging Bagging` obtains when applied to `Hoeffding Trees`, it barely increases performance when applied to k `Nearest Neighbour`. Figure 2 shows the difference in performance of both algorithms. Note that, although `Leveraged Bagging` k-NN performs slightly better, the difference is minuscule. This may be due to the fact that even in a stream setting, k-NN is extremely stable; Bagging exploits the variation in the predictions of classifiers. In [5], it was already shown that `Bagging` will not improve k-NN in the batch setting, due to its stability.

Discovery 3. The data streams on which k-NN or `Leveraging Bagging` k-NN performs best all have a negative Mean Skewness of attributes. Skewness is a measure of the asymmetry in the distribution of a range of values. A negative Skewness indicates that there are some outliers with low values. This was already observed in the first interval of 1,000 instances, the `Decision Stump` extensively uses this feature to distinguish between predicting k-NN and `Hoeffding trees`. This is surprising, since it has been reported that simple, statistical and information theoretic meta-features do not add much predictive power to land-markers [18].

Discovery 4. NaiveBayes works well when the meta-feature measuring the number of changes detected by an ADWIN-equipped Hoeffding Tree has a high value. The NaiveBayes algorithm generally needs only relatively few observations to achieve good accuracy compared to more sophisticated algorithms such as Hoeffding Trees. Assuming that a high number of changes detected by this landmarker indicates that the concept of the stream is indeed changing quickly, this could explain why a classifier like NaiveBayes outperforms more sophisticated learning algorithms that need more observations of the same concept to perform well.

7   Conclusions

We have performed an extensive experiment on meta-learning on data streams, running a wide range of steam mining algorithms over a large number of data streams, and published all results online in OpenML [15], so that others can verify, reproduce and build upon these results. Containing more than 1,000 experiments on data streams, with extensive meta-information calculated over data (windows), this now forms a rich source of meta-learning experiments on data streams. In order to obtain a good number of data streams, the Bayesian Network Generator was introduced, a new data stream generator used to generate a comprehensive set of data streams describing various concepts.

Our approach to perform meta-learning on these streams seems promising: Meta-features calculated on a small interval at the start of the data stream already provide information about which classifier will outperform others. Beyond the classical setting of the algorithm selection problem, we can even use the meta-models obtained from earlier experiments to improve the current state of the art classifiers. We have sketched an abstract algorithm that uses multiple classifiers and a voting schema based on meta-models that outperforms the performance of the individual classifiers in its ensemble, but also is very competitive with state of the art ensembles, measured over 49 data streams spanning more that 46,000,000 instances. In addition, we discussed interesting patterns that emerged from the meta-dataset, which will form a basis for future work.

Moreover, in this work we have treated the algorithm selection problem as a classification task. In future work we will focus on ranking or regression tasks. We will also include more data streams containing concept drift, and study the effect on classifier performance. Due to the use of data generators this study may potentially be biased towards this kind of generated data. We hope that making our meta-dataset publicly available will persuade others to share more data streams, eventually enabling much larger studies on even more diverse data.

Acknowledgments. This work is supported by grant 600.065.120.12N150 from the Dutch Fund for Scientific Research (NWO).

References

1. Bache, K., Lichman, M.: UCI machine learning repository (2013), http://archive.ics.uci.edu/ml
2. Bifet, A., Gavalda, R.: Learning from Time-Changing Data with Adaptive Windowing. In: SDM, vol. 7, pp. 139–148. SIAM (2007)
3. Bifet, A., Holmes, G., Kirkby, R., Pfahringer, B.: MOA: Massive Online Analysis. J. Mach. Learn. Res. 11, 1601–1604 (2010)
4. Bifet, A., Holmes, G., Pfahringer, B.: Leveraging Bagging for Evolving Data Streams. In: Balcázar, J.L., Bonchi, F., Gionis, A., Sebag, M. (eds.) ECML PKDD 2010, Part I. LNCS, vol. 6321, pp. 135–150. Springer, Heidelberg (2010)
5. Breiman, L.: Bagging Predictors. Machine Learning 24(2), 123–140 (1996)
6. Domingos, P., Hulten, G.: Mining high-speed data streams. In: Proceedings of the Sixth ACM SIGKDD International Conference on Knowledge Discovery and Data Mining, pp. 71–80 (2000)
7. Gama, J., Brazdil, P.: Cascade Generalization. Machine Learning 41(3), 315–343 (2000)
8. Gama, J., Medas, P., Castillo, G., Rodrigues, P.: Learning with Drift Detection. In: Bazzan, A.L.C., Labidi, S. (eds.) SBIA 2004. LNCS (LNAI), vol. 3171, pp. 286–295. Springer, Heidelberg (2004)
9. Hall, M., Frank, E., Holmes, G., Pfahringer, B., Reutemann, P., Witten, I.H.: The WEKA Data Mining Software: An Update. ACM SIGKDD Explorations Newsletter 11(1), 10–18 (2009)
10. Hansen, L., Salamon, P.: Neural network ensembles. IEEE Transactions on Pattern Analysis and Machine Intelligence 12(10), 993–1001 (1990)
11. Oza, N.C.: Online Bagging and Boosting. In: 2005 IEEE International Conference on Systems, Man and Cybernetics, vol. 3, pp. 2340–2345. IEEE (2005)
12. Pfahringer, B., Bensusan, H., Giraud-Carrier, C.: Tell me who can learn you and I can tell you who you are: Landmarking various learning algorithms. In: Proceedings of the 17th International Conference on Machine Learning, pp. 743–750 (2000)
13. Read, J., Bifet, A., Pfahringer, B., Holmes, G.: Batch-Incremental versus Instance-Incremental Learning in Dynamic and Evolving Data. In: Hollmén, J., Klawonn, F., Tucker, A. (eds.) IDA 2012. LNCS, vol. 7619, pp. 313–323. Springer, Heidelberg (2012)
14. Rice, J.R.: The Algorithm Selection Problem. Advances in Computers 15, 65–118 (1976)
15. van Rijn, J.N., et al.: OpenML: A Collaborative Science Platform. In: Blockeel, H., Kersting, K., Nijssen, S., Železný, F. (eds.) ECML PKDD 2013, Part III. LNCS, vol. 8190, pp. 645–649. Springer, Heidelberg (2013)
16. van Rijn, J.N., Holmes, G., Pfahringer, B., Vanschoren, J.: The Bayesian Network Generator: A data stream generator. Tech. Rep. 03/2014, Computer Science Department, University of Waikato (2014)
17. Schapire, R.E.: The Strength of Weak Learnability. Machine Learning 5(2), 197–227 (1990)
18. Sun, Q., Pfahringer, B.: Pairwise meta-rules for better meta-learning-based algorithm ranking. Machine Learning 93(1), 141–161 (2013)
19. Vanschoren, J., Blockeel, H., Pfahringer, B., Holmes, G.: Experiment databases. A new way to share, organize and learn from experiments. Machine Learning 87(2), 127–158 (2012)
20. Wang, H., Fan, W., Yu, P.S., Han, J.: Mining Concept-Drifting Data Streams using Ensemble Classifiers. In: KDD, pp. 226–235 (2003)

Sparse Coding for Key Node Selection over Networks

Ye Xu[1] and Dan Rockmore[1,2,3]

[1] Department of Computer Science, Dartmouth College, Hanover, NH 03755
[2] Department of Mathematics, Dartmouth College, Hanover, NH 03755
[3] The Santa Fe Institute, 1399 Hyde Park Rd., Santa Fe, NM 87501

Abstract. The size of networks now needed to model real world phenomena poses significant computational challenges. *Key node selection in networks*, (KNSIN) presented in this paper, selects a representative set of nodes that preserves the sketch of original nodes in the network and thus, serves as a useful solution to this challenge. KNSIN is accomplished via a sparse coding algorithm that efficiently learns a basis set over the feature space defined by the nodes. By executing a stop criterion, KNSIN automatically learns the dimensionality of the node space and guarantees that the learned basis accurately preserves the sketch of the original node space. In experiments, we use two large scale network datasets to evaluate the proposed KNSIN framework. Our results on the two datasets demonstrate the effectiveness of the KNSIN algorithm.

Keywords: Node Selection, Sparse Coding, Social Networks, Reconstruction Cost.

1 Introduction

The network paradigm has proved to be a crucial mathematical framework for the articulation and understanding of a wide range of phenomena (see e.g., [16,29]). In particular, social networks have emerged as among the most important mathematical frameworks for the analysis of the structure of societal interactions [17,22,26,7,6]. The size of many of the networks that are part of the modern "data deluge" often poses challenges for various forms of analysis (e.g., [8,15]). Perhaps, nowhere is this more true than in the case of social networks [20].

One way to address this obstacle is by selecting a set of representative nodes that preserves the characteristics of the original network. Such a set and the induced network derived from original network is called a *sketch* [9]. Such an idea is directly analogous to the various kinds of sampling done in signal processing [32] or statistics [3]. In the context of network analysis we call this the problem of *key node selection*. It is a problem that remains almost untouched in the network setting and is addressed in this paper.

Our approach to the key node selection problem is inspired by the technique of *sparse coding*, an idea originally discovered in computer vision [30]. Sparse coding derives a collection of vectors optimized to produce a sparse representation (in

S. Džeroski et al. (Eds.): DS 2014, LNAI 8777, pp. 337–349, 2014.

terms of the number of vectors used) of any representative element of a given kind of phenomenon. For example, in image processing, given a collection of images of some fixed size, a sparse coding methodology will produce a set of basis vectors (they need not be linearly independent) that efficiently represent (i.e., require only a small number of nonzero elements for a good approximation) a random image patch [33].

Sparse coding has a number of important attributes. First, it provides sparse and efficient representation for original data. This in turn enables the efficient interpretation of the original data. Second, sparse coding naturally offers an index scheme that facilitates quick data reduction and retrieval [5]. Sparse coding may learn an overcomplete basis. This allows for more stable and powerful representations compared with original data [36]. Due to its effectiveness, sparse coding has been widely used in various domains such as speech recognition [21], image restoration [24], and objection detection [19].

In this paper, we use a form of sparse coding to produce a node selection framework that we call *key node selection in networks* (KNSIN). The key nodes are determined by using sparse coding as applied to the feature space (the so-called "node space") that we assume is a part of the network data. Specifically, by considering the dependence relation between nodes, KNSIN incrementally extracts an independent basis set. We also make use of a novel efficient *data skipping* technique, so that KNSIN learns basis sets using only a portion of observed nodes rather than all the nodes of the network. By designing an adaptive acceptance threshold technique, KNSIN is capable of automatically deciding the suitable dimensionality for the node space. Finally, using the derived basis for the node space, we are then able to go back to the original set of nodes to determine the subset of key nodes that make up the sketch. In summary, the contributions of the paper are as follow: (1) We design a node selection framework over network data that is driven by sparse coding in node space. As far as we know, it is the first key node selection framework via coding in network scenario. (2) We develop an effective sparse coding algorithm to learn the basis for the node space. Our proposed KNSIN automatically estimates the dimensionality of the node space and learns an independent basis set. It deserves noting that the learned dimensionality for node space is completely dependent on the data. (3) We give a theoretical analysis showing that the learned basis set guarantees a small reconstruction cost [35]. Here, reconstruction cost is a measure to evaluate if the basis vectors accurately preserve the sketch of original data [9,47]. (4) We validate the method on two large-scale email network datasets from diverse sources: a large university and a large IT corporation.

2   Related Work

Due to its powerful modeling ability, networks have drawn much attention in various domains. Prominent examples include the use of networks in social sciences [11,12], biology [40,38], and finance [1]. A wide range of algorithms have been proposed and implemented in order to uncover various aspects of the structure of networks. Examples relevant to our paper include work on node interaction [22,48], graphlet analysis [2,37], and node classification [28,39]. Many of these algorithms are challenged when the network size is large. For instance, in [8] we learn that analysis on networks might be prohibitively expensive as the number of nodes exceeds more than a few thousand. It is mentioned in [2] that as the scale of the network becomes large, the computation cost is excessively large for even simply counting the frequency of each type of graphlet. Examples like these show that it is essential for us to develop a scheme to select key nodes from original networks. A few attempts [20] to solve the challenge were made using statistic techniques such as random sampling. However, the performance is not good.

Our work is inspired by sparse coding. Sparse coding learns basis vectors in order to capture high-level semantics from original data. Increasingly, ideas derived from the sparse coding methodology are finding applications in many areas [19,21,24,43,13]. Generally speaking, these methods can be divided into several classes. The first class are greedy algorithms [25,41]. For this, basis vectors are chosen sequentially based on certain criteria. Greedy methods are simple and straightforward, addressing the NP-hard l_0-*decomposition problem* directly [4]. The *homotopy method* represents another important class of techniques [10,31], wherein a set of solutions are indexed by a particular *convergence-control* parameter in the constructed *homotopy*. Homotopy algorithms are effective for relatively small networks, but may fail as the scale of networks increases [10]. Several soft-thresholding based methods [18,42] have also been proposed to solve the optimization problem of sparse coding. For example, a feature-sign search method was proposed in [18] to reduce the non-differentiable $L1$-norm problem in an $L1$-regularized least square method, and accelerate the optimization procedure. Another example is the work of Wright et al. [42] making use of a Barzilai-Borwein step length [34] in the negative gradient direction which is then applied to the soft-thresholding mechanism to learn basis vectors.

3   Sparse Coding for Networks

3.1   Problem Description

Before presenting the proposed algorithm in detail, we give a formal description of key node selection in networks. Let $G = (V, E)$ denote a network, where V is the set of nodes and E is the set of edges. The task of key node detection is formulated as follows: given a graph $G = (V, E)$ with original *node set* $V = \{x_1, x_2, ..., x_N\}$ (x_i represents the feature vector for node i), select a subset V^* from V such that V^* accurately represents the original node set, i.e., V^* is capable of preserving the sketch [9] of the original node set. The notion of "sketch" depends upon a measure of distance between the original node set and learned basis set. Often this is evaluated in terms of reconstruction cost [35] which is defined as follows:

$$\mathcal{E}(\mathbf{M}) = \frac{1}{N} \sum_{j=1}^{N} \|x_j - span\{\mathbf{w_1}, \mathbf{w_2}, ..., \mathbf{w_d}\}\|_2. \tag{1}$$

Here $\mathbf{M} = \{\mathbf{w_1}, \mathbf{w_2}, ..., \mathbf{w_d}\}$ is the learned basis set and $\mathcal{E}(\mathbf{M})$ is the reconstruction cost of basis $\mathbf{M}$.

3.2 Learning the Basis Set

In what follows, we describe the sparse coding algorithm KNSIN used to select "key" nodes over networks. Initially, we assume $V = V^{(0)} = [\boldsymbol{x}_1^{(0)}, \boldsymbol{x}_2^{(0)}, ..., \boldsymbol{x}_N^{(0)}]$, where $\boldsymbol{x}_i \in \mathcal{R}^D$. (Here D is the dimensionality of each node features.) The learned basis $\mathbf{M}$ starts as empty.

For every iteration, we conduct column pivoting to exchange the current node vector with the one that has the largest 2-norm as follows: (Suppose the first k pivot columns have been found and the interchange processes have been performed.)

$$j = \arg \max_{k < i \le N} \|\boldsymbol{x}_i^{(k)}\|_2. \tag{2}$$

Then we interchange the pivot column j with column $k+1$, where k is the current number of components. After the process of column pivoting and exchange, we calculate the norm of the $(k + 1)^{th}$ column vector $r_{k+1,k+1}$ by

$$r_{k+1,k+1} = \|\boldsymbol{x}_{k+1}^{(k)}\|_2. \tag{3}$$

And we learn the potential basis vector $\mathbf{w_{k+1}}$ by

$$\mathbf{w_{k+1}} = \frac{\boldsymbol{x}_{k+1}^{(k)}}{r_{k+1,k+1}}. \tag{4}$$

By "potential basis vector" we mean the learned basis vectors that may be accepted or rejected determined by a threshold policy that is discussed later. If $\mathbf{w_{k+1}}$ is accepted, the node matrix $V^{(k)}$ is updated using orthogonal transformation:

$$\boldsymbol{x}_j^{(k+1)} \leftarrow \boldsymbol{x}_j^{(k)} - r_{k+1,j} \mathbf{w_{k+1}}, j = k + 2, ..., N. \tag{5}$$

Therein, $r_{k+1,j}$ is computed by

$$r_{k+1,j} = \mathbf{w_{k+1}}^\top \boldsymbol{x}_j^{(k)}, j = k + 2, ..., N. \tag{6}$$

We then obtain node matrix $V^{(k+1)} = [\mathbf{w_1}, ..., \mathbf{w_{k+1}}, \boldsymbol{x}_{k+2}^{(k+1)}, ..., \boldsymbol{x}_N^{(k+1)}]$. If the threshold condition is not satisfied, we will stop the whole algorithm.

It deserves noting that many sparse coding methods work on all training data to learn the basis vectors. This can be very computationally expensive. In KNSIN, a "training data skipping" scheme is designed to overcome this shortcoming (see Section 3.2.3). In the rest of the section, we analyze KNSIN in detail.

3.2.1 Measuring Independence between Nodes and Learned Basis

To incrementally learn representative basis vectors and reduce the redundancy in basis, we need to consider the independence between an incoming node and the basis set that is already learned [44,47]. Suppose we have learned basis vectors $\{\mathbf{w_1}, \mathbf{w_2}, ..., \mathbf{w_k}\}$. If a node $\boldsymbol{x_{k+1}}$ is dependent on the learned basis, the corresponding computed $\mathbf{w_{k+1}}$ is unsuitable to be accepted as a new basis vector. It is because that if a basis vector is transformed from a node that has strong dependence on learned basis set, the independence among the whole basis set would be affected [14]. Thus it is necessary to consider the independence between the current set of basis vectors and subsequent input vectors.

In KNSIN we use the projection distance of $\boldsymbol{x_{k+1}}$ onto the subspace $span\{\mathbf{w_1}, \mathbf{w_2}, ..., \mathbf{w_k}\}$ to measure the degree of independence between $\boldsymbol{x_{k+1}}$ and the learned basis for node space ($\mathbf{w_1}, \mathbf{w_2}, ..., \mathbf{w_k}$ are the learned basis vectors for node space):

$$\|span\{\mathbf{w_1}, \mathbf{w_2}, ..., \mathbf{w_k}\} - \boldsymbol{x_{k+1}}\|_2. \tag{7}$$

We define the matrix $\mathbf{W_k} = [\mathbf{w_1}, \mathbf{w_2}, ..., \mathbf{w_k}] \in \mathbb{R}^{D \times k}$. The column vectors in $\mathbf{W_k}$ are orthogonal. Thereby, based on $\mathbf{w_k}$, we easily construct an orthogonal square matrix $\mathbf{Q_2}$ in the following form: $\mathbf{Q_2} = [\mathbf{W_k}, \mathbf{Q_1}]$. Therein, $\mathbf{Q_1} \in \mathbb{R}^{D \times (D-k)}$ is a rectangular matrix whose columns are orthogonal to each column of $\mathbf{W_k}$. It is straightforward to see that the value of (7) equals $\|\mathbf{Q_1}^\top \boldsymbol{x_{k+1}}\|_2$.

On the other hand, from (3) and (5), we know $r_{k+1,k+1}$ is the projection of the original data $\boldsymbol{x_{k+1}}$ onto the orthogonal complement space $span\{\mathbf{w_1}, \mathbf{w_2}, ..., \mathbf{w_k}\}^\perp$, thus $r_{k+1,k+1} = \|\mathbf{Q_1}^\top \boldsymbol{x_{k+1}}\|_2$. Therefore, $r_{k+1,k+1}$ (computed by (3)) measures the independence between the learned basis and incoming node $\boldsymbol{x_{k+1}}$.

3.2.2 Estimating Intrinsic Dimension by Threshold Policy

As discussed above, the dependence between the learned basis $\mathbf{M}$ and node $\boldsymbol{x_{k+1}}$ is measured by $r_{k+1,k+1}$. If $r_{k+1,k+1}$ is small, the vector $\boldsymbol{x_{k+1}}$ is almost entirely or entirely in the space $span\{\mathbf{w_1}, \mathbf{w_2}, ..., \mathbf{w_k}\}$. In this case, $\boldsymbol{x_{k+1}}$ has a strong dependence on the space $span\{\mathbf{w_1}, \mathbf{w_2}, ..., \mathbf{w_k}\}$. As a result, it is not reasonable to add the potential base vector $\mathbf{w_{k+1}}$ into the basis $\mathbf{M}$ because $\mathbf{w_{k+1}}$ is learned from $\boldsymbol{x_{k+1}}$. On the contrary, if $r_{k+1,k+1}$ is large, the computed vector $\mathbf{w_{k+1}}$ is an ideal potential basis vector. Due to the fact that $r_{1,1}$ is not smaller than $r_{i,i}$ for all $i > 1$ (this will be demonstrated in Theorem 2), for the purposes of convenient comparison, we use the value of $\frac{r_{k+1,k+1}}{r_{1,1}}$ as a criterion to make the final decision whether the computed potential vector $\mathbf{w_{k+1}}$ should be accepted as a basis vector or not.

Besides obtaining an independent basis, we need to guarantee that the learned basis preserves the sketch (á la [9]) of the original node space. Namely, the reconstruction cost $\mathcal{E}(\mathbf{M})$ indicated in Eq. 1 is small.

To satisfy the above constraints, we propose an adaptive threshold as follows: we accept the computed potential basis vector $\mathbf{w_{k+1}}$ if

$$\frac{r_{k+1,k+1}}{r_{1,1}} \geq T = f\left(\frac{Dim(\mathbf{M})}{N}\right) \tag{8}$$

Therein, N is the number of original nodes, and $f(t)$ is a monotonically increasing function with $0 \leq f(t) \leq 1$ for $0 \leq t \leq 1$.

Now we specify that the designed threshold technique accurately preserves the sketch of original data space via the following Theorem.

Theorem 1. $\mathcal{E}(\mathbf{M}) < r_{1,1} f(\frac{Dim(\mathbf{M})}{N})$.

Proof. Suppose at the end of the algorithm, we obtain d basis vectors $\mathbf{w}_1, \mathbf{w}_2, ...,$ $\mathbf{w_d}$ (i.e., $Dim(\mathbf{M}) = d$). Therefore, for nodes $\mathbf{x_j}$ ($j \leq d$), we have $\|\boldsymbol{x}_j - span\{\mathbf{w}_1,$ $\mathbf{w}_2, ..., \mathbf{w_d}\}\|_2 = \sqrt{\|\boldsymbol{x}_j\|_2^2 - \sum_{i=1}^d (\mathbf{w_i}^\top \boldsymbol{x}_j)^2} = 0$. And for those nodes $\boldsymbol{x}_j$ ($j > d$), due to the non-increasing monotonicity of $r_{i,i}$, (see Theorem 2), we have $\|\boldsymbol{x}_j -$ $span\{\mathbf{w}_1, \mathbf{w}_2, ..., \mathbf{w_d}\}\|_2 = \|\boldsymbol{x}_j - \sum_{i=1}^d (\mathbf{w_i}^\top \boldsymbol{x}_j)\mathbf{w_i}\|_2 \leq r_{d+1,d+1}$. As a result, $\mathcal{E}(\mathbf{M}) \leq \frac{1}{N}(N - d)r_{d+1,d+1}$. Hence, we obtain

$$\mathcal{E}(\mathbf{M}) < \frac{N-d}{N} r_{1,1} f(\frac{Dim(\mathbf{M})}{N}) < r_{1,1} f(\frac{Dim(\mathbf{M})}{N}),$$

which completes the proof. ∎

Since $f(t) \leq 1$ is monotonically increasing and $Dim(\mathbf{M}) << N$, the value of $r_{1,1} f(\frac{Dim(\mathbf{M})}{N})$ will be small. It means that the adaptive threshold technique guarantees that basis vectors obtained by KNSIN accurately preserves the sketch of the original node space.

3.2.3 Learning the Basis Set by Data Skipping

In KNSIN, we design a data skipping policy to learn basis vectors from a portion of rather than all data based on the following theorem.

Theorem 2. $\{r_{i,i}\}_{i=1}^N$ *achieved in every iteration of KNSIN is monotone nonincreasing, i.e., if $j \leq k$ then $r_{j,j} \geq r_{k,k}$.*

Proof. Suppose currently we have learned k basis vectors $\mathbf{w}_1, \mathbf{w}_2, ..., \mathbf{w_k}$, and data matrix $A^{(k)} = [\mathbf{w}_1, ..., \mathbf{w_k}, \boldsymbol{x}_{k+1}^{(k)}, ..., \boldsymbol{x}_N^{(k)}]$. From (3) and (4), we have $\boldsymbol{x}_j^{(k+1)} = \boldsymbol{x}_j^{(k)} - r_{k+1,j} \frac{\boldsymbol{x}_{k+1}^{(k)}}{r_{k+1,k+1}}$ for $j > k + 1$. Because $\boldsymbol{x}_j^{(k+1)}$ is orthogonal to $\mathbf{w_{k+1}}$, we obtain

$$\|\boldsymbol{x}_j^{(k+1)}\|_2^2 = \boldsymbol{x}_j^{(k)^\top} \boldsymbol{x}_j^{(k)} - r_{k+1,j}^2 \leq \|\boldsymbol{x}_j^{(k)}\|_2^2.$$

Due to the column pivoting policy, $\|\boldsymbol{x}_j^{(k)}\|_2^2 \leq r_{k+1,k+1}^2$ for $j > k + 1$ and $r_{k+2,k+2} = \max_{j \geq k+2} \|\boldsymbol{x}_j^{(k+1)}\|_2$. Therefore, we conclude that $r_{k+1,k+1} \geq r_{k+2,k+2}$, which completes the proof. ∎

In KNSIN, we skip training data by a stop policy: we stop the algorithm if we arrive at the first potential basis vector $\mathbf{w_k}$ satisfying the following inequality:

$$\frac{r_{k,k}}{r_{1,1}} < T = f(\frac{Dim(\mathbf{M})}{N}). \tag{9}$$

Suppose $\mathbf{w_k}$ is the first potential basis vector that does not satisfy the threshold condition, i.e., $\frac{r_{k,k}}{r_{1,1}} < T$. $\frac{r_{k,k}}{r_{1,1}}$ is monotone nonincreasing (in k). Therefore, if the integer k does not satisfy $\frac{r_{k,k}}{r_{1,1}} \geq T$, all the integers $j > k$ does not satisfy the threshold condition either. This means that even if we continue to consider the remaining nodes, the potential basis vectors learned later ($\{\mathbf{w_j}\}_{j=k+1}^{N}$) also will not satisfy the threshold condition defined in (8). Therefore, those potential basis vectors will be rejected. It implies that the learned basis set remains the same even if we consider all nodes.

Therefore, we need not execute the KNSIN any longer if we meet a potential basis vector that does not satisfy the threshold condition. Based on this stop policy (9), KNSIN is able to effectively skip the training data that are numerically dependent on the learned basis, which in turn reduces computing complexity. Without this data skipping scheme, KNSIN has to process all the data; the time complexity for learning basis vectors would have reached $O(N^2 D)$. By the skipping scheme, the time complexity of KNSIN reduces to $O(NDd)$. Here, N is the size of training set, D is the dimensionality of original data, and d is number of learned basis vectors. $d << N$, thus the skipping scheme reduces the computation complexity to linear.

3.3   Summary of KNSIN

In Algorithm 1 we give pseudocode for KNSIN. The algorithm generates independent basis vectors and learns the proper dimension for node space. The learned basis guarantees that the sketch of original node space is accurately preserved.

Input: Node matrix $V^{(0)} = [\boldsymbol{x}_1^{(0)}, \boldsymbol{x}_2^{(0)}, ..., \boldsymbol{x}_N^{(0)}]$.

1. Initialize the basis $\mathbf{M} = \emptyset$ and its dimensionality $k = Dim(\mathbf{M}) = 0$.
2. **repeat**
3.   Column pivoting and interchange.
4.   Compute $r_{k+1,k+1}$ by (3) and $\mathbf{w_{k+1}}$ by (4).
5.   Accept $\mathbf{w_{k+1}}$ as a basis vector and add it into $\mathbf{M}$.
6.   Update $\boldsymbol{x}_j^{(k)}$ by (5), for $j = k + 2 : N$.
7.   Update basis dimensionality $k \leftarrow k + 1$.
8. **until** Threshold condition (8) is not satisfied.
9. Obtain key node set according to learned basis.

Output: Key node set and basis $\mathbf{M}$ for node space

Algorithm 1. Key Node Selection In Networks

4   Experiments

In what follows, we introduce two real-world network datasets to evaluate the proposed KNSIN Algorithm. The first dataset was collected in a major university over 6 semesters (three years). Another dataset was collected from a large IT corporation over three years. We first verify whether the sketch of the original data [9] is preserved by the learned basis vectors by calculating the reconstruction cost $\mathcal{E}(\mathbf{M})$ indicated in Eq. (1). Then to evaluate whether the encoded subgraph is close to the original network, we evaluate the link prediction accuracy under the selected key node based subgraph (i.e., the subgraph is constructed using the selected nodes along with the edges among them.). We use the statistic based network sampling algorithm Random Node Sampling (RNS) introduced in [20], a typical Sparse Coding algorithm (SC) [18] and Principal Component Analysis (PCA), a classic method to learn principal components from original data to compare with KNSIN. Meanwhile, the original representation (OR) of the un-sampled network is applied as another baseline.

Table 1. The reconstruction cost $\mathcal{E}(\mathbf{M})$ using KNSIN, SC, and PCA under the university email network datasets

$\mathcal{E}(\mathbf{M})$	KNSIN	SC	PCA
Semester1	**0.153 ± 0.012**	0.323 ± 0.024	0.322 ± 0.031
Semester2	**0.200 ± 0.016**	0.326 ± 0.027	0.317 ± 0.028
Semester3	**0.254 ± 0.021**	0.413 ± 0.049	0.314 ± 0.034
Semester4	**0.076 ± 0.008**	0.392 ± 0.037	0.310 ± 0.041
Semester5	**0.119 ± 0.011**	0.366 ± 0.033	0.301 ± 0.038
Semester6	**0.222 ± 0.019**	0.430 ± 0.052	0.270 ± 0.045

Table 2. The link prediction accuracy AC (%) over subgraph generated using KNSIN, SC,PCA, RNS, and OR under the university email network datasets

AC	KNSIN	SC	PCA	RNS	OR
Semester1	**86.8 ± 1.7**	84.2 ± 2.0	76.9 ± 2.1	72.9 ± 3.0	85.1 ± 1.3
Semester2	**91.0 ± 0.6**	84.3 ± 2.1	80.5 ± 2.4	77.2 ± 2.7	87.5 ± 1.7
Semester3	**92.0 ± 1.0**	86.7 ± 1.7	78.3 ± 2.9	80.3 ± 3.2	87.9 ± 1.4
Semester4	**87.8 ± 1.8**	83.8 ± 2.4	78.9 ± 2.3	72.2 ± 2.9	85.6 ± 1.9
Semester5	**89.3 ± 1.1**	84.0 ± 1.6	80.2 ± 1.9	75.4 ± 2.5	87.8 ± 1.2
Semester6	**92.0 ± 0.5**	81.3 ± 2.2	80.0 ± 1.8	75.2 ± 3.1	89.0 ± 1.4

4.1   University Email Network Dataset

In this section, we use a university email network dataset. The dataset contains email messages delivered to users via the university email system over six separate semesters. The email user population is a mix of students, faculty members, staff, and "affiliates" (a category including postdocs, visiting scholars, and alumni) in a major university in the United States. Every email record is composed of date, time, sender, and list of recipients. Out of privacy and security concerns, the contents of email messages are discarded and the email addresses are encrypted. However, in this email system, we are allowed to access an email user table that describes the personal information of each user, namely occupation, birth, gender, home country, postal code, years at the university, academic department (for student and faculty), division (for student only), and dormitory building (for student only). Besides those observable features, we calculate latent features for each node. A few models have been proposed to extract latent features for network, such as Latent Feature Relational Model [27] etc. In our experiments, we extract latent features by Multidimensional Scaling (MDS) – a widely used technique to learn latent features in networks (see e.g., [46]). Each person is represented by a node in the network, and we say there is an edge between two nodes (persons) if and only if the two persons have an email communication. The average number of email users for each dataset is $67,736$, and the average number of edges is $14,253,468$. Detailed information about the dataset can be found in [45].

First we evaluate the quality of learned basis set. We execute KNSIN to obtain the basis vectors under each of the six datasets respectively. Then we calculate the reconstruction cost $\mathcal{E}(\mathbf{M})$ of the learned basis vectors based on Eq.(1), and list the results in Table 1. To compare with KNSIN, we also list the reconstruction cost computed using basis vectors learned from SC and PCA respectively. Note that in KNSIN, the dimensionality of basis $\mathbf{M}$ is automatically learned by the algorithms themselves. But for SC and PCA, it necessitates that users predetermine the basis dimension as a learning parameter. The predetermined dimensionality affects the reconstruction cost of the learned basis. Therefore, for SC and PCA, we adopt the same basis dimensionality as KNSIN to guarantee that the comparison is fair.

Compared with SC and PCA, KNSIN achieves the smallest reconstruction cost under all datasets, which demonstrates that the basis learned by KNSIN is able to accurately preserve the sketch of original data better than two baselines.

To further verify if the selected key nodes preserve the characteristic of original node space, we perform the link prediction task under the selected subgraph, i.e., the graph containing selected nodes along with the edges among selected nodes. For the subgraph obtained using each algorithm, we apply the supervised link prediction algorithm described in [23], and employ the 10-fold cross validation scheme to calculate the average prediction accuracy. Besides SC and PCA, the link prediction accuracy under the subgraph derived using RNS, and the original graph (OR) is calculated as a baseline. We list the link prediction accuracy results under all methods in Table 2.

The results in Table 2 indicates that the subgraph generated by KNSIN achieves better link prediction accuracy than SC, PCA, and RNS under all the datasets. Compared with OR, KNSIN still obtains comparable or even better accuracy. It demonstrates that KNSIN can effectively select key nodes that accurately preserve the characteristic of original networks.

4.2   IT Company Email Network Dataset

In what follows, an email network dataset collected from a large information technology company is employed to evaluate the proposed KNSIN framework. The dataset contains the complete record, as drawn from the company's servers, of email communications among 30,328 employees from 2006 to 2008. The employees in the company are located in 289 different offices around 50 states in United States and collectively comprise about one quarter of the company's employee population. Each email record comprises the timestamp, sender, lists of recipients, and the size of the message. Privacy laws and corresponding company policies preclude the collection of the content of messages. However, some personal information for each employee is accessible from the HR department of the company, namely work years in the company, employee's job function, office location code, the state of the office, and employee's group ID. Email messages from each of the three years are treated as a separate dataset. Each person is regarded as a node in the network while all people appearing in one email message are regarded as vertices associated with the edge. The personal information for each employee obtained from HR department is treated as observed features. The average number of vertices for each network is 28, 199, and the average number of links is 846, 882.

For each of the three datasets, we run KNSIN. Then reconstruction cost $\mathcal{E}(\mathbf{M})$ is computed and listed in Table 3. We also compare KNSIN with SC and PCA. Similarly to last subsection, the dimensionality of basis learned by KNSIN is automatically determined, while the dimensionality of basis learned by SC and PCA is set up as the same as KNSIN. The results in Table 3 show that the basis learned by KNSIN obtains the smallest reconstruction cost under each condition, indicating that KNSIN is able to accurately preserve the sketch of original data.

Table 3. The reconstruction cost $\mathcal{E}(\mathbf{M})$ using KNSIN, SC, PCA, RNS, under the IT company email network datasets

$\mathcal{E}(\mathbf{M})$	KNSIN	SC	PCA
2006	**0.217 ± 0.068**	0.302 ± 0.072	0.514 ± 0.078
2007	**0.149 ± 0.056**	0.227 ± 0.053	0.512 ± 0.074
2008	**0.211 ± 0.107**	0.243 ± 0.123	0.511 ± 0.157

Table 4. The link prediction accuracy AC (%) over subgraph generated using KNSIN, SC,PCA, RNS, and OR under the IT company email network datasets

AC	KNSIN	SC	PCA	RNS	OR
2006	**76.1 ± 1.8**	70.9 ± 2.8	71.9 ± 3.2	67.8 ± 3.1	74.6 ± 2.9
2007	**76.8 ± 1.6**	70.8 ± 3.1	69.8 ± 3.4	66.9 ± 3.3	72.3 ± 2.8
2008	**76.7 ± 2.1**	71.8 ± 3.2	72.3 ± 2.7	70.2 ± 3.5	75.5 ± 2.2

In this Company Email Network dataset, we still verify the effectiveness of our proposed node selection framework by the link prediction task. Similar to the last subsection, we construct a subnetwork (subgraph) based on the selected nodes using KNSIN, and calculate the link prediction accuracy over the subnetwork. SC, PCA, and RNS are still used to compare with the proposed KNSIN. Meanwhile, the link prediction accuracy over the original network (OR) is computed as a baseline. All link prediction results are listed in Table 4. From Table 4, we can conclude that KNSIN selects a set of nodes that accurately preserves the characteristics of original network.

5 Conclusion

In this paper, we propose a node selection framework for networks, KNSIN, to learn the basis for the node space via a sparse coding algorithm. By designing a stop criterion technique, KNSIN automatically determines the dimensionality of node space and extracts a set of independent basis vectors. The learned basis vectors guarantee that the sketch of original node space can be accurately preserved. By proposing a "data skipping scheme", KNSIN learns the basis efficiently from a portion rather than all of the data. The results from experiments run on two large email networks from diverse sources demonstrate the effectiveness of KNSIN.

References

1. Allen, F., Babus, A.: Networks in finance. In: Network-based Strategies and Competencies, pp. 367–382 (2009)
2. Bhuiyan, M.: Guise: Uniform sampling of graphlets for large graph analysis. In: ICDM 2012 (2012)
3. Bishop, C.: Pattern Recognition and Machine Learning. Springer (2006)
4. Bonnans, J., Gilbert, J., Lemarechal, C., Sagastizabal, C.: Numerical optimization: theoretical and practical aspects. Springer-Verlag New York Inc. (2006)
5. Chen, B.-C., Kuo, Y.-H., Chen, Y.-Y., Chu, K.-Y., Hsu, W.: Semi-supervised face image retrieval using sparse coding with identity constraint. In: ACM MM 2011, pp. 1369–1372 (2011)
6. Chen, Z., et al.: Inferring social contextual behavior from bluetooth traces. In: UbiComp 2013 (2013)
7. Chen, Z., Zhou, J., Chen, Y., Chen, X., Gao, X.: Deploying a social community network in rural areas based on wireless mesh networks. In: IEEE YC-ICT 2009 (2009)
8. Dimitropoulos, X.A., Riley, G.F.: Creating realistic BGP models. In: IEEE/ACM MASCOTS 2003 (2003)
9. Dong, W., Charikar, M., Li, K.: Asymmetric distance estimation with sketches for similarity search in high-dimensional spaces. In: SIGIR 2008, pp. 479–490 (2008)
10. Efron, B., Hastie, T., Johnstone, I., Tibshirani, R.: Least angle regression. Annals of Statistics 32(2), 407–419 (2004)
11. Gao, H., Tang, J., Hu, X., Liu, H.: Modeling temporal effects of human mobile behavior on location-based social networks. In: ACM CIKM (2013)
12. Gao, H., Tang, J., Liu, H.: Exploring social-historical ties on location-based social networks. In: AAAI (2012)
13. Gao, X., et al.: SOML: Sparse online metric learning with application to image retrieval. In: AAAI 2014 (2014)
14. Golub, G.H., Loan, C.F.V.: Matrix Computations, 3rd edn. Johns Hopkins University Press (1996)
15. Hasan, M.A., Zaki, M.J.: Output space sampling for graph patterns. In: VLDB 2009 (2009)

16. Kadushin, C.: Understanding social networks: Theories, concepts, and findings. Oxford University Press (2012)
17. Lakkaraju, H., McAuley, J., Leskovec, J.: What's in a name? Understanding the interplay between titles, content, and communities in social media. In: AAAI 2013 (2013)
18. Lee, H., Battle, A., Rajat, R., Ng, A.Y.: Efficient sparse coding algorithms. In: NIPS 2007 (2007)
19. Leordeanu, M., Hebert, M., Sukthankar, R.: Beyond local appearance: Category recognition from pairwise interactions of simple features. In: CVPR 2007 (2007)
20. Leskovec, J., Faloutsos, C.: Sampling from large graphs. In: SIGKDD 2006 (2006)
21. Lewicki, M.S., Sejnowski, T.J.: Learning overcomplete representations. Neural Computation 12(2), 337–365 (2000)
22. Liben-Nowell, D., Kleinberg, J.: The link prediction problem for social networks. In: CIKM 2003 (2003)
23. Lichtenwalter, R.N., Lussier, J.T., Chawla, N.V.: New perspectives and methods in link prediction. In: SIGKDD 2010 (2010)
24. Mairal, J., Bach, F., Ponce, J., Sapiro, G., Zisserman, A.: Non-local sparse models for image restoration. In: ICCV 2009 (2009)
25. Mallat, S., Zhang, Z.: Matching pursuit in a time-frequency dictionary. IEEE Trans. Signal Processing 41(12), 3397–3415 (1993)
26. McPherson, M., Smith-Lovin, L., Cook, J.M.: Birds of a feather: Homophily in social networks. Annual Review of Sociology 27(1), 415–444 (2001)
27. Miller, K.T.: Bayesian nonparametric latent feature models. Ph.D. Thesis, University of California, Berkeley (2011)
28. Neville, J., Jensen, D., Friedland, L., Hay, M.: Learning relational probability trees. In: SIGKDD 2003 (2003)
29. Newman, M.: The structure and function of complex networks. SIAM Review 45(1), 167–256 (2003)
30. Olshausen, B.A., Field, D.J.: Sparse coding with an overcomplete basis set: A strategy employed by V1? Vision Research 37, 3311–3325 (1997)
31. Osborne, M.R., Presnell, B., Turlach, B.A.: On the lASSO and its dual. Journal of Computational and Graphical Statistics 9(2), 319–337 (2000)
32. Proakis, J.G.: Digital Signal Processing: Principles, Algorithms and Applications. Prentice-Hall Press (2000)
33. Protter, M., Elad, M.: Image sequence denoising via sparse and redundant representations. IEEE Trans. Image Processing 18(1), 27–36 (2009)
34. Raydan, M.: On the barzilai and borwein choice of the step length for the gradient method. IMA Journal on Numerical Analysis 13, 321–326 (1993)
35. Roweis, S.T., Saul, L.K.: Nonlinear dimensionality reduction by locally linear embedding. Science 290, 2323–2326 (2000)
36. Sallee, P., Olshausen, B.A.: Learning sparse multiscale image representations. In: NIPS 2002 (2002)
37. Shervashidzea, N.: Efficient graphlet kernels for large graph comparison. In: AISTATS 2009 (2009)
38. Sporns, O.: Network analysis, complexity, and brain function. Complexity 8(1), 56–60 (2002)
39. Tang, L., Liu, H.: Relational learning via latent social dimensions. In: SIGKDD 2009 (2009)
40. Uetz, P., et al.: A comprehensive analysis of protein-protein interactions in saccharomyces cerevisiae. Nature 403, 623–627 (2000)

41. Weisberg, S.: Applied Linear Regression. Wiley, New York (1980)
42. Wright, J., Yang, A., Ganesh, A., Sastry, S., Ma, Y.: Robust face recognition via sparse representation. IEEE Trans. PAMI 31(2), 210–227 (2009)
43. Xiang, S., Shen, X., Ye, J.: Efficient sparse group feature selection via nonconvex optimization. In: ICML 2013 (2013)
44. Xu, Y.: Orthogonal component analysis for dimensionality reduction, Nanjing Univerisity (2010)
45. Xu, Y., Rockmore, D.: Feature selection for link prediction. In: PIKM 2012 (2012)
46. Xu, Y., Rockmore, D., Kleinbaum, A.M.: Hyperlink prediction in hypernetworks using latent social features. In: Fürnkranz, J., Hüllermeier, E., Higuchi, T. (eds.) DS 2013. LNCS, vol. 8140, pp. 324–339. Springer, Heidelberg (2013)
47. Xu, Y., Shen, F., Ping, W., Zhao, J.: Takes: a fast method to select features in the kernel space. In: CIKM 2011 (2011)
48. Zhu, J.: Max-margin nonparametric latent feature models for link prediction. In: ICML 2012 (2012)

Variational Dependent Multi-output Gaussian Process Dynamical Systems

Jing Zhao and Shiliang Sun

Department of Computer Science and Technology, East China Normal University
500 Dongchuan Road, Shanghai 200241, P.R. China
jzhao2011@gmail.com, slsun@cs.ecnu.edu.cn

Abstract. This paper presents a dependent multi-output Gaussian process (GP) for modeling complex dynamical systems. The outputs are dependent in this model, which is largely different from previous GP dynamical systems. We adopt convolved multi-output GPs to model the outputs, which are provided with a flexible multi-output covariance function. We adapt the variational inference method with inducing points for approximate posterior inference of latent variables. Conjugate gradient based optimization is used to solve parameters involved. Besides the temporal dependency, the proposed model also captures the dependency among outputs in complex dynamical systems. We evaluate the model on both synthetic and real-world data, and encouraging results are observed.

Keywords: Gaussian process, variational inference, dynamical system, multi-output modeling.

1 Introduction

Dynamical systems are widespread in machine learning applications. Multi-output time series such as motion capture data and video sequences are typical examples of these systems. Modeling complex dynamical systems has a number of challenges such as only time as inputs, nonlinear mapping from time to observations, large data sets and possible dependency among multiple outputs. Gaussian processes (GPs) provide an elegant method for modeling nonlinear mappings in the Bayesian nonparametric learning framework [15]. Some extensions of GPs have been developed in recent years, which aim to solve these challenges.

Lawrence [9, 10] proposed the GP latent variable model (GP-LVM) as a nonlinear extension of the probabilistic principal component analysis [18]. GP-LVM can provide a visualization of high dimensional data by optimizing the latent variables with the maximum a posterior (MAP) solution. To overcome the difficulty of time and storage complexities for large data sets, some approximate methods, e.g., sparse GP [11] have been proposed for learning GP-LVM. By adding a Markov dynamical prior on the latent space, GP-LVM is extended to the GP dynamical model (GPDM) [21, 22] which is able to model nonlinear dynamical systems. GPDM captures the variability of outputs by constructing the variance of outputs with different parameters.

S. Džeroski et al. (Eds.): DS 2014, LNAI 8777, pp. 350–361, 2014.

Instead of seeking a MAP solution for the latent variables as in the former methods, Titsias and Lawrence [20] introduced a variational Bayesian method for training GP-LVM. This method computes a lower bound of the logarithmic marginal likelihood by variationally integrating out the latent variables that appear nonlinearly in the inverse kernel matrix of the model. It was built on the method of variational inference with inducing points [19, 16]. This Bayesian GP-LVM was later adapted to multi-view learning [5] through introducing a softly shared latent space. Similarly, Damianou et al. [6] extended the Bayesian GP-LVM by imposing a dynamical prior on the latent space to the variational GP dynamical system (VGPDS). Park et al. [14] developed an almost direct application of VGPDS to phoneme classification. Besides variational approaches, expectation propagation based methods [7] are also capable of conducting approximate inference in Gaussian process dynamical systems (GPDS).

However, all the models mentioned above for GPDS ignore the dependency among multiple outputs, which usually assume that the outputs are conditionally independent. Actually, modeling the dependency among outputs is necessary in many applications such as sensor networks, geostatistics and time-series forecasting, which helps to make better predictions. Indeed, there are some recent works that explicitly considered the dependency of multiple outputs in GPs [3, 2, 23]. Latent force models (LFMs) [3] are a recent state-of-the-art modeling framework, which can model multi-output dependencies. Later, a series of extensions of LFMs were presented such as linear, nonlinear, cascaded and switching dynamical LFMs [1]. People also gave sequential inference methods for LFMs [8]. Álvarez and Lawrence [2] employed convolution processes to account for the correlations among outputs to construct a convolved multiple outputs GP (CMOGP) which can be regarded as a specific case of LFMs. Wilson et al. [23] combined neural networks with GPs to construct a GP regression network (GPRN). However, CMOGP and GPRN are neither introduced nor directly suitable for dynamical system modeling. When a dynamical prior is imposed, marginalizing over the latent variables is needed, which can be very challenging.

This paper proposes a variational dependent multi-output GP dynamical system (VDM-GPDS). The convolved process covariance function [2] is employed to capture the dependency among all the data points across all the outputs. To learn VDM-GPDS, we first approximate the latent functions in the convolution processes, and then variationally marginalize out the latent variables in the model. This leads to a convenient lower bound of the logarithmic marginal likelihood, which is then maximized by the scaled conjugate gradient method to find out the optimal parameters. Our model is applicable to general dependent multi-output dynamical systems rather than being specially tailored to a particular application. We adapt the model to different applications and obtain promising results.

2   The Proposed Model

Suppose we have multi-output time series data $\{\mathbf{y}_n, t_n\}_{n=1}^{N}$, where $\mathbf{y}_n \in \mathbb{R}^D$ is an observation at time $t_n \in \mathbb{R}^+$. We assume that there are low dimensional latent variables that govern the generation of the observations and a GP prior for the latent variables conditional on time captures the dynamical driving force of the observations, as in Damianou et al. [6]. However, a large difference compared with their work is that we explicitly model the dependency among the outputs through convolution processes [2].

Our model is a four-layer GP dynamical system. Here $t \in \mathbb{R}^N$ represents the input variables in the first layer. Matrix $X \in \mathbb{R}^{N \times Q}$ represents the low dimensional latent variables in the second layer with element $x_{nq} = x_q(t_n)$. Similarly, matrix $F \in \mathbb{R}^{N \times D}$ denotes the latent variables in the third layer, with element $f_{nd} = f_d(\mathbf{x}_n)$ and matrix $Y \in \mathbb{R}^{N \times D}$ denotes the observations in the fourth layer whose nth row corresponds to $\mathbf{y}_n$. The model is composed of an independent multi-output GP mapping from t to X, a dependent multi-output GP mapping from X to F, and a linear mapping from F to Y.

Specifically, for the first mapping, $\mathbf{x}$ is assumed to be a multi-output GP indexed by time t similarly to Damianou et al. [6], that is $x_q(t) \sim \mathcal{GP}(0, \kappa_x(t, t'))$, $q = 1, ..., Q$, where individual components of the latent function $\mathbf{x}(t)$ are independent sample paths drawn from a GP with a certain covariance function $\kappa_x(t, t')$ parameterized by $\boldsymbol{\theta}_x$. There are several commonly used covariance functions such as the squared exponential covariance function (RBF) and Matern 3/2 function [6]. Given the above assumption, we have

$$p(X|t) = \prod_{q=1}^{Q} p(\mathbf{x}_q|t) = \prod_{q=1}^{Q} \mathcal{N}(\mathbf{x}_q|0, \mathbf{K}_{t,t}), \tag{1}$$

where $\mathbf{K}_{t,t}$ is the covariance matrix. The covariance matrix may be constructed with any of the above covariance functions according to different applications.

For the second mapping, we assume that $\mathbf{f}$ is another multi-output GP indexed by $\mathbf{x}$, whose outputs are dependent, that is $f_d(\mathbf{x}) \sim \mathcal{GP}(0, \kappa_{f_d, f_{d'}}(\mathbf{x}, \mathbf{x}'))$, $d, d' = 1, ..., D$, where $\kappa_{f_d, f_{d'}}(\mathbf{x}, \mathbf{x}')$ is a convolved process covariance function which can capture the dependency among all the data points across all the outputs with parameters $\boldsymbol{\theta}_f = \{\{\Lambda_k\}, \{P_d\}, \{S_{d,k}\}\}$. The detailed formulation of $\kappa_{f_d, f_{d'}}(\mathbf{x}, \mathbf{x}')$ will be given in Sect. 2.1. From the conditional dependency among the latent variables $\{f_{nd}\}_{n=1, d=1}^{N, D}$, we have

$$p(F|X) = p(\mathbf{f}|X) = \mathcal{N}(\mathbf{f}|0, \mathbf{K}_{\mathbf{f}, \mathbf{f}}), \tag{2}$$

where $\mathbf{f}$ is a shorthand for $[\mathbf{f}_1^\top, ..., \mathbf{f}_D^\top]^\top$ and $\mathbf{K}_{\mathbf{f}, \mathbf{f}}$ sized $ND \times ND$ is the covariance matrix in which the elements are calculated by $\kappa_{f_d, f_{d'}}(\mathbf{x}, \mathbf{x}')$.

The third mapping, which is from f_{nd} to the observation y_{nd} can be written as $y_{nd} = f_{nd} + \epsilon_{nd}$, where $\epsilon_{nd} \sim \mathcal{N}(0, \beta^{-1})$. Thus, we get

$$p(Y|F) = \prod_{d=1}^{D} \prod_{n=1}^{N} \mathcal{N}(y_{nd}|f_{nd}, \beta^{-1}). \tag{3}$$

Given the above setting, the graphical model for the proposed VDM-GPDS on the training data $\{\mathbf{y}_n, t_n\}_{n=1}^{N}$ can be depicted as Fig. 1. From (1), (2) and (3), the joint probability distribution for the VDM-GPDS model is given by

$$p(Y, F, X|\mathbf{t}) = p(\mathbf{f}|X) \prod_{d=1}^{D} \prod_{n=1}^{N} p(y_{nd}|f_{nd}) \prod_{q=1}^{Q} p(\mathbf{x}_q|\mathbf{t}). \tag{4}$$

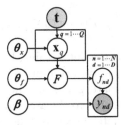

Fig. 1. The graphical model for VDM-GPDS

2.1 Convolved Process Covariance Function

Since the outputs in our model are dependent, we need to capture the correlations among all the data points across all the outputs. Bonilla et al. [4] and Luttinen and Ilin [12] used a Kronecker product covariance matrix, which is very limited and actually a special case of some general covariances when covariances calculated from output dimensions and inputs are independent. In this paper, we use a more general and flexible model in which these two covariances are not separated. In particular, the convolution processes [2] are employed to model the latent function $F(X)$.

Now we introduce how to construct the convolved process covariance functions. Using latent functions $\{u_k(\mathbf{x})\}_{k=1}^{K}$ and smoothing kernels $\{G_{d,k}(\mathbf{x})\}_{d=1,k=1}^{D,K}$, $f_d(\mathbf{x})$ is supposed to be expressed through a convolution integral,

$$f_d(\mathbf{x}) = \sum_{k=1}^{K} \int_X G_{d,k}(\mathbf{x} - \tilde{\mathbf{x}}) u_k(\tilde{\mathbf{x}}) d\tilde{\mathbf{x}}. \tag{5}$$

The smoothing kernel is assumed to be Gaussian and formulated as $G_{d,k}(\mathbf{x}) = S_{d,k} \mathcal{N}(\mathbf{x}|\mathbf{0}, P_d)$, where $S_{d,k}$ is a scalar value that depends on the output index d and the latent function index k, and P_d is assumed to be diagonal. The latent process $u_k(\mathbf{x})$ is assumed to be Gaussian with covariance function

$$\kappa_k(\mathbf{x}, \mathbf{x}') = \mathcal{N}(\mathbf{x} - \mathbf{x}'|\mathbf{0}, \Lambda_k). \tag{6}$$

Thus, the covariance between $f_d(\mathbf{x})$ and $f_{d'}(\mathbf{x}')$ is

$$\kappa_{f_d, f_{d'}}(\mathbf{x}, \mathbf{x}') = \sum_{k=1}^{K} S_{d,k} S_{d',k} \mathcal{N}(\mathbf{x}|\mathbf{x}', P_d + P_{d'} + \Lambda_k). \tag{7}$$

The covariance between $f_d(\mathbf{x})$ and $u_k(\mathbf{x}')$ is

$$\kappa_{f_d, u_k}(\mathbf{x}, \mathbf{x}') = S_{d,k} \mathcal{N}(\mathbf{x} - \mathbf{x}'|\mathbf{0}, P_d + \Lambda_k). \tag{8}$$

These covariance functions will be used for approximate inference in Sect. 3.

3   Inference and Optimization

The fully Bayesian learning for our model requires maximizing the logarithm of the marginal likelihood

$$p(Y|\mathbf{t}) = \int p(Y|F)p(F|X)p(X|\mathbf{t})dXdF. \tag{9}$$

Note that the integration w.r.t X is intractable, because X appears nonlinearly in the inverse of the matrix $\mathbf{K_{f,f}}$. We attempt to make approximations for (9).

To begin with, we approximate $p(F|X)$ which is constructed by convolution process $f_d(\mathbf{x})$ in (5). Similarly to Álvarez and Lawrence [2], a generative approach is used to approximate $f_d(\mathbf{x})$ as follows. We first draw a sample, $\mathbf{u}_k(Z) = [u_k(\mathbf{z}_1), ..., u_k(\mathbf{z}_M)]^\top$, where $Z = \{\mathbf{z}_m\}_{m=1}^M$ are introduced as a set of input vectors for $u_k(\tilde{\mathbf{x}})$ and will be learned as parameters. We next sample $u_k(\tilde{\mathbf{x}})$ from the conditional prior $p(u_k(\tilde{\mathbf{x}})|\mathbf{u}_k)$. According to the above generating process, $u_k(\tilde{\mathbf{x}})$ in (5) can be approximated by the expectation $\mathcal{E}(u_k(\tilde{\mathbf{x}})|\mathbf{u}_k)$. Let $U = \{\mathbf{u}_k\}_{k=1}^K$ and $\mathbf{u} = [\mathbf{u}_1^\top, ..., \mathbf{u}_K^\top]^\top$. We get the probability distribution of $\mathbf{f}$ conditional on $\mathbf{u}, X, Z$ as follows

$$p(\mathbf{f}|\mathbf{u}, X, Z) = \mathcal{N}(\mathbf{f}|\mathbf{K_{f,u}}\mathbf{K_{u,u}^{-1}}\mathbf{u}, \mathbf{K_{f,f}} - \mathbf{K_{f,u}}\mathbf{K_{u,u}^{-1}}\mathbf{K_{u,f}}), \tag{10}$$

where $\mathbf{K_{f,u}}$ is the cross-covariance matrix between $f_d(\mathbf{x})$ and $u_k(\mathbf{z})$ with element $\kappa_{f_d, u_k}(\mathbf{x}, \mathbf{x}')$ in (8), block-diagonal matrix $\mathbf{K_{u,u}}$ is the covariance matrix between $u_k(\mathbf{z})$ and $u_k(\mathbf{z}')$ with element $\kappa_k(\mathbf{x}, \mathbf{x}')$ in (6), and $\mathbf{K_{f,f}}$ is the covariance matrix between $f_d(\mathbf{x})$ and $f_{d'}(\mathbf{x}')$ with element $\kappa_{f_d, f_{d'}}(\mathbf{x}, \mathbf{x}')$ in (7). Therefore, $p(F|X)$ is approximated by $p(\mathbf{f}|X, Z) = \int p(\mathbf{f}|\mathbf{u}, X, Z)p(\mathbf{u}|Z)d\mathbf{u}$ and $p(Y|\mathbf{t})$ is converted to

$$p(Y|\mathbf{t}) = \int p(\mathbf{y}|\mathbf{f})p(\mathbf{f}|\mathbf{u}, X, Z)p(\mathbf{u}|Z)p(X|\mathbf{t})dFdUdX, \tag{11}$$

where $p(\mathbf{u}|Z) = \mathcal{N}(\mathbf{0}, \mathbf{K_{u,u}})$ and $\mathbf{y} = [\mathbf{y}_1^\top, ..., \mathbf{y}_D^\top]^\top$. It is worth noting that (11) is still intractable as the integration w.r.t X remains difficult.

Then, we introduce a lower bound of the $\log p(Y|\mathbf{t})$. We construct a variational distribution $q(F, U, X|Z)$ to approximate the distribution $p(F, U, X|Y, \mathbf{t})$ and compute the Jensen's lower bound on the $\log p(Y|\mathbf{t})$ as

$$\mathcal{L} = \int q(F, U, X|Z) \log \frac{p(Y, F, U, X|\mathbf{t}, Z)}{q(F, U, X|Z)} dXdUdF. \tag{12}$$

The variational distribution is assumed to be factorized as

$$q(F, U, X|Z) = p(\mathbf{f}|\mathbf{u}, X, Z)q(\mathbf{u})q(X). \tag{13}$$

$p(\mathbf{f}|\mathbf{u}, X, Z)$ in (13) is the same as the second term in (11), which will be eliminated during the variational computation. $q(\mathbf{u})$ is an approximation to $p(\mathbf{u}|X, Y)$, which is arguably Gaussian by maximizing the variational lower bound [6, 20]. $q(X)$ is an approximation to $p(X|Y)$, which is assumed to be a product of independent Gaussian distributions $q(X) = \prod_{q=1}^Q \mathcal{N}(\mathbf{x}_q|\boldsymbol{\mu}_q, S_q)$.

After some calculations and simplifications, the optimal lower bound becomes

$$\mathcal{L} = \log\left[\frac{\beta^{\frac{ND}{2}}|\mathbf{K_{u,u}}|^{\frac{1}{2}}}{(2\pi)^{\frac{ND}{2}}|\beta\psi_2 + \mathbf{K_{u,u}}|^{\frac{1}{2}}}\exp\{-\frac{1}{2}\mathbf{y}^\top W\mathbf{y}\}\right] \tag{14}$$
$$-\frac{\beta\psi_0}{2} + \frac{\beta}{2}\mathrm{Tr}(\mathbf{K_{u,u}^{-1}}\psi_2) - \mathbf{KL}[q(X)\|p(X|\mathbf{t})],$$

where $W = \beta I - \beta^2\psi_1(\beta\psi_2 + \mathbf{K_{u,u}})^{-1}\psi_1^\top$, $\psi_0 = \mathrm{Tr}(\langle\mathbf{K_{f,f}}\rangle_{q(X)})$, $\psi_1 = \langle\mathbf{K_{f,u}}\rangle_{q(X)}$ and $\psi_2 = \langle\mathbf{K_{u,f}K_{f,u}}\rangle_{q(X)}$. $\mathbf{KL}[q(X)\|p(X|\mathbf{t})]$ defined by $\int q(X)\log\frac{q(X)}{p(X|\mathbf{t})}dX$ is

$$\mathbf{KL}[q(X)\|p(X|\mathbf{t})] = \frac{Q}{2}\log|\mathbf{K_{t,t}}| - \frac{1}{2}\sum_{q=1}^{Q}\log|S_q| \tag{15}$$
$$+ \frac{1}{2}\sum_{q=1}^{Q}[\mathrm{Tr}(\mathbf{K_{t,t}^{-1}}S_q) + \mathrm{Tr}(\mathbf{K_{t,t}^{-1}}\boldsymbol{\mu}_q\boldsymbol{\mu}_q^\top)] + const.$$

Note that although the lower bound in (14) and the one in VGPDS [6] look similar, they are essentially distinct and have different meanings. In particular, the variables U in this paper are the samples of the latent functions $\{u_k(\mathbf{x})\}_{k=1}^{K}$ in the convolution process while in VGPDS they are samples F. Moreover, the covariance functions of F involved in this paper are multi-output covariance functions while VGPDS adopts single-output covariance functions. As a result, our model is more flexible and challenging.

3.1 Computation of ψ_0, ψ_1, ψ_2

Recall that the lower bound (14) requires computing the statistics $\{\psi_0, \psi_1, \psi_2\}$. We now detail how to calculate them. ψ_0 is a scalar that can be calculated as

$$\psi_0 = \sum_{n=1}^{N}\sum_{d=1}^{D}\int\kappa_{f_d,f_d}(\mathbf{x}_n, \mathbf{x}_n)\mathcal{N}(\mathbf{x}_n|\boldsymbol{\mu}_n, S_n)d\mathbf{x}_n = \sum_{d=1}^{D}\sum_{k=1}^{K}\frac{NS_{d,k}S_{d,k}}{(2\pi)^{\frac{Q}{2}}|2P_d + \Lambda_k|^{\frac{1}{2}}}. \tag{16}$$

ψ_1 is a $V \times W$ matrix whose elements are calculated as[1]

$$(\psi_1)_{v,w} = \int\kappa_{f_d,u_k}(\mathbf{x}_n, \mathbf{z}_m)\mathcal{N}(\mathbf{x}_n|\boldsymbol{\mu}_n, S_n)d\mathbf{x}_n = S_{d,k}\mathcal{N}(\mathbf{z}_m|\boldsymbol{\mu}_n, P_d + \Lambda_k + S_n), \tag{17}$$

where $V = N \times D$, $W = M \times K$, $d = \lfloor\frac{v-1}{N}\rfloor + 1$, $n = v - (d-1)N$, $k = \lfloor\frac{w-1}{M}\rfloor + 1$ and $m = w - (k-1)M$. Here the symbol "$\lfloor\rfloor$" means rounding down. ψ_2 is a

---

[1] We borrow the density formulations to express ψ_1 as well as ψ_2.

$W \times W$ matrix whose elements are calculated as

$$(\psi_2)_{w,w'} = \sum_{d=1}^{D} \sum_{n=1}^{N} \int \kappa_{f_d,u_k}(\mathbf{x}_n, \mathbf{z}_m)\kappa_{f_d,u_{k'}}(\mathbf{x}_n, \mathbf{z}_{m'})\mathcal{N}(\mathbf{x}_n|\boldsymbol{\mu}_n, S_n)d\mathbf{x}_n$$
(18)
$$= \sum_{d=1}^{D} \sum_{n=1}^{N} S_{d,k}S_{d,k'}\mathcal{N}(\mathbf{z}_m|\mathbf{z}_{m'}, 2P_d + \Lambda_k + \Lambda_{k'})\mathcal{N}(\frac{\mathbf{z}_m + \mathbf{z}_{m'}}{2}|\boldsymbol{\mu}_n, \Sigma_{\psi_2}),$$

where $k = \lfloor \frac{w-1}{M} \rfloor + 1$, $m = w - (k-1)M$, $k' = \lfloor \frac{w'-1}{M} \rfloor + 1$, $m' = w' - (k'-1)M$ and $\Sigma_{\psi_2} = (P_d + \Lambda_k)^{\top}(2P_d + \Lambda_k + \Lambda_{k'})^{-1}(P_d + \Lambda_{k'}) + S_n$.

3.2   Conjugate Gradient Based Optimization

The parameters involved in (14) include the model parameters $\{\beta, \boldsymbol{\theta}_x, \boldsymbol{\theta}_f\}$ and the variational parameters $\{\{\boldsymbol{\mu}_q, S_q\}_{q=1}^{Q}, Z\}$. In order to reduce the variational parameters to be optimized and speed up convergence, we reparameterize the variational parameters $\boldsymbol{\mu}_q$ and S_q as $\bar{\boldsymbol{\mu}}_q$ and $\bar{\Lambda}_q$, respectively, as in Opper and Archambeau [13] and Damianou et al. [6]. The corresponding transformations are $S_q = (\mathbf{K}_{t,t}^{-1} + \bar{\Lambda}_q)^{-1}$ and $\boldsymbol{\mu}_q = \mathbf{K}_{t,t}\bar{\boldsymbol{\mu}}_q$. All the parameters are jointly optimized by the scaled conjugate gradient method to maximize the lower bound in (14).

4   Prediction

4.1   Prediction with Only Time

In the Bayesian framework, we need to compute the posterior distribution of the predicted outputs $Y_* \in \mathbb{R}^{N_* \times D}$ on some given time instants $\mathbf{t}_* \in \mathbb{R}^{N_*}$. With the parameters and time $\mathbf{t}_*$ omitted, the posterior density is given by

$$p(Y_*|Y) = \int p(Y_*|F_*)p(F_*|X_*, Y)p(X_*|Y)dF_*dX_*,$$
(19)

where $F_* \in \mathbb{R}^{N_* \times D}$ denotes the set of latent variables (the noise-free version of Y_*) and $X_* \in \mathbb{R}^{N_* \times Q}$ denotes the latent variables in the low dimensional space. The distribution $p(F_*|X_*, Y)$ is approximated by the variational distribution

$$q(\mathbf{f}_*|X_*) = \int p(\mathbf{f}_*|\mathbf{u}, X_*)q(\mathbf{u})d\mathbf{u},$$
(20)

where $\mathbf{f}_*^{\top} = [\mathbf{f}_{*1}^{\top}, ..., \mathbf{f}_{*D}^{\top}]$, and $p(\mathbf{f}_*|\mathbf{u}, X_*)$ is Gaussian expressed as $\mathcal{N}(\mathbf{f}_*|\mathbf{K}_{\mathbf{f}_*,\mathbf{u}}\mathbf{K}_{\mathbf{u},\mathbf{u}}^{-1}\mathbf{u}, \mathbf{K}_{\mathbf{f}_*,\mathbf{f}_*} - \mathbf{K}_{\mathbf{f}_*,\mathbf{u}}\mathbf{K}_{\mathbf{u},\mathbf{u}}^{-1}\mathbf{K}_{\mathbf{u},\mathbf{f}_*})$. Since the optimal setting for $q(\mathbf{u})$ is Gaussian, $q(\mathbf{f}_*|X_*)$ is Gaussian that can be computed analytically.

The distribution $p(X_*|Y)$ is approximated by the variational distribution $q(X_*)$ formulated as $q(X_*) = \mathcal{N}(\boldsymbol{\mu}_{X_*}, \Sigma_{X_*})$, where $\boldsymbol{\mu}_{X_*}$ is composed of column vector $\boldsymbol{\mu}_{\mathbf{x}_{*q}}$ with $\boldsymbol{\mu}_{\mathbf{x}_{*q}} = \mathbf{K}_{t_*,t}\mathbf{K}_{t,t}^{-1}\boldsymbol{\mu}_q$ and block-diagonal matrix Σ_{X_*} has diagonal element $\Sigma_{\mathbf{x}_{*q}}$ with $\Sigma_{\mathbf{x}_{*q}} = \mathbf{K}_{t_*,t_*} - \mathbf{K}_{t_*,t}\mathbf{K}_{t,t}^{-1}(\mathbf{K}_{t,t} - S_q\mathbf{K}_{t,t}^{-1}\mathbf{K}_{t,t_*})$.

However, the integration of $q(\mathbf{f}_*|X_*)$ w.r.t $q(X_*)$ is not analytically feasible. Following Damianou et al. [6], we give the expectation of $\mathbf{f}_*$ as $\mathcal{E}(\mathbf{f}_*)$ and its element-wise autocovariance as vector $\mathcal{C}(\mathbf{f}_*)$ whose $(\tilde{n} \times d)$th entry is $\mathcal{C}(f_{\tilde{n}d})$ with $\tilde{n} = 1, ..., N_*$ and $d = 1, ..., D$.

$$\mathcal{E}(\mathbf{f}_*) = \psi_{1*}\mathbf{b}, \tag{21}$$

$$\mathcal{C}(f_{\tilde{n}d}) = \mathbf{b}^{\top}(\psi_{2\tilde{n}}^{d} - (\psi_{1\tilde{n}}^{d})^{\top}\psi_{1\tilde{n}}^{d})\mathbf{b} + \psi_{0*}^{d} - \mathrm{Tr}\left[(\mathbf{K}_{\mathbf{u},\mathbf{u}}^{-1} - (\mathbf{K}_{\mathbf{u},\mathbf{u}} + \beta\psi_2)^{-1})\psi_{2*}^{d}\right],$$

where $\psi_{1*} = \langle\mathbf{K}_{\mathbf{f}_*,\mathbf{u}}\rangle_{q(X_*)}$, $\mathbf{b} = \beta(\mathbf{K}_{\mathbf{u},\mathbf{u}} + \beta\psi_2)^{-1}\psi_1^{\top}\mathbf{y}$, $\psi_{1\tilde{n}}^{d} = \langle\mathbf{K}_{f_{\tilde{n}d},\mathbf{u}}\rangle_{q(\mathbf{x}_{\tilde{n}})}$, $\psi_{2\tilde{n}}^{d} = \langle\mathbf{K}_{\mathbf{u},f_{\tilde{n}d}}\mathbf{K}_{f_{\tilde{n}d},\mathbf{u}}\rangle_{q(\mathbf{x}_{\tilde{n}})}$, $\psi_{0*}^{d} = \mathrm{Tr}(\langle\mathbf{K}_{\mathbf{f}_{*d},\mathbf{f}_{*d}}\rangle_{q(X_*)})$, $\psi_{2*}^{d} = \langle\mathbf{K}_{\mathbf{u},\mathbf{f}_{*d}}\mathbf{K}_{\mathbf{f}_{*d},\mathbf{u}}\rangle_{q(X_*)}$. Since Y_* is the noisy version of F_*, the expectation and element-wise auto-covariance of Y_* are $\mathcal{E}(\mathbf{y}_*) = \mathcal{E}(\mathbf{f}_*)$ and $\mathcal{C}(\mathbf{y}_*) = \mathcal{C}(\mathbf{f}_*) + \beta^{-1}\mathbf{1}_{N_*D}$, where $\mathbf{y}_*^{\top} = [\mathbf{y}_{*1}^{\top}, ..., \mathbf{y}_{*D}^{\top}]$.

4.2 Prediction with Time and Partial Observations

In this case which is referred as reconstruction, we need to predict Y_*^{m} which represents the outputs on missing dimensions, given Y_*^{pt} which represents the outputs observed on partial dimensions. The posterior density of Y_*^{m} is given by

$$p(Y_*^{m}|Y_*^{pt}, Y) = \int p(Y_*^{m}|F_*^{m})p(F_*^{m}|X_*, Y_*^{pt}, Y)p(X_*|Y_*^{pt}, Y)dF_*^{m}dX_*. \tag{22}$$

$p(X_*|Y_*^{pt}, Y)$ is approximated by $q(X_*)$ whose parameters need to be optimized for the sake of considering the partial observations Y_*^{pt}. This requires maximizing a new lower bound of $\log p(Y_*^{pt}, Y)$ which can be computed analogously to (14). Moreover, parameters of the new variational distribution $q(X, X_*)$ are jointly optimized because of the coupling of X and X_*. Then the marginal distribution $q(X_*)$ is obtained from $q(X, X_*)$. Note that multiple sequences where X_* and X are independent, only the separated variational distribution $q(X_*)$ is optimized.

5 Experiment

5.1 Synthetic Data

In this section, we evaluate our method on synthetic data generated from a complex dynamical system. The latent variables X are independently generated by the Ornstein-Uhlenbeck (OU) process

$$dx_q = -\gamma x_q dt + \sqrt{\tilde{\sigma}^2}dW, \quad q = 1, ..., Q. \tag{23}$$

The outputs Y are generated through a multi-output GP

$$y_d(\mathbf{x}) \sim \mathcal{GP}(0, \kappa_{f_d,f_{d'}}(\mathbf{x}, \mathbf{x}')), \quad d, d' = 1, ..., D, \tag{24}$$

where $\kappa_{f_d,f_{d'}}(\mathbf{x}, \mathbf{x}')$ employs the convolution process with one latent function. In this paper, the number of the latent functions in (5) is set to one, i.e., $K = 1$, which is also the common setting used in Álvarez and Lawrence [2]. We sample the synthetic data by two steps. First we use the differential equation with parameters $\gamma = 0.5$, $\sigma = 0.01$ to sample $N = 200$, $Q = 2$ latent variables at time interval $[-1, 1]$. Then we sample $D = 4$ dimensional outputs, each of which has 200 observations through the multi-output GP with parameters $S_{1,1} = 1$, $S_{2,1} = 2$, $S_{3,1} = 3$, $S_{4,1} = 4$, $P_1 = [5, 1]^\top$, $P_2 = [5, 1]^\top$, $P_3 = [3, 1]^\top$, $P_4 = [2, 1]^\top$ and $\Lambda = [4, 5]^\top$. In addition, white Gaussian noise is added to each output.

Prediction. Here we evaluate the performance of our method for predicting the outputs given only time compared with CMOGP, GPDM and VGPDS. We randomly select 50 points from each output for training with the remaining 150 points for testing. This is repeated for ten times. The latent variables X in VGPDS and VDM-GPDS with two dimensions are initialized by using principal component analysis on the observations. Moreover, the Matern 3/2 covariance function and 30 inducing points are used in VGPDS and VDM-GPDS.

Table 1 presents the averaged root mean square error (RMSE) with the standard deviation (std) for predictions. The best results are shown in bold. Since the data in this experiment are generated from a complex dynamical system that combines two GP mappings, CMOGP which consists of only one GP mapping can not capture the complexity well. Moreover, VDM-GPDS models the explicit dependency among the multiple outputs while GPDM and VGPDS does not. Therefore, our model gives the best performance among the four models as expected. Besides highest accuracies, VDM-GPDS also has the smallest variances. In addition, to verify the flexibility of VDM-GPDS, we do experiments on the independent output data which are generated analogously to Sect. 5.1. GPDM and VGPDS which do not make the assumption of output dependency is included as comparisons. The results are given in Table 2 where we can see that our model performs as well as VGPDS and significantly better than GPDM.

Table 1. Averaged RMSE (%) with std (%) for predictions on the dependent output data

	CMOGP	GPDM	VGPDS	VDM-GPDS
y_1	1.75±0.38	1.70±0.18	1.51±0.31	**1.43 ± 0.23**
y_2	3.46±0.67	3.32±0.27	2.99±0.53	**2.82 ± 0.35**
y_3	5.19±0.99	4.83±0.28	4.24±0.85	**4.09 ± 0.59**
y_4	7.50±0.94	5.98±0.55	5.16±0.92	**5.00 ± 0.60**

Reconstruction. In this part, we compare VDM-GPDS with the k-nearest neighbor best (k-NNbest) method which chooses the best k from $\{1, \ldots, 5\}$, CMOGP and VGPDS for recovering missing points given time and partially observed outputs. Here, we do not include the results of GPDM because that GPDM is not directly suitable for reconstructing some dimensions given data of other. We set $S_{4,1} = -4$ to generate data in this part, which makes that the

Table 2. Averaged RMSE (%) with std (%) for predictions on the independent output data

	GPDM	VGPDS	VDM-GPDS
y_1	3.82±1.55	**2.18 ± 0.06**	2.21±0.06
y_2	3.45±1.70	2.06±0.19	**2.05 ± 0.13**
y_3	3.57±1.71	**1.68 ± 0.09**	1.72±0.12
y_4	7.10±1.28	4.48±0.23	**4.45 ± 0.20**

output y_4 be negatively correlated with the others. We remove all outputs y_1 or y_4 at time interval $[0.5, 1]$ from the 50 training points, resulting in 35 points as training data. Note that CMOGP considers all the present outputs as the training set while VGPDS and VDM-GPDS only consider the outputs at time interval $[-1, 0.5)$ as the training set. Table 3 shows the results with four methods for reconstructions on the missing points for y_1 and y_4. It indicates the superior performance of our model for the reconstruction task.

Table 3. Averaged RMSE (%) with std (%) for reconstructions on y_1 and y_4

	k-NNbest	CMOGP	VGPDS	VDM-GPDS
y_1	1.87±0.62	1.90±0.31	1.49±0.94	**0.98 ± 0.34**
y_4	13.51±2.54	9.31±0.87	6.79±6.07	**5.56 ± 1.88**

5.2   Human Motion Capture Data

Here the sequences of runs/jogs from subject 35 in the CMU motion capture database are employed for the reconstruction task. We preprocess the data as in Lawrence [11], which leads to nine independent training sequences and one testing sequence. The average length of each sequence is 40 frames and the output dimension is 59.

The RBF kernel is adopted in this set of experiments to construct $\mathbf{K}_{t,t}$ which is a block-diagonal matrix because the sequences are independent. We compare our model with the nearest neighbor in the angle space (NN) and the scaled space (NN sc.) [17] and VGPDS. For parameter optimization of VDM-GPDS and VGPDS, the maximum numbers of iteration steps are set to be identical.

Table 4 gives results of four methods. LS and LA correspond to the reconstructions on the right leg in the scaled space and angle space. Similarly, BS and BA correspond to the upper body in the same two spaces. Clearly, our model outperforms the other approaches. We conjecture that this is because VDM-GPDS effectively considers both the dynamical characteristics and the dependency among the outputs in the complex dynamical system. Since GPDM cannot reconstruct the missing outputs on some dimensions given the others as explained in Sect. 5.1. We do experiments according to Wang et al. [22] to reconstruct the missing frames $21 - 43$ on all dimensions of the test data. We get the RMSE for reconstruction: 0.7323 with VDM-GPDS versus 0.9448 with GPDM and 5.1099 with VDM-GPDS versus 7.8984 with GPDM in the scaled space and angle space, respectively. It turns out that our model also defeats GPDM.

Table 4. The RMSE for reconstructions on the motion capture data

	NN sc.	NN	CMOGP	VGPDS	VDM-GPDS
LS	0.8170	0.8493	1.1468	0.6502	**0.6379**
LA	6.7495	7.9441	13.5338	5.5356	**5.3026**
BS	1.0027	1.4018	3.5564	0.6569	**0.5961**
BA	5.6332	9.5748	5.0171	2.8108	**2.6033**

6　Conclusion

In this paper we have proposed a dependent multi-output GP for modeling complex dynamical systems. The convolved process covariance function is employed to model the dependency among all the data points across all the outputs. We adapt the variational inference method involving inducing points to our model so that the latent variables are variationally integrated out.

Modeling the possible dependency among multiple outputs can help to make better predictions. The effectiveness of the proposed model is empirically demonstrated. However, when the dimensionality of the output is very high, our model may take a long time to converge. This opens the possibility for future work to accelerate training for high dimensional dynamical systems.

Acknowledgments. This work is supported by NNSFC Projects 61370175 and 61075005, and Shanghai Knowledge Service Platform Project (No. ZF1213).

References

1. Álvarez, M.A., Luengo, D., Lawrence, N.D.: Linear latent force models using Gaussian processes. IEEE Transactions on Pattern Analysis and Machine Intelligence 35, 2693–2705 (2013)
2. Álvarez, M.A., Lawrence, N.D.: Computationally efficient convolved multiple output Gaussian processes. Journal of Machine Learning Research 12, 1459–1500 (2011)
3. Álvarez, M.A., Luengo, D., Lawrence, N.D.: Latent force models. In: Proceedings of the 12th International Conference on Articicial Intelligence and Statistics, pp. 9–16 (2009)
4. Bonilla, E.V., Chai, K.M., Williams, C.K.I.: Multi-task Gaussian process prediction. In: Advances in Neural Information Processing Systems, vol. 18, pp. 153–160 (2008)
5. Damianou, A.C., Ek, C.H., Titsias, M.K., Lawrence, N.D.: Manifold relevance determination. In: Proceedings of the 29th International Conference on Machine Learning, pp. 145–152 (2012)
6. Damianou, A.C., Titsias, M.K., Lawrence, N.D.: Variational Gaussian process dynamical systems. In: Advances in Neural Information Processing Systems, vol. 24, pp. 2510–2518 (2011)
7. Deisenroth, M.P., Mohamed, S.: Expectation propagation in Gaussian process dynamical systems. In: Advances in Neural Information Processing Systems, vol. 25, pp. 2618–2626 (2012)

8. Hartikainen, J., Särkkä, S.: Sequential inference for latent force models (2012), http://arxiv.org/abs/1202.3730
9. Lawrence, N.D.: Gaussian process latent variable models for visualisation of high dimensional data. In: Advances in Neural Information Processing Systems, vol. 17, pp. 329–336 (2004)
10. Lawrence, N.D.: Probabilistic non-linear principal component analysis with Gaussian process latent variable models. Journal of Machine Learning Research 6, 1783–1816 (2005)
11. Lawrence, N.D.: Learning for larger dataset with the Gaussian process latent variable model. In: Proceedings of the 11th International Workshop on Artificial Intelligence and Statistics, pp. 243–250 (2007)
12. Luttinen, J., Ilin, A.: Efficient Gaussian process inference for short-scale spatiotemporal modeling. In: Proceedings of the 15th International Conference on Artificial Intelligence and Statistics, pp. 741–750 (2012)
13. Opper, M., Archambeau, A.: The variational Gaussian approximation revisited. Neural Computation 21, 786–792 (2009)
14. Park, H., Yun, S., Park, S., Kim, J., Yoo, C.D.: Phoneme classification using constrained variational Gaussian process dynamical system. In: Advances in Neural Information Processing Systems, vol. 22, pp. 2015–2023 (2012)
15. Rasmussen, C.E., Williams, C.K.I.: Gaussian Process for Machine Learning. MIT Press (2006)
16. Sun, S.: A review of deterministic approximate inference techniques for Bayesian machine learning. Neural Computing and Applications 23, 2039–2050 (2013)
17. Taylor, G.W., Hinton, G.E., Roweis, S.: Modeling human motion using binary latent variables. In: Advances in Neural Information Processing Systems, vol. 17, pp. 1345–1352 (2007)
18. Tipping, M.E., Bishop, C.M.: Probabilistic principal component analysis. Journal of the Royal Statistical Society 61, 611–622 (1999)
19. Titsias, M.K.: Variational learning of inducing variables in sparse Gaussian processes. In: Proceedings of the 12th International Conference on Artificial Intelligence and Statistics, pp. 567–574 (2009)
20. Titsias, M.K., Lawrence, N.D.: Bayesian Gaussian process latent variable model. In: Proceedings of the 13th International Conference on Artificial Intelligence and Statistics, pp. 844–851 (2010)
21. Wang, J.M., Fleet, D.J., Hertzmann, A.: Gaussian process dynamical models. In: Advances in Neural Information Processing Systems, vol. 19, pp. 1441–1448 (2006)
22. Wang, J.M., Fleet, D.J., Hertzmann, A.: Gaussian process dynamical models for human motion. IEEE Transactions on Pattern Analysis and Machine Intelligence 30, 283–398 (2008)
23. Wilson, A.G., Knowles, D.A., Ghahramani, Z.: Gaussian process regression networks. In: Proceedings of the 29th International Conference on Machine Learning, pp. 599–606 (2012)

Author Index